A Collection of Essays on
Complexity and Management

A Collection of Essays on
Complexity and Management

The Summer School on Managerial Complexity

Granada, Spain July 11 – 25 1998

Editor

Walter Baets
Nyenrode University,
The Netherlands Business School

World Scientific
Singapore • New Jersey • London • Hong Kong

Published by

World Scientific Publishing Co. Pte. Ltd.

P O Box 128, Farrer Road, Singapore 912805

USA office: Suite 1B, 1060 Main Street, River Edge, NJ 07661

UK office: 57 Shelton Street, Covent Garden, London WC2H 9HE

British Library Cataloguing-in-Publication Data
A catalogue record for this book is available from the British Library.

A COLLECTION OF ESSAYS ON COMPLEXITY AND MANAGEMENT
The Summer School on Managerial Complexity

ISBN 981-02-3714-6

This book is printed on acid-free paper.

Printed in Singapore by Uto-Print

FOREWORD

The Summer School on Managerial Complexity has discussed the use and consequences of dynamic non-linear systems behaviour for managerial problems and companies. It has created some understanding of the role which complexity plays. It has illustrated how important it can be for companies to have some insight in these issues, in order to acquire a better understanding of some of the chaos presently known on many of our markets.

The Summer School has equally given attention to ways to deal with complexity in real life managerial problems. In practice, a lot of applications on knowledge management and learning/adaptive systems have been discussed. Recent trends in managerial theory have been considered against the knowledge about complexity theory.

The main trends that were present during the summer school were the following: organisational learning; the virtual organisation and knowledge; chaos in emerging markets; chaotic management in practice; post-modern views on chaos management.

A few interesting emerging areas were identified during the Summer School. Post-modern management theory is little known and caught some interest. There is not yet a lot of material available, but we have organised a guest lecture about the topic (which is also included in the proceedings). Another interesting observation made is that most participants are only vaguely aware of the content of the (mathematical) chaos theory. It sounds strange, but it confirms the subdivision of the research practice in universities.

Most researchers in the social sciences (and I count management amongst the social sciences) have a lack of knowledge of the more mathematical developments of interest for their discipline. But probably more problematic is that it proved very difficult to gather a number of participants of multidisciplinary backgrounds. Multidisciplinary research is not really appreciated in the academic world, which only reinforces the need for this kind of summer schools.

This limited communication between disciplines results in a rather different jargon used, even amongst the business school people. Some hype seems to be present and again this makes summer schools like this one an interesting event for getting some parts of the field progressing. In general, research projects are very limited in scope and focus and the summer school has particularly opened some those projects to a wider horizon.

Those participants having a corporate experience were clearly more apt to discuss complexity in management. It remains a problem to discuss with scientists applications for management, though the approach chosen for this summer school was appreciated by both parties in order to improve that dialogue. By the end of the two weeks, a lot of mutual understanding was created.

I trust and hope that the success of this Summer School leads to the creation of a small network of researchers in the area and I hope this works out. I trust and hope that the proceedings of this Summer School are a stimulus in that direction and attract attention both from practitioners and academics alike.

I would like to thank a number of people who have all invested in this Summer School and without whom this Summer School would never have been possible. For sure I forget some and I already apologise for this. Hector Gachs Sanchez and Erna Oldenboom have contributed a lot of time and effort to the recruitment and the early set up of the event. Eva Ripolles Iturralde was the perfect course coordinator, appreciated both by the participants and myself. The guest faculty during this Summer School were: Jacques-Andre Bartoli, Juan Julian Merelo, Erna Oldenboom, Gert Van der Linden, Frans van Eijnatten, V. Venugopal and Fons Wijnhoven.

This summer school has received financial support from the Training and Mobility of Researchers Programme (TMR) of DG XII of the European Commission.

It was my pleasure of serving as the Scientific Co-ordinator of this Summer School.

Breukelen (The Netherlands), November 1998

Prof Dr Walter Baets
Philips Chair in Information and Communication Technology
Universiteit Nyenrode
The Netherlands Business School
Straatweg 25
3621 BG Breukelen
The Netherlands
E-mail: w.baets@nyenrode.nl

CONTENT

EVOLUTIONARY PATHS TOWARDS THE FUTURE: A SCENARIO APPROACH

PETER HOFMEISTER
Department of Operations Research, RWTH Aachen, Templergraben 64, 52056 Aachen, Germany
E-mail: peter@or.rwth-aachen.de

Companies have to formulate long-term strategies and therefore, rely on long-term forecasts. Since traditional forecasting methods cannot cope with disturbances and regard them as structural breaks, scenario analysis is a method to describe the future as a general picture in alternative views by quantitative and qualitative factors. Traditional scenario techniques are static since they focus on one future point in time and cannot reveal the dynamic development. In addition they need condensed estimates of the factors' complex interrelations which are very demanding for the user. The new scenario approach presented here is dynamic following the evolutionary path from the present to the future. This results in an even higher complexity of the problem but can be relieved by the use of an approximate reasoning algorithm. The algorithm evaluates linguistic rules which describe the relationship between the factors as a (mainly) causal relation in a human way of thinking to reproduce the mental model of the user.

1 Introduction

Due to the increasing globalization companies recognize a stronger competition in their markets. There are more competitors, more substitutes for their products, and new production processes threatening their once unbeatable positions. Opposite to these risks there are chances as new supply and demand markets arise and new niches can be exploited. On the one hand, these aspects seem to indicate an increase of the complexity of the firms' environments. On the other hand, the worldwide standardization, the tearing down of trade barriers, the simplification of legal affairs etc. simplify production and trade.

The effects of these changes are difficult to quantify and even qualitative statements remain doubtful. To maintain or strengthen their position in the market the companies have to formulate appropriate strategies. To deal with the uncertainty they can rise their flexibility to counter unexpected events. Being flexible is always at the expense of efficiency because one cannot choose the action which is most efficient in the actual situation but has to take into account future changes of the environment as well. Therefore, one has to decide which position to take on the trade-off curve of flexibility and efficiency.

This trade-off curve is not fixed but can be changed by new technologies. Information technology is the outstanding example for shifting production processes simultaneously to more efficiency and flexibility by, for example, CAD, CAE, and CIM. This shift allows to act more efficiently keeping the flexibility constant (see Figure 1a).

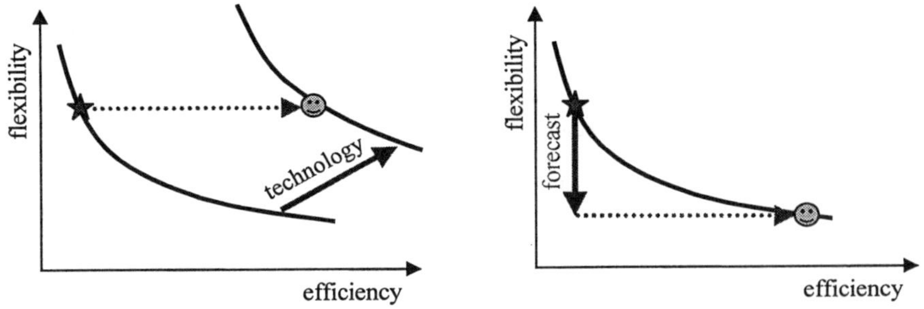

a) technology b) forecast

Figure 1: Gains in efficiency by technology shifts or forecasting effects

Forecasting effects influence the trade-off position on the line instead of altering the line. Better forecasting methods allow to reduce flexibility because the forecasts are more accurate and reliable and, therefore, lessen the demand for flexibility (see Figure 1b). Leaving technological aspects aside we focus on forecasting methods as tools for reinforcing competitiveness in today's business environment. Since the uncertainty is risky for all competitors, companies whose forecasting abilities are superior outperform others by a gain in efficiency.

2 Forecasting Methods

There are several methods how to catch the future based on different philosophical views. Some people who deny more or less their influence on their future, consider a higher being or fate to be responsible for the course of life. These god- or fate-fearing people look for revelations, wait for predictions at holy places, read holy books, pray to god, or sacrifice valuable goods to mysterious deities characters. Others believe in chance and suppose a random process to rule their life. The random walk method is a typical forecasting method which assumes this kind of random process and is used e.g. in the stock market analysis [18, p. 60]. From another point of view one can be convinced that the will of people forms the future.

In the former socialist countries in Eastern Europe the future could be regarded as a manifestation of will because their plans for a period of 5 years covered most of the social life and the economy. The task of prediction consisted of reading the plan (and estimating the deviation from reality). Since the last century cause-effect relations are increasingly regarded to determine the future. Consequently, forecasting methods based on cause-effect relations become more significant. The progress of science enables to discover these interrelations and to unveil former mysterious processes.

At the end of the last century the demon of Laplace, who could foresee the future exactly if he exactly knows the starting position and all the interrelations, seemed to become true and just a matter of time and technical progress [21, pp. 29-30]. Two scientific findings shattered this hope: at first, Poincaré discovered the phenomenon of the sensitivity of nonlinear processes and their chaotic behavior [15, p. 28]. This sensitivity causes minor differences of starting positions or relations to become unexpectedly large differences after some iterations, so any approximation will fail. Consequently, this so called deterministic chaos prevents from predicting a trajectory if we do not know the exact starting positions and the relations. Secondly, Heisenberg stated that it is technically and theoretically impossible to determine the actual state of a physical system arbitrary exactly [21, p. 95]. From the combination of both statements we conclude that we cannot predict the future by cause-effect relationships because we never know our actual position exactly and chaotic effects prevent from approximations.

Despite these insights people use statistical means supposing cause-effect relations to predict the future. These forecasts are based on the analysis of time-series, which are supposed to contain all data necessary to describe the past (and the future). As long as the underlying processes are stable predictions are possible. But time-series consist of a limited number of data and, therefore, cannot represent the underlying processes of the past arbitrary exactly. To remedy this objection the data are approximated by mathematical functions like regression curves. This adds information like, for example, the order of the curve, which is not undeniable included in the data. Whenever this forecasting method fails a structural break in the underlying relations is assumed [4, p. 291]. These breaks, which are apparently impossible to avoid, are less frequent and less substantial in shorter periods. Consequently, one predicts for the short periods where the errors caused by approximations and a lack of stability has only small effects.

Since structural breaks, i.e. changes in the underlying processes, are responsible for the quantitative unpredictability of the long-term development, qualitative techniques are recommended like, for example, the Delphi technique [2, pp. 116-120] or the scenario analysis [25, p. 73]. These techniques traditionally focus less on the path (which is supposed to be hard to follow due to the breaks) but on the end of the development. The scenario approach is noteworthy because it redefines the forecasting problem. Whereas the other techniques try to predict the future, the scenario technique describes the future by more than one forecast. The emphasis is on covering the possible evolution to prepare for and to make aware of future developments instead of predicting one - possibly wrong - long-term forecast

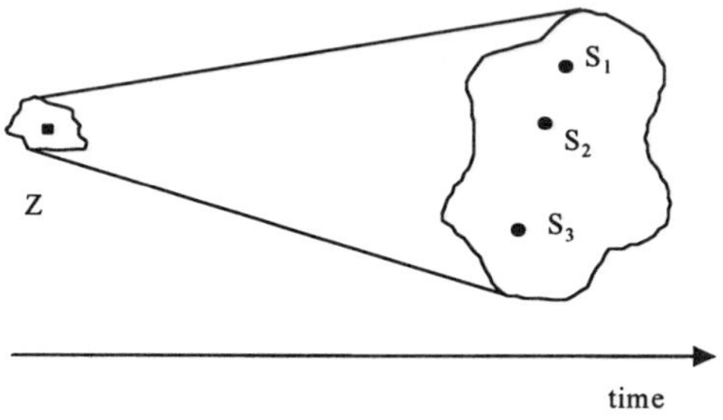

Figure 2: Scenario approach

triggering a strategy which could ruin the company. Starting from the actual state Z one looks for 3 to 5 alternative descriptions S_i of the future (see Figure 2).

3 Scenario Techniques

Although there are several variations of the scenario analysis, they share the same phases (see Figure 3) [23, p. 197]. Each scenario technique process starts with the definition of the issues to describe the scope of the analysis, the time frame etc. In the next steps the factors which influence the situation under consideration are identified and specified in detail. Then a construction process calculates all possible future states which are evaluated and selected with regard to their importance or plausibility, respectively. If too many possible states are left, the number is reduced by a clustering procedure which fuses similar states. The centers of these clusters are regarded as raw scenarios. Thereafter they are more precisely interpreted and embroidered to get an embellished story for each final scenario. In the last step the consequences for the companies' strategies, which can be derived from the scenarios, are investigated and strategic actions are designed.

Typically quantitative procedures support the construction by combinatorial means. Supposed we have two factors like, for example economic growth and unemployment rate both having three possible qualitative values (low, medium, high). Then there exist nine (three times three) combinations of these values. Since this easy procedure does not take into account the interdependencies between the factors, some combinations might be implausible (e.g. the combination of a high economic growth and a high unemployment rate seems less sound).

The plausibility - sometimes called consistence - of these combinations has to be assessed with regard to their relevance in the evaluation step. This assessment enables to select only those combinations for the fusion, which are ranked above a certain level of consistence to maintain plausibility.

1. DEFINITION
2. IDENTIFICATION
3. SPECIFICATION
4. CONSTRUCTION
5. EVALUATION + SELECTION
6. FUSION
7. INTERPRETATION
8. FORMULATION

Figure 3: Phases of the scenario analysis

This traditional scenario technique has some drawbacks. At first, the evaluation is often done by estimating two by two to reduce the number of comparisons [3, pp. 35-36]. Therefore, interdependencies of higher order (e.g. three, four factors influencing each other at the same time) are neglected which results in a wrong representation of the interrelations. Secondly, these estimates refer to probabilities [5, 1984], correlation numbers [24, p. 32], or consistence numbers [3, pp. 36-41], which are hard to determine reliably for human beings and have only a low validity. Some algorithms support the user to avoid estimates which contradict the laws of probability theory or statistics [22] but nevertheless, these estimates remain rather mysterious. Thirdly, the traditional method is focused on the ends and does not adequately take into account the paths from the present into the future. The dynamic evolution included in the paths could give valuable insights for the strategic planning and underlines the function of forecasts to warn far ahead of impending danger.

Our last objection refers to the fundamental concept of interpreting sudden extreme behavior of processes. We agree that this extreme behavior exists but we deny the random character expressed by the term "structural break" mistakenly used by the majority of forecasters. In our view these breaks are deterministically fixed and – according to Poincaré - induced by nonlinear cause-effect relations. It is not a change of or a break in the underlying structure of a system but a new and up to now unobserved behavior of the same stable and continuous structure. In addition, traditionalists implicitly deny the possibility to follow the path because the relations change (at random) and are unknown. We only admit that following these interrelations is a strain but believe in the existence and stability of the interdependencies.

4 Scenario Paths

The idea of following the paths raises objections referring not only to the costs but also to the result. Independent of the modeling technique the relations guiding the paths have to be cause-effect relations and, therefore, the findings of Heisenberg and Poincaré contradict the intention of an exact prediction. Since scenario analysis is not used to forecast *exactly one* future but suppose a *few roughly* defined *alternative* states, this argument can be rejected.

The cost argument refers to the demanding modeling of the relations. To forecast the behavior of the reality partially, we use a kind of a system dynamics view [9]. We comprehend reality as a combination of objects which are linked and get feedback from each other. Therefore, objects influence other objects and themselves directly and indirectly. This combination can be characterized by a system view which models the objects as elements and their mutual influences as interrelations. Traditional scenario techniques also describe the future by objects and relations. They adopt the system view by taking factors for objects and measures of dependency (probabilities, consistence numbers, correlation numbers) for relations. Therefore, the system view is plausible and widespread in the scenario analysis.

Our approach adapts the concept of system dynamics but does not cling to the technical realization. In our view, the technical modeling of system dynamics is too rigid and demanding. For example, in its intention to guarantee the behavior being driven by the deterministic structure and not at random, it rigidly bans stochastic elements. Inputs from outside the system are only tolerated as noise of peripheral influence. But sometimes, especially if it is difficult to demarcate the boundaries of the system to specify the required closed system, stochastic elements have to be used to get a realistic and not too extensive model of reality. Another strong objection refers to the formulation of the nonlinear equations. The debate on the validity of system dynamics models points to its weak spot (see [1], [10], [11], [19] and more recently [8, pp. 78-82]) because without an appropriate support it is too demanding to formulate exact mathematical equations which represent the interrelations. Therefore, we extend the capabilities of system dynamics to relieve users of too demanding tasks and import techniques from Artificial Intelligence and Fuzzy Set Theory to model complex systems. These techniques are aimed at reducing the complexity and maintaining the validity of the model simultaneously.

4.1 *How to handle complexity*

Complexity is not defined unequivocally and it is often confused with its frequent concomitant, the difficulty to handle it. But difficulty is only one aspect of complexity and is emphasized by the definition of computational complexity, which indicates the amount of resources needed to solve a class of problems by a certain algorithm. The definition of complexity by Kolmogorov in information theory stresses the problem of representing information [21, p. 146]. If we generate a certain pattern with a minimal Turing machine program, the length of this program gives us the Kolmogorov complexity. In system theory complexity refers to the number, variety and characteristics of the elements and interrelations of that system and, therefore, depends on the easiness to describe a system (see [7] and [12, p. 4]). With regard to this definition dynamic systems are often very complex, because time is another dimension of the problem that adds a lot of interrelations between the elements at each point in time.

These and other definitions are not comprehensive and have to be used with caution outside their areas. They contradict each other partially which becomes evident if we compare, for example, the system theoretic complexity and the Kolmogorov complexity: the increase in the number of interrelations between system elements amplifies the system theoretic complexity. On the other hand, these interrelations indicate that the elements are related and, therefore, each element carries information about others. There is a redundancy and, consequently, the possibility to compress the length of the minimal program and to reduce the complexity.

Since there is no clear and comprehensive definition of complexity, this term remains subjective. Complexity is not an absolute measure but a relative degree of a subjective impression of uneasiness. The relative complexity of two objects depends on the level of comparison because what we measure by complexity is not the complexity of the objects themselves. We rather calculate the complexity of the modeling relation between the object and a formal system [20]. Taking a formal system which can be adapted to an object only with a lot of troubles we discriminate this object against another one. Just as much we will possibly favor an object if we change the precision of the view. Objects with confusing structures at a micro-level might look very simple at a macro-level. Applying the most precise level does not solve the problem of measuring complexity because at the atomic level all objects are infinite complex and any distinction is neither proper nor possible.

To summarize complexity is a subjective attribute and depends on the intention we apply it for. From a system analysis view we are interested neither in the quantity of resources needed to solve a problem nor in the amount of information contained in the pattern. We look for a proper description of the problem to model reality as a system and, consequently, search for tools which reduce the number, variety, and characteristics of the elements and the interrelations of the system.

The last resort to reduce complexity is to ignore it. Ignorance is a widespread method bearing high risks. In most situations it is used unconsciously and sometimes intentionally but without realizing the consequences accurately. For applications which seek to quantify risks like scenario analysis this method is not suited. But there are three more strategies to cope with complexity: aggregation, decomposition and approximation.

Aggregation in the literal sense refers to the melting of similar elements. Macroeconomics, for instance, do not look at individuals but at aggregated units like households, industries, or the public sector. This reduces the number not only of elements but also of interrelations because all relations inside an aggregated unit are left aside. Aggregation does not change the problem but our view at it. Similarity is not an absolute attribute but depends relatively on our intentions. If we investigate e.g. the trade balance of two countries, we combine such different units like tourists and importers without loss of insights with respect to this specific task.

Instead of handling one complex problem one can decompose the problem in a few smaller and less complex problems. This method is used for example in sensitivity analysis where the effects on the solutions of changes of each single uncertain parameter is investigated. This relieves from the examination of the changes of all parameters at once. The disadvantage of this strategy is the well-known disregard of synergetic effects that can seriously change the outcome. From a systems perspective the elements are directly or indirectly connected to each other. Therefore the decomposition results in a loss of information but is unavoidable in real life applications.

Approximations change not only the representation but also the possibilities to gain insights into a system for a specific purpose. In a nonlinear optimization problem, for example, we look for an optimal solution and use partially linear approximations to simplify the search. The result is not necessarily optimal due to the loss of precision. This kind of approximation changes the characteristics of the system, in this particular example from nonlinear to partially linear. Another type of approximation transforms the variety of the system by a coarser precision. If we run calculations, the accuracy of the computations can be limited. Then we take only those solutions into account which can be expressed by this level of accuracy and all others are left aside. Sometimes approximations add information to a system. If there are no facts about certain states of the system, approximating techniques like interpolating or extrapolating estimate information. These methods are used in regression analysis and in simple forecasting models [2].

The types of approximations, decomposition and aggregation are related. Simplifying the characteristics of a system can change the variety because the loss of special characteristics reduces the possibility to differentiate. This decrease in differentiation is accompanied by an increase in similarity which induces a higher potential for aggregation. On the other hand, approximations can raise the number of elements if, for example, an exceptional interrelation is decomposed in a few parts, each represented by simple relations. Exchanging number for specialty can reduce complexity if the advantage by standardizing is not overpowered by the increase of size.

The reason for the difficulties to formulate the equations of system dynamics refers to the limitations of human abilities with respect to rigid mathematics. Mathematics offer a very abstract formal system to describe reality. In contrast, humans are used to act concretely in a real system. Therefore, to formulate mathematical equations is not an intuitive way of coping with reality and humans are overtaxed if the mathematical formulation demands too many and unfamiliar operations. Successful applications of mathematics are characterized by an adequate modeling of the problem by separation, the use of simple and familiar constructs of mathematics, and their automated combination. In the next section a concept from the Artificial Intelligence will be presented, which models the interrelations by simple linguistic rules instead of using too demanding mathematical equations.

4.2 Reproduction of interrelations by approximate reasoning

To replace the differential equations of system dynamics we divide the representation into parts and use linguistic expressions to approximate the equations. To exemplify we use a part of the Logistic map

$$x_t = 4 \cdot x_{t-1} \cdot (1 - x_{t-1}) \quad \textit{for } x_{t-1} \in [0,1]$$

which forms a parabola in the interval [0,1] (Figure 4).

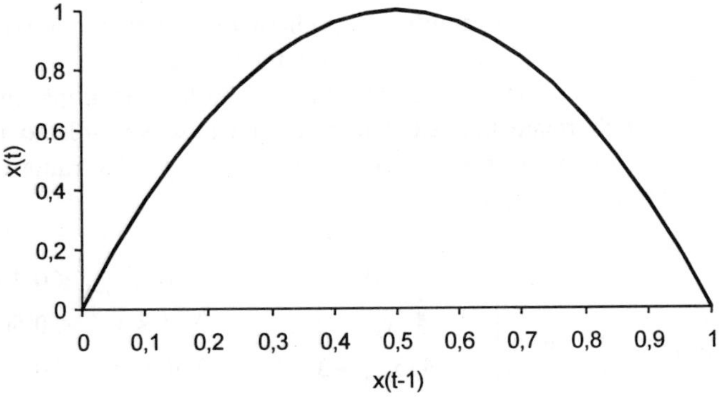

Figure 4: Logistic map

A rough linguistic description of the Logistic map would be:

$$\text{if} \quad x_{t-1} \quad \text{is} \quad \begin{cases} \textit{very small} \\ \textit{small} \\ \textit{medium} \\ \textit{large} \\ \textit{very large} \end{cases} \quad \text{then} \quad x_t \quad \text{is} \quad \begin{cases} \textit{very small} \\ \textit{large} \\ \textit{very large} \\ \textit{large} \\ \textit{very small} \end{cases}$$

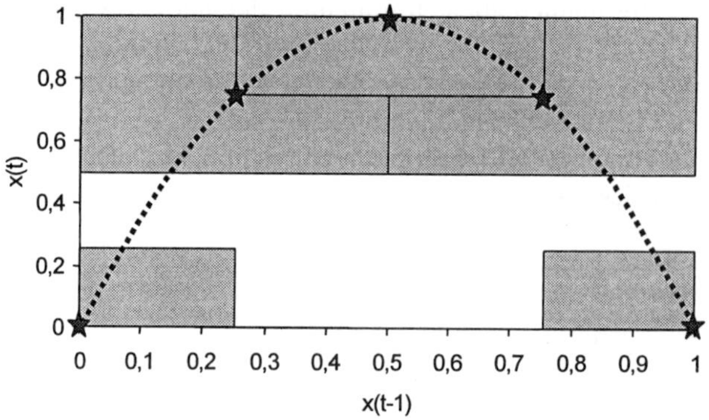

Figure 5: Granulation

The granulated information contained in this approximation is presented graphically by the rectangles which partially cover the function (Figure 5) [26, p. 122]. The linguistic description can be transformed in an exact mathematical expression by using the concept of linguistic variables of Fuzzy Set Theory and an approximate reasoning calculus [27, pp. 131, 144].

A linguistic variable (for example x_{t-1}) is characterized by a name (here x_{t-1}), the linguistic values (here {very small, small, medium, large, very large}), its range (here [0,1]), the grammar, and the semantic rule. The last are given by degrees of membership μ, which relate the real values of the base variable to its linguistic values. To give an example, the degrees of membership of a value of x_{t-1} to the linguistic term "medium" could be

$$\mu_{medium}(x_{t-1}) = \begin{cases} 0 & & 0 \le x_{t-1} \le 0.25 \\ 4 \cdot x_{t-1} - 1 & \textit{for} & 0.25 \le x_{t-1} \le 0.50 \\ -4 \cdot x_{t-1} + 3 & & 0.50 \le x_{t-1} \le 0.75 \\ 0 & & 0.75 \le x_{t-1} \le 1 \end{cases}$$

For example, the definitions of all terms of x_{t-1} are given graphically in Figure 6. The value 0.5 of x_{t-1} is only related to the term "medium" by a degree of 1. In comparison, the value 0.6 of x_{t-1} is associated to the term "medium" by a degree of 0.6 and to the term "large" by a degree of 0.4. From a set-theoretic point of view the element x_{t-1}=0.6 belongs to the set "medium" and "large". This is consistent with traditional set theory. The relaxation of traditional set theory by fuzzy set theory is expressed by the possibility for an element to belong to a set partially. Traditionally an element may be either a member of a set or not. This is modeled by a degree of membership taking its values from the set {0,1}. In contrast, the degree of member-

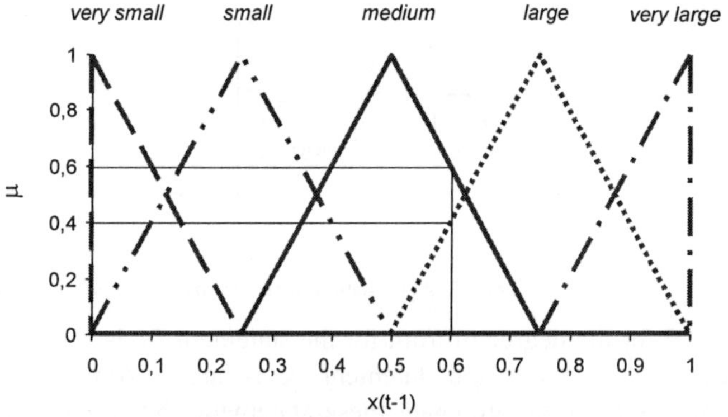

Figure 6: Linguistic variable x_{t-1}

ship in fuzzy set theory takes its values in the interval [0,1]. This gives the advantage of a finer differentiation and a better approximation to the human way of classifying and making inferences. The information contained in the granules of Figure 5 is not uniformly distributed. The stars represent the centers of information where the degrees of membership take the value of 1. These centers do not coincide with the centers of the granules if the linguistic terms are asymmetrically defined.

Approximate reasoning is a generalization of the traditional modus ponens inference [17, p. 231]. In traditional logic a conclusion is true if the premise is true as well. For example, from the rule "if A then B" in combination with the fact "A is true" follows "B is true". Since a linguistic variable forms the premises in our rules of the Logistic map, there are some linguistic values of xt-1 which are only partially true. Therefore, we have to define a method which takes into account the partial truth of premises. Approximate reasoning offers a family of operators to calculate the degree of truth of a conclusion. Typically the minimum (maximum) operator is used, i.e. the minimum (maximum) of the truth values of AND- (OR-) connected premises determines the truth of the conclusion. If we take the value 0.6 of xt-1, two linguistic terms of xt-1 have a non-zero degree of membership. Medium and large are supported by the value of 0.6 and 0.4, respectively. Therefore, two rules can be applied but with different degrees of support since their premises are only partially true (see Table 1).

Table 1: Applied rules and their degrees of support

rule	degree of support
if x_{t-1} is *medium* then x_t is *very large*	0.6
if x_{t-1} is *large* then x_t is *large*	0.4

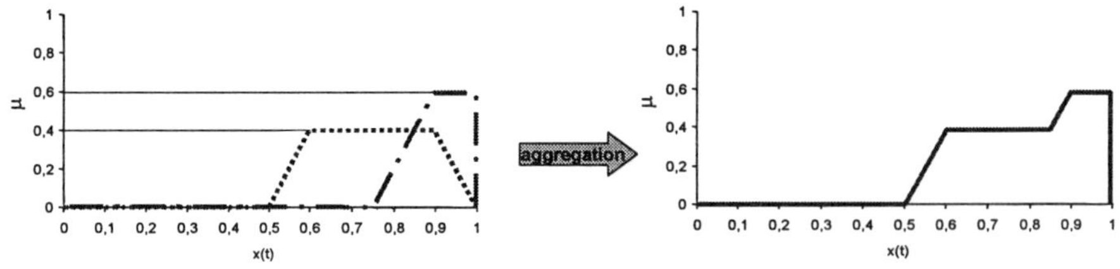

Figure 7: Aggregation of two terms

We calculate 0.6 for the degree of truth for the statement "x_t is very large" and give a value of evidence of 0.4 to the statement "x_t is large". Approximate reasoning offers a lot of operators to aggregate these statements but usually the maximum operator is applied. For each value of x_t it takes the term with the highest degree of membership. The graphical example illustrates the combination of the membership functions (cut at the 0.6 and 0.4 level of truth respectively) of both conclusions (Figure 7).

To get a crisp value for x_t we have to defuzzify the result of the aggregation. The numerous methods of defuzzification differ with regard to computational complexity and plausibility [6, p. 144]. An often applied procedure which simplifies the aggregation and defuzzification is the center-of-maximum method. It calculates the crisp value for each rule before the aggregation by taking the value x_t with the maximal degree of membership or if it is ambiguous, it takes the mean of these x_t's. In the second step the crisp values of rules are combined weighted by their degrees of truth.

Figure 8 gives three approximations of the Logistic map, the first one takes 5 terms per each linguistic variable and 5 rules, the second one 7 terms and 11 rules, and the third one 9 terms and 17 rules. The finer grid improves the fit of the approximation significantly resulting in an ignorable error of the (9_17)-approximation.

The approximate reasoning method replaces the task of formulating one exact equation, which is difficult for humans to quantify in exact mathematical terms, by a few less demanding tasks. At first, we define linguistic expressions, then we formulate rules, which are calculated individually and afterwards combined and interpreted. The calculation, the aggregation, and the interpretation are executed automatically by the approximate reasoning algorithm.

This method reduces the complexity by, for example, aggregating the infinite numbers in the unit interval by 5, 7 or 9 linguistic variables. In addition it decomposes the complexity since the mathematical formula is represented by smaller pieces of knowledge: the linguistic variables, the rule base, the operators etc. As a result one approximates the function and gets a close representation of the functional dependence.

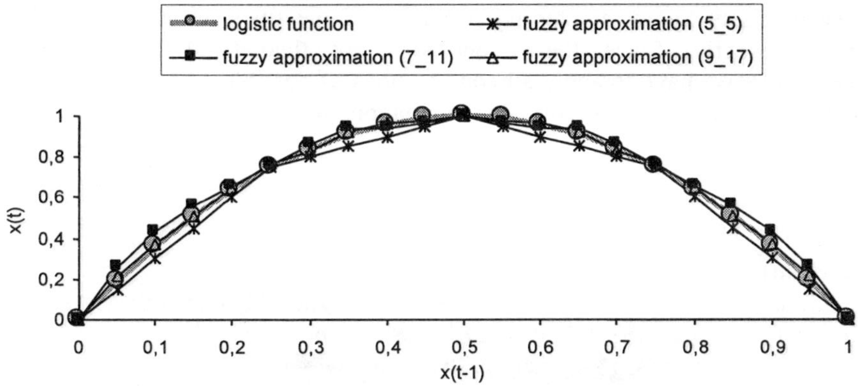

Figure 8: Fuzzy approximation of the Logistic map

4.3 The Simulation of the future

As is demonstrated in section 0, the transition from a state x_{t-1} to the next state x_t can be expressed by linguistic rules instead of mathematical equations. Therefore, we can calculate a sequence of crisp states by taking the actual crisp value x_{t-1}, fuzzifying it, applying the rules by an inference process, aggregating the consequences and defuzzifying them to one crisp result x_t (Figure 9). Thus, the concept of system dynamics to calculate mathematically the consequences of a system following its states step by step is kept but the equations (dotted line in Figure 9) are approximated by rules with linguistic components.

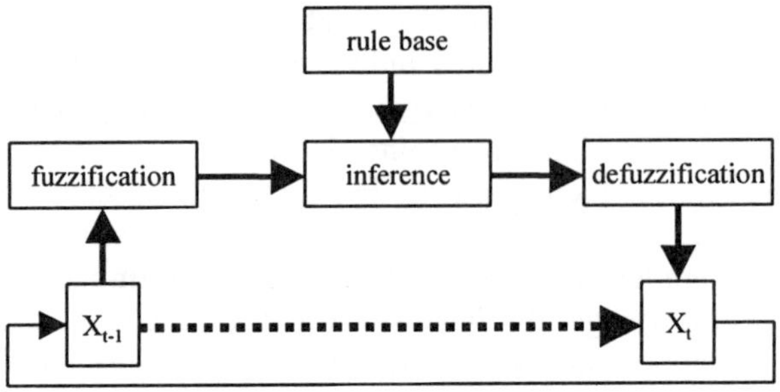

Figure 9: Simulation by rules

If we precisely know the starting position and the rules, we would be able to forecast exactly. Since these conditions are theoretically and technically unrealizable, we have to apply all combinations of possible positions and rules. In accordance with Chaos Theory this simulation of the future will result in very different ends. The set

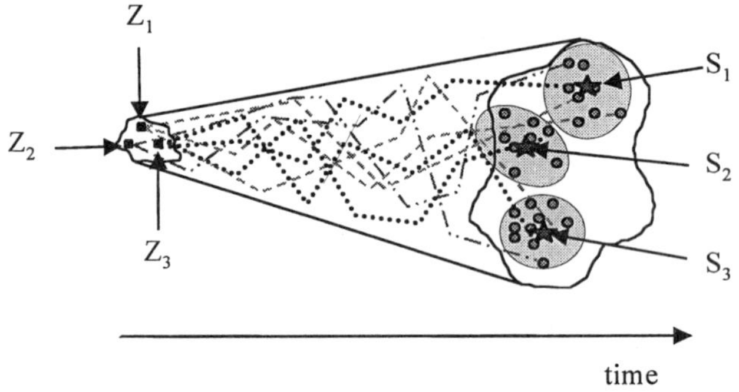

Figure 10: Scenario paths

of ends can be clustered like in traditional scenario analysis and the centers of the clusters are interpreted as scenarios (Figure 10).

5 Evolutionary Scenario Analysis

In comparison to the static view of traditional scenario techniques the dynamic systems approach is characterized by an evolutionary aspect. The rules of the system under consideration can be interpreted as laws of nature, which have to be followed. From this point of view the different starting points are individuals and the laws rule their growth. Each combination of rules can be interpreted as a different environment, which influences the development of the individuals. Therefore, the simple fundamental laws of evolution - mutation and selection - rule the process.

Mutation is induced by the mutual influences of the elements and selection is carried out according to the rules. Opposite to pure qualitative modeling techniques a state of a system has only one successor because of the aggregation and defuzzification. The calculations are less equivocal though the variety is maintained by the diverse simulation runs.

The phases of the scenario analysis do not alter compared to the traditional technique but their significance and their tasks change according to the system dynamics view (Figure 11). At first, the evolutionary scenario analysis deals with a forecasting problem which is different from the ones solved by traditional forecasting methods due to some particular characteristics of the underlying system. Since we use the causal interrelations as the main driving force of the development, this method can only be applied if we are convinced that deterministic relations rule the system. This is in contrast with traditional scenario techniques which assume causal relations as well but with a different intention. Traditionally, the relations finally reorganize situations perturbed by stochastic effects during the process and the rules do not deterministically determine the process completely.

1. DEFINITION

- scope
- time frame

2. IDENTIFICATION

- factors
- events
- stakeholders
- interrelations

3. SPECIFICATION

- units
- range
- linguistic variables
- rules
- inference parameters
- coarseness of rules and starting positions

4. CONSTRUCTION

- simulation

5. EVALUATION + SELECTION

- structural validation tests
- pattern behavior tests

6. FUSION

- (dynamic) clustering

7. INTERPRETATION

- embroidered stories

8. FORMULATION

- strategic actions
- consequences

Figure 11: Detailed phases

Secondly, the former implicit mental models about the reality have to be declared explicitly and in finer detail in the identification and specification phases to get a realistic model of the system. More factors, events, stakeholders, and their relationships have to be identified to grasp the system as comprehensive as possible. To reduce the numbers of possible combinations of starting points and rules the range of them should not be too broad. In case of doubt one should not restrict the range because this tempts to draw erroneous conclusions from the model. Instead one should increase the granulation by a coarser division or spend more resources and time on the calculations. For documentation and communication purposes we suggest an enriched cognitive map [16, pp. 173-175]. This map includes not only elements and directed qualitative interrelations but also symbols, which indicate exact mathematical functions, the time-lag, the inertia, stochastic effects etc.

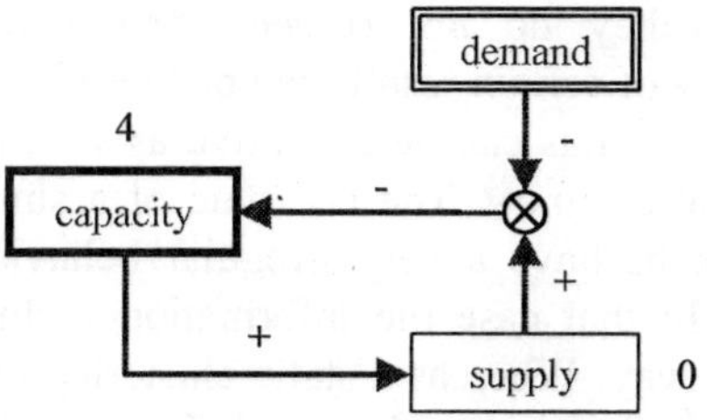

Figure 12: Extract of an enriched cognitive map

The enrichment enables to symbolize a more realistic interplay because some factors can be influenced after a while due to delays of the effects and some can only be changed very slowly due to their persistence. For example, in Figure 12 the difference (⊗) of supply and demand reduces (-) the production capacity. According to the thickness of its frame, adding capacity depends on several observations of the past because this is a decision with a long-term effect and is difficult to change. The double line of the demand's frame indicates that stochastic effects influence the demand, i.e. it is not only determined from within the model itself. When the producer has decided, it takes 4 more iterations to realize the additional capacity. The extra capacity increases (+) the supply immediately without any delay. The cognitive map relieves the specification of the rules and prevents from loosing track of the model which will become broad for realistic applications. Additionally, sources of error are more easily identified and can be backtracked.

Like in the traditional methods the construction is done automatically. Each simulation is defined by a certain starting position and specification of the rules. The simulation of all combinations and the more extensive inference process cause a bigger expenditure compared to the easy combinations of traditional scenario techniques. But this pays off in the evaluation phase when the user is not forced to rate all combinations. A few results of the simulation models might be unrealistic since the model inspects only the consistence taking the rules as fixed for each run. There is no meta-control of the consistence of the rules themselves. Therefore, the validity of the results has to be inspected for outstanding behavior, which seems to be inexplicable. Two explanations can be found. The specification of the model is wrong or the ideas of the forecaster about the system are biased or too narrow. The former should result in a new specification or an exclusion of these results in the selection phase. The latter is a favorable gain of insights in the behavior of reality and triggers learning.

Due to the higher number of ends the fusion of the simulated results takes more computation. The number of ends belonging to a cluster and thus its simulated frequency is a deceptive indicator for the probability of the scenario. Since combinations of the starting points and the rules - as inputs of the simulations - are uniformly distributed, they do not respect their real frequency distribution. Therefore, the frequency of certain results can be biased.

The clustering of the ends can be deceptive as well because the length of the time frame is somewhat arbitrary. The last state of a simulation run might not be representative if the paths have a very irregular behavior at the end due to the absence of attractors. In that case the information included in the paths can be analyzed in a dynamic way. Whereas a static clustering considers only the ends of the development, the similarity can also be defined on the correspondence of the paths. This increases the computational expenditures and raises questions of the appropriate dynamic clustering method (see [13], [14]) but it gives hints to similar pattern of behavior.

The interpretation of the scenario is supported by the sounder background, which is obtained by the paths leading to the centers of the clusters. Just as much the consequences of possible strategies can be analyzed. We include them in the model and simulate the outcome for a specific scenario. Opposite to traditional techniques, not only a qualitative but also a quantitative evaluation of the strategy can be deduced. In addition, one can identify situations where the possibility to favorably influence the development is very easy. Then a steering of the environment at low costs takes place and strategic actions shape the future.

6 Conclusions

The scenario approach presented in this paper is different from existing scenario techniques in several aspects. Compared to the static method the path method underlines the dynamic connection between the present and the future. The underlying ratio is more deterministically oriented because the interdependencies are regarded as a feedback system based on cause-effect relationships. Finally, the calculations are strictly quantitative and qualitative aspects have to be converted into fuzzy quantitative statements.

Comparing traditional techniques and the presented evolutionary scenario analysis we are aware that our approach demands more information. The identification and specification of the model is more extensive because a detailed description of the system under consideration has to be given. On the other hand, the information quality is somewhat less complex since the user does not have to calculate probabilities or other very demanding parameters. The required quantity of information increases but can be given in smaller and less ambitious packages. Therefore, the susceptibility to wrong specifications is decreased and can be kept under control by the explicit formulation of the model. The validity of the explicit models can be checked faster and more exactly whereas in traditional scenario techniques one had to trust in the users' estimates without almost any possibility of control. In addition, the system-theoretic approach and the use of IF-THEN-rules is very intuitive and does not strain.

To the best of our knowledge there is no easy-to-use fuzzy simulation software to model feedback systems. Nevertheless the method is easy to implement even for users who are not experienced with programming languages. We use a spreadsheet software (Excel©) to model the feedback system and combine it with a fuzzy logic software (FuzzyTech©) which supports the linguistic approximation and the inference. The efficiency of the implementation is not crucial. Since the forecasts are oriented to the long term, the calculations do not have to be carried out in real-time.

The more extensive input of information results in more benefits. At first, the more detailed information contained in the paths can be exploited in the fusion and interpretation phase. Secondly, strategic actions and promising occasions for interventions can be identified and their consequences are quantified in the model. In addition, the paths serve as a monitoring tool and can warn far ahead if the real system proceeds unfavorably.

The explicit formulation raises the transparency and relieves the communication. It is easier to explain and to understand the explicit model compared to the implicit models of traditional approaches. The model can be used to teach the long-term evolution to people inexperienced with the modeling technique and the real system. Finally, former mysterious calculations are unveiled and can be checked in detail, so that confidence in the model is inspired.

The presented new scenario technique has to be considered as a complementary approach due to some limitations. At first, the range of its application is limited to systems which have boundaries. From a system's view this claim can never be met absolutely since every part of reality has an influence on the other parts. The demarcation of the system, therefore, has to make a compromise between taking into account all elements and the size of the model. Secondly, the elements and their relations have to be known sufficiently exactly. Otherwise the number of simulation runs needed to cover the combinations of possible specifications is too large and the deterministic approach runs the risk to become a stochastic method like the Monte Carlo simulation.

To sum it up, specification and construction becomes harder but evaluation and selection is easier by using the evolutionary approach. The quantitative effort is more extensive, but the forecaster is relieved from too demanding qualitative tasks. The former implicit calculations and estimations are avoided and the user can devote his efforts to aspects not practicable for computers like identification or specification of factors.

Acknowledgements

The author would like to thank L. Mikenina, K. Palczynski, and K. Allen for their help and constructive criticism during the preparation of this paper. The presentation of the topic benefited a lot from their valuable hints.

References

1. Ansoff, H. I. and Slevin, D. P., An Appreciation of Industrial Dynamics, *Management Science* **14** (1968) pp. 383-397.
2. Armstrong, J. S., *Long-Range Forecasting* (John Wiley & Sons, New York, 1985).
3. Brauers, J. and Weber, M., A New Method of Scenario Analysis for Strategic Planning, *Journal of Forecasting* **7** (1988) pp. 13-47.
4. Bunn, D. W. and Salo, A. A., *Forecasting with Scenarios,* European Journal of Operational Research **68** (1993) pp. 291-303.
5. DeKluyver, C. A. and Moskowitz, H., Assessing Scenario Probabilities via Interactive Goal Programming, *Management Science* **30** (1984) pp. 273-278
6. Driankov, D., Hellendoorn, H. and Reinfrank, M., *An Introduction to Fuzzy Control* (Springer, Berlin, 1993).
7. Edmond, B. *What is Complexity?* www.cpm.mmu.ac.uk/~bruce/evolcomp/evolcomp-1.html (16.04.98).
8. Flood, R. L. and Jackson, M. C., *Creative Problem Solving* (John Wiley & Sons, New York, 1991).
9. Forrester, J. W., *Industrial Dynamics* (Productivity Press, Cambridge, 1961).
10. Forrester, J. W., Industrial Dynamics - A Response to Ansoff and Slevin, *Management Science* **14** (1968) pp. 601-618.
11. Forrester, J. W. / Low, G. W. and Mass, N. J., The Debate on World Dynamics: A Response to Nordhaus, *Policy Sciences*, **5** (1974) pp. 169-190.
12. Heylighen, F., The Growth of Structural and Functional Complexity during Evolution. In *The Evolution of Complexity* ed. by F. Heylighen and D. Aerts (Kluwer, Dordrecht, 1997) [in print].
13. Joentgen, A., Mikenina, L., Weber, R. and Zimmermann, H.-J., Dynamic Data Analysis: Problem Description and Solution Approaches. In *Fuzzy-Neuro Systems '98 – Computational Intelligence* ed. by. W. Brauer (Infix, Sankt Augustin, 1998) pp. 282-289.
14. Joentgen, A., Mikenina, L., Weber, R. and Zimmermann, H.-J., Dynamic Data Analysis: Similarity Between Trajectories. In *Fuzzy-Neuro Systems '98 – Computational Intelligence* ed. by. W. Brauer (Infix, Sankt Augustin, 1998) pp. 98-105.

15. Kaplan, D. and Glass, L., *Understanding Nonlinear Dynamics* (Springer, Berlin, 1995).
16. Klayman, J. and Schoemaker, P. J. H., Thinking about the Future: A Cognitive Perspective, *Journal of Forecasting* **12** (1993) pp. 161-186.
17. Klir, G. J. and Yuan, B., *Fuzzy Sets and Fuzzy Logic* (Prentice Hall, Upper Saddle River, 1995).
18. Makridakis, S. and Wheelwright, S. C., *Forecasting Methods for Management* (John Wiley & Sons, New York, 1989).
19. Nordhaus, W. D., World Dynamics: Measurement without Data, *Economic Journal* **83** (1973) pp. 1156-1183.
20. Prideaux, J., *Complexity... What Exactly is Complex Here*? www.cpm.mmu.ac.uk/~uce/PRNCYB-L/0724.html (16.04.98).
21. Ruelle, D., *Chance and Chaos* (Princeton University Press, Princeton, 1991).
22. Sarin, R. K., A Sequential Approach to Cross-Impact Analysis, *Futures* **10** (1978) pp. 53-62.
23. Schoemaker, P. J. H., Multiple Scenario Development: Its Conceptual and Behavioral Foundation, *Strategic Management Journal* **14** (1993) pp. 193-213.
24. Schoemaker, P. J. H., Scenario Planning: A Tool for Strategic Thinking, *Sloan Management Review*, **Winter** (1995) pp. 25-40.
25. Wack, P., Scenarios: Uncharted Waters Ahead, *Harvard Business Review*, **September-October** (1985) pp. 73-89.
26. Zadeh, L., Toward a Theory of Fuzzy Information Granulation and its Centrality in Human Reasoning and Fuzzy Logic, *Fuzzy Sets and Systems* **90** (1997) pp. 111-127.
27. Zimmermann, H.-J., *Fuzzy Set Theory – and its Applications*, (Kluwer, Boston, 1996) 3rd Edition.

MODELLING AND SIMULATION OF VARIABLE STRUCTURE SYSTEMS: A WAY TO HANDLE COMPLEXITY

KOSTIS CHRISTODOULOU

City University Business School, Frobisher Crescent, Barbican Centre, London EC2Y 8HB, UK

E-mail:K.Christodoulou@city.ac.uk

This paper describes a variable structure, object oriented platform used for strategic business simulation. The underlying motivation is to model industries and markets that can endogenously change their own structure and thus provide a tool that helps in dealing with complexity. This is achieved by treating individual players as agents and providing the means for addition and deletion of agents as well as exchange of information among them. Today, most industries are characterised by constant change; in addition, the trends towards enhanced competition, liberalisation and deregulation lead us to consider scenarios of an industry's dynamic taxonomies and the role that economic agents play in defining these changing structures.

We discuss our implementation and use the resulting platform to build a case study that proves the new methodology's technical feasibility and also addresses the issues of market structure in a highly dynamic industry characterised by an evolutionary economics model. Our concern is not to consider the states that will finally emerge, but rather to describe the original system, to ascertain how agents act and react to current circumstances and thus understand the actual process of market evolution.

1 Introduction

Contemporary enterprises inhabit a complex interconnected environment of constant change. Today's markets (consumer, investment, capital and knowledge) are highly dynamic and competitive and a firm's successful management needs to be able to adapt to the continuous change and learn from its environment. "The ability to learn faster than our competitors may be the only sustainable competitive advantage", noted de Geus (1988), emphasising the importance of institutional learning which is "the process whereby management teams change their shared mental models of their company, their markets and their competitors" (ibid.). The trend towards enhanced competition, liberalisation and deregulation leads us to consider industry structures with a multiplicity of interacting actors and objectives. In addition, firms need to consider and prepare for change that transforms in a fundamental way the economic-industrial system that they inhabit. This necessity highlights the importance of modelling tools than can play out a number of scenarios of management's view of the market, competition and evolution of the industry.

Modelling and simulation is widely accepted as an experimental science that offers the means to study large complex systems that evolve over time (Campbell et al. 1985, Casti 1997, Allen and McGlade 1987). Simulation forces the researcher "[...] to be precise about the relationship among entities, to make implicit assumptions explicit, and to describe in detail the mechanisms by which entities and relationships change" (Carley and Pietrula, 1994). Simulation is also increasingly used as a strategy exploration tool, as it addresses the shortcomings of a static analysis that most established frameworks offer (see figure 1).

	Five Forces (S-C-P)	*Core Competence* (RBV)	Game Theory
Assumptions	Stable industry structure	Firm as a bundle of competencies	Industry as a dynamic oligopoly
Goal	Defend position	Sustain advantage	Temporary advantage
Performance Driver	Industry structure	Unique firm competencies	Right 'strategies' (moves)
Strategy	Pick industry, a strategic position, fit organisation	Create a vision, build and exploit competencies	Make right competitive and cooperative moves
Success	Profits	Long term dominance	Short-term win

Figure 1: Models of strategy

Specifically, the most widely used framework for modelling and analysing competitive industry dynamics is Porter's (1980) Five Forces model. The structure-conduct-performance paradigm (Jacquemin, 1987) upon which Porter's five forces is based, assumes that each player adopts a single, well-defined and unchanging role. In addition, it only assumes that industry structure is fixed, only considers differences in average industry profitability and does not take into account rivals' choices as to whether to compete or not. The core competencies framework originating from the resource-based view of the firm, treats the firm as a bundle of resources. Firms try to build unique tangible and intangible resources that define their competencies and the long-term dominance of a firm is the result of exploiting these competencies. Finally, game theory models strategic decisions and offers

valuable insight on matters of competition and bargaining, but addresses the issue of dynamic situations through a collection of static equilibria. In addition, game theory assumes optimal behaviour and perfect rationality on the part of the players.
These traditional approaches to strategy assume clear industry structure and boundaries, economic agents with perfect rationality, a more-or-less 'predictable' competition and an easily foreseeable future.

However, actual human decision-making is often far from optimal and perfectly rational and researchers have argued about economic agents' *bounded rationality* (see Simon 1979 and Conlisk 1996 for an extensive review of the literature). Given the highly dynamic markets that firms inhabit today, the problem we are faced with is how to model and simulate markets and economic systems that exhibit a particularly complex behaviour. The aim is to model the evolution of market and its participants over time, to gain a deeper understanding of how the market under consideration works and finally learn how best to manage change.

2 Structural Change

Schumpeter (1934) was the first to recognise the dynamic relation between market structure and competition and we concentrate our efforts in using simulation to model and examine the dynamics of industries and markets over time. Various computer modelling and simulation techniques to represent the structure and dynamics of business and social systems have been proposed. *Strategic business simulation*, also referred to as *industry simulation*, allows the modelling of firms, their policies and interactions in the marketplace and is used for the development of business strategy models at the industry level (Morecroft and van der Heidjen 1994). Participants think about the existing market structure and the behaviour of the main players, a model based on the shared mental model of how the industry functions is developed and a number of scenarios are examined.

System Dynamics (SD) (Forrester 1961) and OO/DEVS (Ninios 1994) are two platforms used for strategic business simulation and their focus is primarily on the dynamic behaviour that results from a given market structure. More specifically, in both of these methodologies, it is the fixed structure of the social system under consideration that brings about the dynamic behaviour of the system. However, in exploring alternative policies and strategies in both SD and in OO/DEVS, one effectively undertakes sensitivity analysis by experimenting with the value of the parameters that define the system. Once the structure and policy rules have been defined, the model will describe the dynamic behaviour of the system, but system components and relations among components do not vary. The model might accurately capture the structure of a complex system, but this will be valid for as long as the current taxonomy holds. In addition, the model will not be able to account for, say, entry or exit of firms from an industry or change in a firm's or person's decision rules.

However, if we think of industries and organisations we often see that the structure of such social systems does not remain static, but changes over time. New firms appear, mergers occur, inefficient companies 'die out' or are overtaken by more efficient or aggressive competitors; thus one could argue that the original description of the system modelled is valid for as long as its taxonomy remains unchanged.

These fixed structure models cannot anticipate and cannot account for changes that might occur in the composition of the system, leading to incorrect causal relations and models that need to be reformulated. In addition, traditional, population-level simulation techniques assume that events occur at their average rate and that all individuals are identical and of the same, 'average' type. The argument works well for mechanical systems, but cannot be easily transferred to social ones where people are not very clear about their choices and are usually "boundedly rational", as defined by Simon (1951) and (1957). Aggregate (macro) behaviour cannot be thought of as simply being the sum of behaviours of 'average' individuals. In addition, the taxonomy of the system studied does not have to remain fixed, but should have the ability to evolve over time as well. The *complex systems* we will be studying can be defined as a set of simple, heterogeneous, interacting agents capable of exchanging information with their environment and capable of adapting their internal structure as a consequence of this interaction. The behaviour of the (economic) agents at the micro level could bring changes at the aggregate market level and, more importantly, new agents can, if appropriate, enter the market.

Our motivation is to model industries and markets that can endogenously change their own structure. This is achieved by treating individual players as agents and providing the means for entry and exit of economic agents in a system (say a marketplace) as well as exchange of information among them. Our concern is not to consider the states that will finally emerge, but rather to describe the original system, to ascertain how agents act and react to current circumstances and thus understand the actual process of market evolution and the role that economic agents play in defining these changing structures. We do not aim to provide exact forecasts, but rather to build exploratory simulation models that will help us examine the *possible* and not the probable. We aim to explore industries and markets where we have a multiplicity of actors that individually make assumptions about what will happen in the market and, as a result, bring about changes both at the individual (firm) and at the aggregate (market) level of observation. We thus aim to understand economic organization as a dynamic, adaptive process of interactions among individuals ('agents') in realistic and complex environments. This sort of models, as noted by Lesourne (1992) offer a useful way of looking at how market structure and organisation emerge.

We are thus facing the 'structural limitations' of traditional modelling and simulation techniques and to explore the issue of changing industry structure by considering how the economic agents' behaviour can alter the structure of a system, as shown in figure 2.

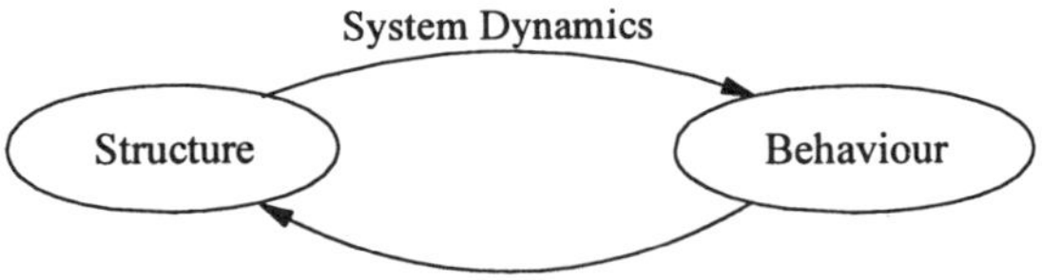

Figure 2: Closing the gap in strategic business simulation: from dynamic behaviour to dynamic structure

By doing this we can understand market dynamics at the population level by considering the actions and interactions of individual agents and look at dynamic models of endogenous mergers and acquisitions, investment, entry and exit using descriptive models based on decentralized agents whose collective interactions produce emergent behaviour.

3 VS/DEVS

We propose VS/DEVS, an object oriented framework that allows for changing model structure; in other words, a model could dynamically adapt its structure during a simulation run when necessary. One of the main issues is how knowledge about the system is structured and incorporated in a systematic way and what sort of deduction scheme is to be employed. The essential distinction between our approach and system dynamics (SD) is that SD and 'traditional' simulation is equation-driven, whereas our approach is rule- and knowledge-driven, which is essentially an alternative representation.

We model economic agent as objects (in object oriented programming terms) that interact with other agents through a series of messages. Each of them has a series of attributes and procedural methods with which it accomplishes things. Most economic models assume agent behaviour to be a procedural one; i.e., agents behave by using an algorithm of constrained optimisation, by calculating game theoretic best response strategies, or by some form of Bayesian update. In addition to this, our agents have a declarative representation of knowledge, i.e., a specification of the implications of a given set of conditions. Using procedural techniques we could determine the best course of action given a state of affairs, whereas declarative modelling will help us analyse the processes that emerge from an initial state. Most of our agents' behaviour dealing with which sort of action an agent should pursue will be declarative, i.e., rule based. Once an agent has chosen a certain mode of behaviour, this will be defined in a procedural way.

The agents of the system do not simply react to external messages and data, but given the 'knowledge' they have they can act as evolving, interrelated entities with the capability of changing their own environment. Our agents' actions depend not only on the messages they receive, but also on the declarative knowledge agents have about the outside world.

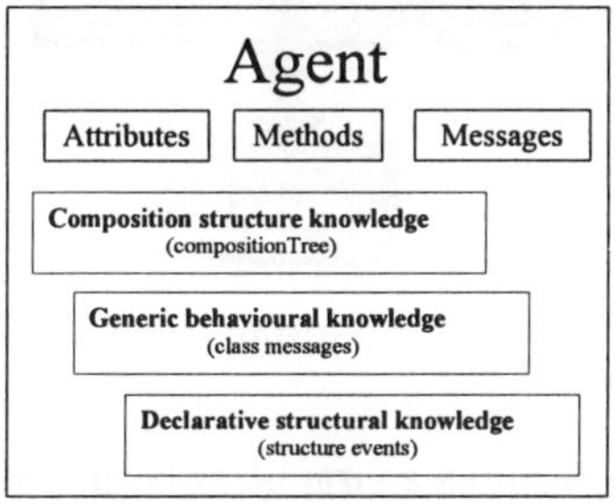

Figure 3: What an *agent* looks like

An agent's declarative knowledge consists of a series of *structure events,* a construct which is similar to Zeigler's (1990) *forward models.* These events affect both the composition of the entities and their behaviour and we can think of them as an agent's declarative structural knowledge. As our implementation for structural change allows a balance between control and autonomy, we also need to equip agents with generic behavioural knowledge that will spell out how new types of agents can be incorporated into the modelled system.

The declarative structural knowledge and the generic behavioural knowledge define an agent's *knowledge base.* Thus, the agents that we model are cognitive, autonomous ones that react to changes in their environment, each with knowledge about how to react to these changes and, hence, *intelligent.* Structural knowledge is defined as a set of rules (*if...then...*), whereas generic behavioural knowledge is defined as a set of messages which can also be thought of as a set of *if...then...* rules. As seen in figure 4, the 'blend' of intelligent agents and rule-based simulation gives us VS/DEVS, the technology we are looking for.

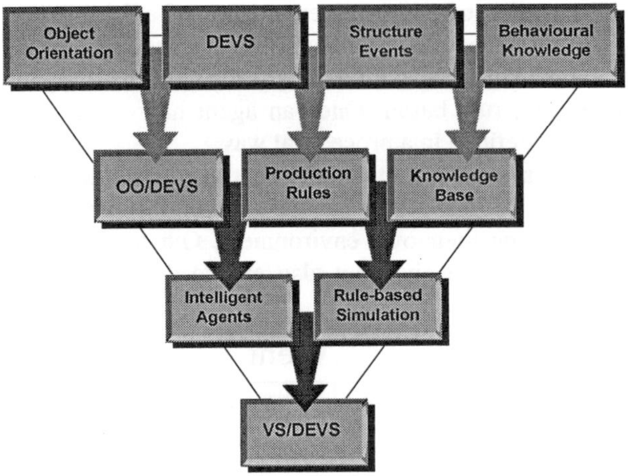

Figure 4: Roadmap to VS/DEVS

The technology also features a Graphical User Interface (GUI) that aids model creation by hiding away from the modeller a lot of the implementation details. Figure 5 shows how to define a structure event using the GUI.

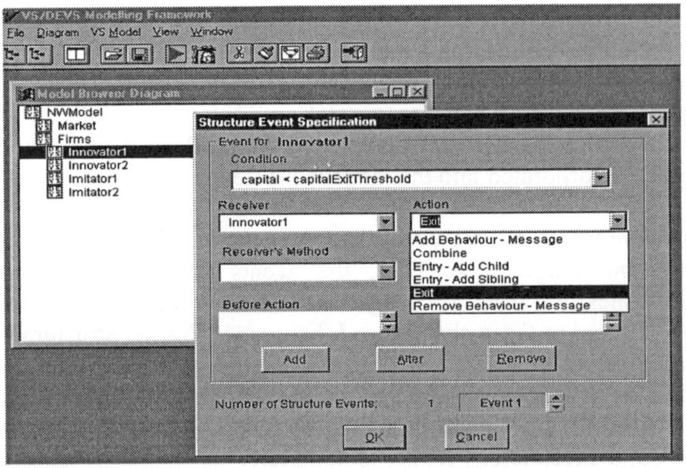

Figure 5: Using the GUI to define a structure event

4 Market dynamics in a fast-paced industry

In the last twenty years, industrial organisation economics (Tirole 1988, Scherer 1980) has produced a number of theories to explain competition and to account for a number of empirical regularities, namely profit rates, entry and exit rates, relation between age and size and growth and survival and, finally, firm size distribution. Originally, most of the work carried out was through a static framework for analysing economic organisation and primarily focused on game theoretic analysis of oligopoly competition and, more recently, on using repeated games to address the same issue. These analytical techniques, when applicable, allow observation of the dynamics only around the equilibrium point whereas simulation techniques allow the evolution of the system to be observed.

The feedback between firm size, innovation and market structure has caused econometric studies to face a number of problems, as testified by the wide range of inconclusive results on the subject reported by Kamien and Schwartz (1975) and Cohen and Levin (1989). As noted by the last authors,

> "A methodological problem common to almost all the studies of the relationship between size and innovation is that they overlook the effect of innovation on firm growth (and hence, ultimately, firm size). It is curious that the endogeneity of firm size, central to Schumpeter's notion of creative destruction has been neglected, while the simultaneity associated with creative destruction has been recognized in some studies of the relationship between innovation and market concentration"

A series of 'structural' models have been proposed that try to provide a description for all, or at least most, of the empirical regularities outlined earlier. In these models one finds an (almost) complete description of the firm's decisions and market processes. These models were introduced with the seminal work of Jovanovic (1982). In this model, there is a competitive industry with a continuum of price taking firms that are characterised by their efficiency □. Jovanovic's main assumtpion is that firms cannot observe their efficiency directly, but know how □ is distributed in the industry. Firms produce based on their efficiency estimate x^* and get to 'learn' their own efficiency through a productivity shock x_t which is a function of the firm's efficiency, i.e, $x_t = f(□+□_t)$. Exit in the industry occurs when a firm's estimated efficiency x_t^* falls below a threshold value $□_t(n)$ where n is the firm's age.

If one considers Jovanovic's model, one can see that there is nothing that a firm can do to improve its own efficiency. The firm will, in the long run, learn its efficiency through the series of productivity shocks that it receives from the market. In general, one can distinguish between passive and active learning models. In the former, see Jovanovic (1982) and Hopenhayn (1992), firms decide whether to enter or exit a market and on an output level. Firms do not engage in any activity that can improve their efficiency/productivity, but either 'learn' their true efficiency through the market (Jovanovic 1982) or draw a Markov expectation about future efficiency based on today's observed productivity shock (Hopenhayn 1992).

In contrast to the above, in active learning models the level of a firm's productivity (efficiency) is the result of the firm's investment in R&D. In these models, firms decide whether to remain in the industry or not and if they do they try to increase their productivity. This is achieved by deciding how much to invest in R&D (Ericson and Pakes 1995). Alternatively, Nelson and Winter (1982) employ the paradigm of biological evolution to describe market dynamics where firms incur costs associated with imitative activities that copy, if succesful, the best available productivity in a market or by undertaking innovative activities that aim to increase a firm's productivity. Both these approaches make market shares endogenous to the process of innovation and thus differ significantly from the structure-conduct-performance paradigm, which is the 'traditional' industrial organisation framework. In the next section we explore an extension to a Nelson and Winter model that accounts for firm entry and exit. Our concern is, primarily, to show the platform's technical feasibility and how a relatively simple model can produce a host of interesting results.

4.1 A Nelson and Winter extension

In their seminal work on evolutionary economics, Nelson and Winter (1982) describe a model of technological change as competition between firms that are innovators or imitators. We will concentrate on the model of "Schumpeterian competition" which deals with the evolution of the production techniques and other behavioural rules of a stylised industry which produces a homogeneous product.

A simple theoretical description of the model is as follows: we have a set of n firms, all of which are explicitly created and at time t firm i is characterised by a capital stock K_{it} and productivity of capital A_{it}. In contrast to Jovanovic's (1982) model we have a discrete number and not a continuum of firms. Each of these firms decides on a production function by simple capacity utilisation rules, i.e., full capacity utilisation. The output of the industry is found by simple aggregation and an exogenously defined demand curve determines the market price per unit of output which, in turn, determines turnover and net profit for each firm.

The next step in the simulation is that firms undertake imitative and/or innovative activities in order to improve their capital productivity. Technology is assumed to improve with time and if an innovating firm is succesful in its R&D activities, it 'gains' this new technology and its efficiency, and therefore output, increase. Not all firms undertake such innovative activities, but they all engage in imitative searches that can provide them with the best possible technology at the time. Imitative activities cost less than innovative ones, but the rewards of the latter can be more significant, in that they provide an advantage, even if temporary, to a firm. Once firms have determined their productivity, they then proceed to determine the amount of capital they will have available in the next period by considering their desired investment and the financial constraints they have. This is a simple model that runs on discrete time and at each period every firm is in search of a better production technique (in terms of capital productivity). Note that this is model determines *probabilistically* what happens in each period. Figure 6 (adapted from Andersen 1994) shows schematically the process described above.

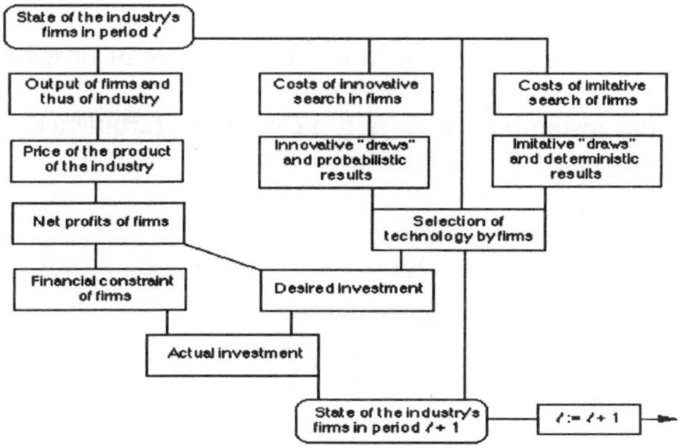

Figure 6: A depiction of the Schumpeterian competition

The extension this exercise considers is to model the entry and exit of firms in the marketplace, a feature not present at Nelson and Winter's original work. The aim is again to consider the evolution of market shares that will prevail in this simple industry in order to gain a better understanding of industry concentration and market share instability in a dynamic market. Besides using the same production and innovation/imitation methods described in the original, we have extended the original work by modelling entry and exit and these processes have been implemented using a set of *structure events*.

More specifically,

1. *Entry* in the market occurs when average profits per unit capital exceed some threshold value x. We assume that there are no barriers to entry and thus a firm immediately enters the market. In our model, there are no entry costs for the new firm and we assume that there is a pool of potential entrants that never dries out. The new firm is similar to existing ones (as they all produce a homogeneous good), with same production, innovation and imitation rules. The values of attributes of the new firm are initalised to what all firms had at the beginning of the simulation and what is determined randomly is whether the new form will be an innovator or imitator, something that will in turn determine its performance over time.
2. The condition for *exit* is met when firm capital falls below some threshold value x. When this occurs, the firm exits the market by deleting itself.

As our concern is to study industry evolution, besides looking at the Herfindahl index, a further issue to explore is market instability. This can be measured through the *Instability index* which was introduced by Hymer and Pashigan (1962) which is defined as $\sum_{i=1}^{n} \left| s_{i,t} - s_{i,t-1} \right|$ where $s_{i,t}$ is the market share of firm i at time t in industry j. Essentially, the instability index is a measure of 'turbulence' in the market that brings out the irregularities and fluctuations in market shares.

4.2 Sample results

In running the model we considered three variants; one with no entry or exit (the 'standard' Nelson and Winter model (Model A), one with entry only (Model B) and one with both entry and exit (Model C). In all models, we started off with 4 firms, two innovators and two imitators (see figure 7).

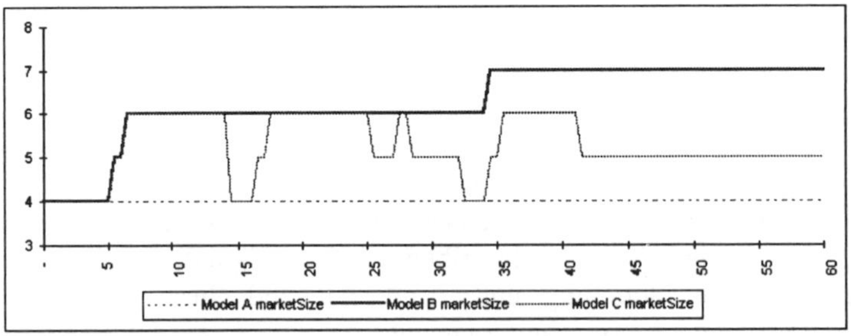

Figure 7: Number of firms in models A, B, C

The arrival in the market of new firms suppresses the average market price which slowly recovers (see figure 8). The same also applies for profits; the higher the entry rate, the further do profits fall.

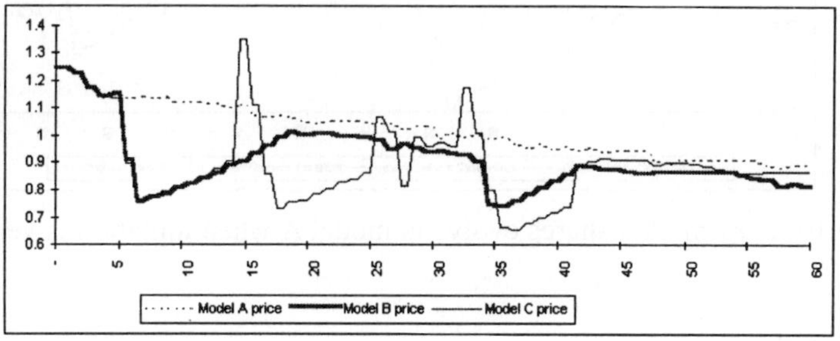

Figure 8: Price evolution in models A, B, C

For model A, we also considered a number of scenarios, by varying the degree of how difficult it is to imitate, and obtained a series of interesting results. The more difficult the imitation, the easier it is for innovating firms to maintain their lead and stay ahead of the game (figures 9 and 10).

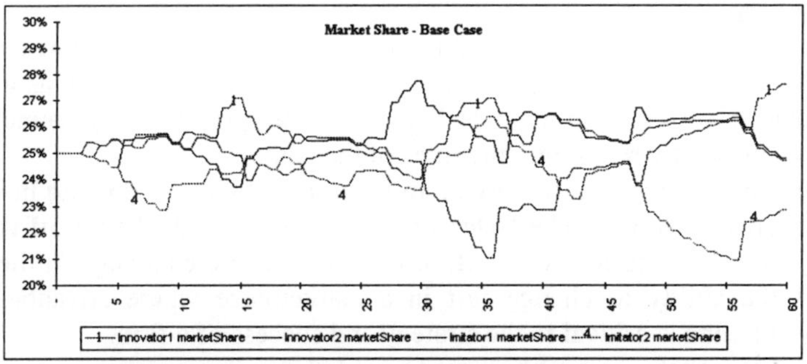

Figure 9: How market shares evolve in model A

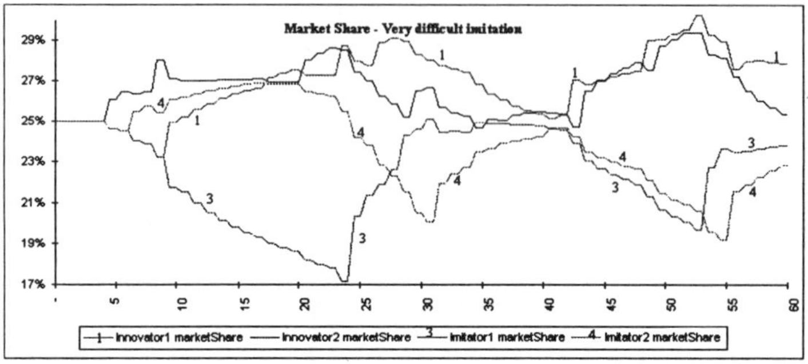

Figure 10: How market shares evolve in model A when imitation is very difficult.

4.3 Conclusion

Our model has successfully shown the technical feasibility of the technology we have developed. We extended the original model of Nelson and Winter by introducing a couple of structure events that bring out the system's rich trajectory. In addition, our model describes a dynamic, fast moving market where there are no real barriers to entry other than regulation and whose environment is too uncertain to plan effectively. The overall system behaviour is dictated by a number of few simple rules that bring out a rich system trajectory, including market entry and exit. An example of such a turbulent market might be some sectors of the telecommunications industry (say, International Simple Resale) or the computer software industry.

Even though we have talked of innovation as a means of improving capital productivity, we could think of innovation as a way of getting ahead, as a means of leading change by creating new technologies and products, by launching new markets, by constantly redefining the rules of the game.

The varying degree of difficulty in imitating might have to do with how original a certain improvement was. The faster a company changes, the better off would it be in forcing other firms to follow it. All in all, the strategic challenge in these simple markets is the ability to change fast in a marketplace where advantage is only temporary and time is critical to the strategy and success of a firm.

Firms that imitate, or simply react to market developments, may perform well but they are always following the leaders. Once their competitors position gets better, they will try to improve too and launch a better product, or repackage what they already offer. Imitation is a defensive technique and is unlikely to create new opportunities in a fast changing world.

5 Discussion

We now have a tool for building models that can dynamically change their structure and can thus study a range of interesting problems. We start by describing how a system works (main agents and their interactions) and introduce a number of structure events to build an exploratory model. This methodology can therefore be used as a learning tool, where management teams can model and test their assumptions and explore some of the possible futures. It is not meant to provide exact forecasts or make any statements about the probability of different outcomes.

Besides the models we can build, there are three issues that need to be addressed: first, how do changes come about? What conditions and/or relationships are capable of creating structural change? Secondly, once the environment has changed, how do individuals and institutions respond to this change? When an industry changes, firms ("individuals") will try to adapt by changing their internal structure. Other firms, though, might choose to try and influence their environment so they do not have to change their internal structure. The final issue is how firms can make better and quicker use of change in their environment. Structural simulation can help us generate scenarios, where processes of structural change are produced by actors operating on the same systems, but with differing information, goals, and learning abilities.

6 Acknowledgments

In carrying out his research, the author was supported by a grant from British Telecommunications plc.

References

1. P.M. Allen and J. M. McGlade, Modelling complex human systems: a fisheries example, *European Journal of Operational Research*, **30**, (1987), 147-167
2. D. Campbell, J. Crutchfield, J. D. Farmer and E. Jen, Experimental mathematics: The role of computation in nonlinear science, *Communications of the ACM*, **28**, (1985), 374-384
3. K. M. Carley and M. J. Prietula, *Computational Organization Theory*, Hillsdale, New Jersey, Lawrence Elbaum Associates, (1994).
4. J. Casti, Would-be worlds: how simulation is changing the frontiers of science, New York, John Wiley & Sons, (1997)
5. W. M. Cohen and R. C. Levin, *Empirical studies of innovation and market structure.* In R. Schmalansee and R. Willig (eds) *Handbook of Industrial Organization*,Amsterdam,North Holland, **2**, (1989), 1059-1107

6. J. Conlisk, Why bounded rationality?, *Journal of Economic Literature*, **34**, (1996), 669-700
7. R. Ericson and A. Pakes, Markov-Perfect Industry Dynamics: A Framework for Empirical Work, *Review of Economic Studies*, **62**, (1995), 53-82
8. J. W. Forrester, *Industrial Dynamics*, Cambridge, MA, MIT Press, (1961)
9. A. de Geus, Planning as learning, *Harvard Business Review*, **66**(2), (1988), 70-74
10. H. A. Hopenhayn, Entry, Exit, and Firm Dynamics in Long Run Equilibrium, *Econometrica*, **60**, (1992), 5, 1127-1150
11. S. Hymer and P. Pashigan, Turnover of firms as a measure of market behavior, *Review of Economics and Statistics*, **44**, (1962), 82-87
12. A. Jacquemin, *The new industrial organization: Market forces and strategic behavior*, Cambridge, MA,The MIT Press, (1987)
13. B. Jovanovic, Selection and the Evolution of Industry, *Econometrica*, **50**, (1982), 3, 649-670
14. M. Kamien and N. Schwartz, Market structure and innovation: a survey, *Journal of Economic Literature*, **13**, (1975), 1-37
15. J. Lesourne, *The economics of order and disorder: The market as organizer and creator*, Oxford, Oxford University Press, (1992)
16. R. Nelson and S. Winter, *An evolutionary theory of economic change*, Cambridge, MA, Harvard University Press, (1982)
17. P. Ninios, *An Object Oriented-DEVS framework for strategic management and industry simulation*, PhD Thesis, London Business School, London, University of London, (1994)
18. M. Porter, *Competitive Strategy: Techniques for analyzing industries and competitors*, New York, Free Press, (1980)
19. F. M. Scherer, *Industrial Market Structure and Economic Performance*, Chicago, IL, Rand McNally, (1980)
20. H. Simon, A formal theory of the employment relation, *Econometrica*, **19**, (1951), 293-305
21. H. Simon, *Administrative behavior*, New York, NY,Free Press, (1957)
22. H. Simon, Rational decision making in business organizations, *American Economic Review*, **69**(4), (1979), 493-513
23. J. Tirole, *The Theory of Industrial Organization*, Cambridge, MA, The MIT Press, (1988)
24. B. P. Zeigler, *Object-Oriented Simulation with hierarchical, modular models-Intelligent Agents and Endomorphic Systems*, London, Academic Press, (1990)

MANAGERIAL LEADERSHIP AND LEARNING ORGANISATION

EVA MOLOCZNIK

MARIA CURIE-SKLODOWSKA UNIVERSITY, Department of Economics, Plac Marii Curie Sklodowskiej 5, 20-031 Lublin, Poland

E-mail: smol@zamoy.2lo.lublin.pl

Great attention paid to increased adaptability leads to new ways of looking at leadership as creative tension and giving up the traditional model of leaders. New definition of leadership implies the need of developing new leaders. In a learning organisation leaders' roles differ from that of traditional decision makers. These are the roles of designer, teacher and steward. To perform all these tasks effectively leaders must acquire numerous specific skills within five critical areas: building shared vision, testing mental models, systems thinking, team dialogue and personal growth. Recently we can observe the return of the emphasis on the significance of developing leaders who can develop organisations. There are some specific features of a learning company, which can be used to identify such organisations as well as "to serve as a guiding star" in the process of becoming a learning company. The purpose of such transformation cannot be effected without highly skilled leaders. Here management learning can be presumed a perfect connection between the two processes of learning - of the organisation and of its leaders. There a role of a manager is not to be a sole actor, but to work with and through others on common objectives. It is assumed that managers will learn only under very restricted conditions which is an important fact for trainers. But a serious problem is managers' resistance to learning. That's where a new challenge emerges - to discover practical solutions, training applications which will make it possible to manage resistance effectively and to help managers internalise the optimal leadership concept. One possible way to help managers become successful leaders can be NEURO LINGUISTIC PROGRAMMING. A promising solution can be accepting NLP training as a part of firm's human resources management strategy and incorporating NLP techniques and attitudes as a pattern in organisational culture. NLP produces numerous advantages and above all, has the power of erasing resistance. It turns "I must" into "I want".

1 New work of a visionary leader

The majority of concepts which concern learning organisations put emphasis on increased adaptability. The impulse to learn goes deeper than desires to respond and adapt more effectively to environmental change. Adaptiveness is only the first stage which leads to focusing on generative learning. This, unlike adaptive learning requires new ways of looking at world, understanding customers and managing business based on creation, not just coping.

The traditional model of leaders has as its origin in individualistic and non-systematic world view.

1.1 Leadership as creative tension

Leadership in a learning organisation starts with the principle of creative tension. It comes from seeing clearly where we want to be, our "vision," and telling the truth about where we are, our "current reality." The gap between the two generates a natural tension [7].

Creative tension can be resolved in two basic ways:

- by raising current reality toward the vision,
- by lowering the vision toward current reality.

Individuals, groups, and organisations who learn how to work with creative tension learn how to use the energy it generates to move reality more effectively toward their visions.

There are certain aspects of the principle of creative tension that have to be recognised by leaders. Without vision there is no creative tension. But creative tension cannot be generated from vision alone. The principle teaches that an accurate picture of current reality is just as important as a picture of a desired future.

There are some significant distinctions between leading through creative tension and solving problems. In problem solving, the energy for change comes from attempting to escape from an undesirable aspect of current reality. With creative tension, the energy for change comes from the vision, from what we aim to create. Many people and organisations find themselves motivated to change only when their problems are bad enough to cause them to change. With problem solving, the motivation for change is extrinsic. With creative tension, it is intrinsic.

2 Developing new leaders

In a learning organisation, leaders' roles differ from that of traditional decision makers. These roles requires new skills.

2.1 *New roles*

2.1.1 Leader as designer

The functions of design (often called "social architecture") are rarely visible. The consequences that appear today are the result of work done long in the past, and work today will show its benefits far in the future [7].

Hanover O'Brien says: "The first task of organisation design concerns designing the governing ideas of purpose, vision, and core values by which people will live. Few acts of leadership have a more enduring impact on an organisation than building a foundation of purpose and *core* values".

The second design task concerns the policies, strategies, and structures that translate guiding ideas into business decisions. Leadership theorist Philip Selznick calls policy and structure the "institutional embodiment of purpose". Jay Forrester adds: "Policy making (the rules that guide decisions) ought to be separated from decision making" [7].

The traditional view that policy making and implementation should be the work of a small number of senior managers is changing. The dynamic business environment and the power of the learning organisation to engage people at all levels now make it clear that this second design task is more subtle. According to Mintzberg successful organisations "craft strategy" as they continually learn about unstable business conditions and balance what is desired against what is possible. The key is not getting the right strategy but fostering strategic thinking. "More important is the need to achieve insight into the nature of the complexity and to formulate concepts for coping with it" [7].

Behind appropriate policies, strategies, and structures are effective learning processes; their creation is the third key design assignment in learning organisations.

2.1.2 Leader as teacher

Leader as teacher does not mean leader as authoritarian expert whose job it is to teach people the "correct" view of reality. It is rather helping every member of the organisation, oneself included, to obtain more clear and insightful views of current reality. This corresponds with a emerging concepts of leaders as coaches, guides, or facilitators.

The role of leader as teacher starts with bringing to the surface people's mental models of important issues[7]. These mental pictures of the world have an important influence on how we perceive threats and opportunities and make decisions.

But tasks of a leader- teacher mean more than just revealing hidden assumptions of their subordinates. "Reality" brings stressful situations that must be dealt with, crises that must be reacted to. Leaders as teachers help people restructure their views of reality to see beyond the superficial conditions into the underlying causes of problems - and therefore to see new possibilities for shaping the future.

Specifically, leaders can influence people to view reality at three distinct levels: events, patterns of behaviour and systemic structure. Contemporary society focuses mainly on events. Pattern-of-behaviour explanations are rarer than event explanations, but they do occur. "Trend analysis" is an example of seeing patterns of behaviour. Structural explanations are the most powerful. They operate at a level ensuring the possibility of causing changes in patterns of behaviour.

Currently, most of leaders focus on events and patterns of behaviour, and, under their influence, their organisations are reactive, or at best responsive - rarely generative. Leaders in learning organisations pay attention to all three levels, but focus especially on systemic structure [7].

2.1.3 Leader as steward

This is the subtlest role of leadership rooted in an attitude critical to learning organisations.

It begins with the natural desire to serve. This conscious choice brings one to aspire to lead. Leaders' sense of stewardship operates on two levels: stewardship for the people they lead and stewardship for the larger purpose or mission of the whole enterprise. The first type has its source in a intense appreciation of the impact one's leadership can have on others. The second type of stewardship arises from a leader's sense of personal purpose and internalisation of the organisation's larger mission. People's impulse to learn is unleashed when they are involved in an effort they consider worthy of their fullest commitment.

2.2 New skills

They must be spread widely throughout the organisation, not just developed by a few individuals. This is one reason that understanding the disciplines of a learning organisation is vital.

Five critical areas of skills (disciplines) are:

2.2.1 Building shared vision - developing a sense of commitment in a group by inventing shared images of the future we desire to create as well as the principles and guiding rules by which we hope to get there [3].

Such procedure includes:

- *Encouraging personal vision.* They are a significant source of shared vision.
- *Communicating and appealing for support.* Leader must be constantly sharing his vision, rather than being the official representative of the corporate vision.

- *Building vision as a continuous process.* It is a never-ending process. At any point there will be a particular image of the future that is predominant, but that image will evolve [7].
- *Mixing extrinsic and intrinsic vision.* Extrinsic visions focus on achieving something relative to an outsider, such as a competitor. But once such vision is achieved, it becomes defensive. To the contrary, intrinsic goals like creating a new product, can call for a new level of innovation. Intrinsic and extrinsic visions need to exist together.
- *Distinguishing positive from negative visions.* Two fundamental sources of energy can motivate organisations: fear and aspiration. Fear, the energy source behind negative visions, can produce extraordinary changes in short periods, but aspiration endures as a continuing source of learning and growth [7].

2.2.2 Testing mental models - deciphering, continually clarifying and improving our internal pictures of the world and observing how they shape our external actions and decisions:

- *Identifying leaps of generalisation.* We tend to confuse our generalisations with the observable data upon which they are based, treating them as if they were data [7].
- *Recognising and removing defensive routines.* According to Chris Argyris "defensive routines" are certain habits that we use to protect ourselves from the embarrassment and threat which come with exposing our thinking. They aggravate all the hardships with revealing hidden mental models and thereby lessen learning. The first challenge is to recognise defensive routines, then to reduce them.
- *Balancing inquiry and advocacy.* Most managers are skilled at presenting their views coherently and persuasively. Advocacy skills can become counterproductive as managers become burdened with greater responsibility and face increasingly complex issues that require collaboration and learning together with different, equally knowledgeable and skilled people.
- *Distinguishing endorsed theory from theory in use.* Recognising gaps between certain views that we declare to hold (e.g. I say that people are trustworthy) and theory in use - deeper views revealed in actions (but I never lend things even to my friends) is all important to deeper learning.

2.2.3 Systems thinking - considering, describing and understanding the interrelationships which shape the behaviour and organisational systems in a way that helps everybody see how to change the systems more effectively.

We all know that leaders should help people see the big perspective. They concentrate less on everyday events and more on underlying long-term trends and forces of change. But they do this almost completely intuitively. Consequently, they are often unable to explain their intuitions to others and feel frustrated when others perceive reality in different way.

Managerial systems thinking as a field of study and practice suggests some key skills for future leaders:

- *Seeing interrelationships, not things, and processes, not snapshots.* Most of us have a tendency to zero in on things and to see the world in static images. This leads us to linear explanations of problems.
- *Refraining from symptomatic and stereotypical solutions.* The pressures to intervene in badly working management systems and the linear thinking that dominating in most organisations cause that we tend to focus on symptomatic and the promptest solutions. It is difficult to avoid using most popular and quick fixes and to force everyone to look for more enduring solutions.
- *Moving beyond blame.* We tend to blame each other or outside circumstances for our problems. But in reality it is badly designed systems or incompetent individuals that cause most problems. Systems thinking shows us that there is no outside - that you and the cause of your problems are part of a single system [7].
- *Concentrating on high leverage areas.* Sometimes small, but well-focused actions can produce significant improvements, if they are in the right place(as the principle of "leverage" says). Handling a difficult problem is often a matter of recognising where the high leverage lies, where a change - with a minimum of effort - would bring considerable effects. The most powerful levers for change are: HRM strategy, organisation, culture management, change management, managing commitment, flexibility, teamworking, quality management, resourcing, human resource development, performance management, reward management, employee relations [1].
- *Distinguishing detail complexity from dynamic complexity.* Some types of complexity are more important strategically than others. Detail complexity arises when there are many variables. Dynamic complexity arises when cause and effect are distant in time and space, and when the consequences over time of interventions are subtle and not obvious to many participants in the system. The leverage in most management situations lies in understanding dynamic complexity, not detail complexity [7].

2.2.4 Team dialogue - transforming individual and collective skills so as to enable a group of people to develop abilities greater than the sum of sole members' talents (synergetic effect).

2.2.5 Personal growth - learning to expand personal capabilities in order to create the most desired effects as well as creating an organisational environment which encourages all its members to achieve goals chosen by them.

3 Developing Learning Organisations - The significance of developing leaders in this process

Currently we can observe the return of emphasis on developing leaders who can develop organisations. Top management is criticised not to be qualified enough to develop culture. Learning organisations represent a potentially significant form of organisational culture. Not surprisingly, such organisations will remain a distant vision until the leadership capabilities they demand are developed. Organisation development and management development are thought to be helpful here under condition that they focus on the roles and skills of leader in learning organisation.
In order to understand fully the significance of developing leaders I should point out the most relevant aspects of learning organisation from the point of view of its leaders.

3.1 What is learning organisation?

It is an organisation which facilitates the learning of all its members and continuously transforms itself - by Pedler,Boydell,Burgoyne (1988).Such a company does not simply engage in a lot of training. The need for the development of individual skills (including leadership skills) is just as important as the necessity of organisational learning.

This is an organisation which means far more than just providing goods and services. It creates the environment in which people spend most of their lives and subsequently, has major impact on their behaviour - by Gibson, Ivancevich, Donelly (1988).

Each organisation manifests its identity through its unique culture, which evolves due to experience and influence. Organisation must be capable of learning defined as "a purposeful activity aimed at the acquisition and development of skills and knowledge and their application. An organisation's skills are found in its accepted behaviour patterns and its collective knowledge in its shared assumptions. The organisational attitude exists in the core values. If these can be developed within an individual, there is no reason why this cannot happen for an organisation. It is more difficult and will take longer to achieve, but these factors do not reduce the possibility of organisational learning" [4].

3.2 What are the features of a learning organisation?

There are some features that can be used to distinguish between a company which is not learning and one which is.
According to Pedler a learning organisation is one which:

- has a climate in which individual members are encouraged to learn and to develop their full potential;
- extends this learning culture to include customers, suppliers and other significant stakeholders;
- makes human resource development strategy central to business policy;
- is in a continuous process of organisational transformation.

The purpose of this process of transformation is to enable the company to search for new ideas, new problems and new opportunities for learning in the environment and in the organisation itself. This "hunting for advantages" in our competitive reality cannot be effected without highly skilled leaders.

3.3 What is it like to be a member of a learning organisation?

Being a part of a learning organisation can be uncomfortable, depending on the individual's views of the world. For entities who find learning and development exciting and look for change and growth, the prospect of continuous learning is very attractive and thus involvement in a learning company is desirable. To those who prefer routine, the idea of change and challenge can be unpleasant and discouraging.

Moreover, there is an incorrect belief among those who are committed to learning and development that this will be beneficial to all. Some people prefer to go to work and do the same job for their whole working lives. They have the right to think that way. Just as people, some organisations prefer stability in their cultures. However, all leaders must do all that is possible to eliminate the tendency to stagnation, which can be detected in statements such as: "We can't do that, we have never done it before", which lead to a systemic inability to change. Therefore it will become even more crucial for organisations to learn how to learn.
There are some characteristic features of membership in learning organisation.

3.3.1 Living with change

Being a part of a learning organisation means living in a state of constant questioning and change. In terms of organisations, this level of change requires people to act and to look at the world in different ways. Furthermore, our evolving environment brings high degrees of uncertainty. Staff may know what the vision statement says because its meaning has been shared. But "what it means to me" may be unclear before the implementation of vision.

All the same, because the situation is evolving and the people and organisation are learning, the plans too will change. The vision, being a long-term dream, will be transformed. As some parts become clearer and move more sharply, into focus, other parts will fade. The vision may not change in essence but how it is seen may. It will be seen by different people from different perspectives at different times and be interpreted in different ways [4].

Since the process of evolution may have contradictory meanings for various people, managers and their subordinates are required to develop special skills if they are to pursue to function effectively. Without such skills people can undergo untenable levels of stress, which create the threat of dysfunctional behaviour or even breakdown [6]. A learning organisation takes pains to help its members develop management skills in ambiguous and uncertain conditions.

3.3.2 Internal harmony

The changing nature of organisation brings into question its internal harmony and balance. Encouraging questions undermines the mutual view and since there are no experts there is no right or wrong. Unless any extreme situation appears, compromise is possible and even healthy conflict is considered to be a favourable mean of working out new, better solution to newly discovered problems.

3.3.3 Managing diversity

Its essence is equalling opportunities. "The new term imposes the need of facing up to contradictions and differences and finding ways of letting them exist together, each with equal validity and significance. This term also expresses the right of every individual to make his contribution"[4]. Managing diversity reveals conflict, which always used to be hidden and not talked about openly. Therefore, managers have to give up all these traditional control procedures.

4 Management learning - a perfect connection between the two processes of learning: of the organisation and of its leaders ?

WHAT CAN A SOLE MANAGER DO? - it is a question asked very often and inspired by the increasing failures of established procedures in contemporary conditions of greater uncertainty. Managers find it more and more difficult to operate in such environment. That is why so much attention is paid to the development of individual managers' personal skills and competence. Thanks to them leaders are to be able to help their subordinates and themselves to be a fully integrated parts of a learning organisation making their contributions to its successful development. This is what management learning has focused on.

Leaders can play a fundamental role in the creation of learning organisation. They have the power of helping people think and act together on "an idea or metaphor that can serve as a guiding star" (Megginson, Pedler). This is nothing more than a vision, which can enable the creation of the conditions in which a learning organisation can be brought about:

- learning strategy;
- participative policy-making;
- informating (i.e. information technology is used to inform and empower people to ask questions and take decisions based on available data);
- formative accounting (i.e. control systems are structured to assist learning from decisions);
- internal exchange;
- reward flexibility;
- enabling structures;
- front-line workers as environmental scanners;
- inter-company learning;
- learning climate;
- self - development for all[4].

But there is a certain obstacle which has to be overcome first. The popular image of a manager is of a person in charge, working alone and burdened with tasks of great complexity and responsibility. Success is possible and can be achieved by long hours and hard work. This picture is certainly false. The real role of a manager is to work with and through others - a group of companions, friends zeroed-in on common objectives that are far too big for one person to achieve alone. This is done by team dialogue (one of a leader's crucial skill areas): the use of appropriate techniques and by treating staff as respected individuals whose contribution is invaluable.

By means of sharing the task team work lessens strain on the individual, improves the quality of the process of work and the outcome achieved. Involvement of appropriate others in appropriate ways can have a number of predictable outcomes:

- Group working avoids tram-line thinking.
- Individual assumptions are questioned by concerned others.
- Debate stimulates the creation of new ideas.
- Group members help each other stand back from the problem to reframe it.
- The available experience, knowledge, skills and brain capacity is increased.
- The vision or purpose can be restated and revalidated as a result of group discussion, challenge and debate.
- Group members can mentor and coach each other, and providing "moral" support and encouragement.
- People are essentially social creatures and generally function better when working with others in conducive settings[4].

Thus, a group working on a common problem can result in better solutions than one person working alone. In the learning company, managers are not sole individuals, but "members of a professional community of co-learners, who help each other and provide support when the going gets tough"[4].

The very interesting technique is the Chinese concept of Wu-wei - the art of non-action, described by Vaill 1n 1989. Too many managers follow "I'm in charge" model. Wu-wei suggests stepping back, observing what is happening, collecting ideas from others, considering options, thinking about the situation and, only then, deciding whether act or not and next-how to act.

4.1 Management Learning

This concept is fairly fresh and popular one. Many managers are seriously undertrained and educated and have a lot of new things to learn. There are various lists of what must be learnt: responsiveness, innovation, participation of all employees, leadership for change, teamwork, quality and flexibility. These behaviours are seen as required by significant changes in organisational environments: complexity, growing competition.

All these necessary behaviours have to be learnt. Managers must be able and willing to focus more on strategic and not operational, modify their management styles and team roles. They must learn to manage in flexible ways encouraging participation. As the message is: learn, this intense in management learning is constantly growing.

4.2 How and why managers learn?

Originally, great interest has been directed to the probable assumptions that managers may resist learning, or may learn in peculiar ways. A relevant statement is that experience of managing in different sorts of organisational environments might shape managers' learning styles. This suggestion has obvious practical importance for trainers.

Trainers, consultants and lecturers have practical reasons for being concerned with how managers learn. The significant conclusion is that managers learn best through doing, through experience, as often as possible. When the subject is highly theoretical it must be supported with numerous examples, case - studies or project work. Most of all, it is presumed that managers will only learn when the subject has obvious and immediate practical application: the words "theory", "academic", "model", are usually perceived as negative [2].

It suggests managers are to be trained and rewarded to learn only under very restricted conditions: lots of variety, professional and immediate presentations in form of short sessions, lots of activities and exercises, obvious relevance to learners' own experience, with maximum of practical techniques and minimum of theory elements.

In some sense this could be seen as active (as it involves doing things) and practical (because of examples and case studies). But it can also be considered passive as a learner does not take pains to discover by him/herself the application and practicability of general knowledge to personal situations or issues. This is done for them. Presumably, as university students managers learnt using conventional academic methods. But once they become managers, they reject them and learn new ways to learn.

"Probably this orientation towards learning was a result of the location of the manager within hierarchic structures, and within cultures which explicitly rewarded certain forms of behaviour (and discouraged, even penalised others)"[2]. The success of training and development programmes could well be confined by the impact of the value/reward structures. Subsequently, not only do management structures define how managers learn, but they may also establish managers' resistance to learning.

4.3 Resistance to learning

Such resistance may originate from various sources: the organisation's structure and culture, from pathologies of teamwork, individuals themselves. At the organisational level, the way in which power is exercised and the behaviours that are rewarded and penalised are the most resistance-creative factors.

The necessity for drastic change in the organisation requires radical alterations towards new behaviour on the part of managers. When they are to be achieved is via training courses, an immediate paradox occurs: "The course is only necessary because the required behaviours are not a normal outcome of existing practice and procedures, otherwise there would be no need to develop them in courses. Yet if they are not a normal part of organisational procedure, managers may sense a dissonance between course content and everyday organisational life, and be cynical about course exhortation"[2].

But, what must be emphasised is that managers learn all the time even when they are not on courses - especially when they aren't on courses. If they hadn't learnt so much away from training, they wouldn't be so reluctant to learn on courses. They learn how to survive and behave within their organisation. This experience may be in opposition to formal learning. "The problem thus is not that managers won't learn, or that they resist learning but that they have learnt too much and too well. These lessons about how their organisation works may obstruct their openness to further learning"[2].

Moreover, the willingness to learn among managers existing within structures of power, reward and evaluation depends on whether they can see that what they are learning will help them do what they are rewarded for and will be valuable within their organisation.

5 Conclusions - in search of practical solutions

The purpose of this paper is to indicate the main ideas of the new concept of leadership but first of all, to draw attention to issues it can generate. These are the real issues which can be observed in numerous companies. Crisis of leadership or even misunderstanding its factual nature is likely to unable such firms to stand the competition of the cleverer enterprises. The example here can often be inexperienced domestic companies in post-communist countries vs. multinational corporations entering those markets.

That's where a new challenge emerges - to discover practical solutions, training applications which will make it possible to manage resistance effectively and to help managers internalise the optimal leadership concept.

5.1 Neuro Linguistic programming as a new tool

One possible way to help managers become successful leaders can be NEURO LINGUISTIC PROGRAMMING. NLP is programming of your neural system and language in order to codify and amplify your perfection. As based on Peter Senge's concept of personal mastery and on John Grinder's and Richard Bandler's analysis of biographies of famous, successful people, NLP supplies the way to raise yourself to the level of perfection in everything you do think and say[4].
The basic rules of NLP are:

- NLP starts with you - if you are able to manage yourself, you can also manage others [3];
- Sub-consciousness has greater power than consciousness and it is better to operate on this level;
- success is generated from the harmony of who you are, what you believe in, what you do and say.

5.1.1 NLP techniques

NLP brings up many techniques useful in different areas of management [5]:

1. Absorbing attitudes which lead to success - your views on yourself and others that shape your behaviour and decisions you make. NLP helps switch to such attitudes:

- Every person is exceptional - it helps understand differences and imagine that in certain situation everyone makes the best choice possible. This moves you from prosecuting to understanding.
- There is no failure, there is reaction – this facilitates learning and prevent resignation.
- In every action there is a positive motive - if your chief is shouting at you for your incompetence it doesn't mean he wants to embarrass you, but to guide you in order to avoid further incompetence.
- The value of communication is shown by its effects - if the reaction is not what you expect, you have to change the way of communicating.
- Every problem has its solution - this opens you to innovation and creativity.
- A person with the highest degree of flexibility in thinking and activity is most likely to win - if one solution doesn't work, try something else.
- Your mind and body are the parts of the same system - this enables you to believe that you can control your emotions and images.

2. Building life scenario (goals) - Technique of a dialogue with yourself about your chances and positive expectations as well as threats and your fears helps you discover what is your life mission:

- Find out what job has a deeper meaning for you;
- Build a thorough image of your ideal post (or the most horrible) and write it down;
- Let this image evolve (make corrections and improvements on what you've written down);
- Take actions to turn your goals into reality.

3. Building the atmosphere of trust and incorporating empathy - the way to effective leadership is through effective communication, which is adjusting yourself to certain communication styles and showing respect to different ones.

5.1.2 The advantages of NLP

The use of NLP brings numerous advantages:

- accelerates the ability to learn;
- facilitates building valuable relationships with others;
- improves empathy and motivation;
- increases flexibility;
- teaches to control emotions;

- guides how to reach to sub-consciousness.

One of the promising solutions can be accepting NLP training as a part of firm's human resources management strategy and incorporating NLP techniques and attitudes as a pattern in organisational culture. This would bring significant advantages for managers - not only would this help them realise their responsibility and impact their behaviour has on the whole organisation, but also, what is more important, would make them discover the real value of human potential. Reaching to deeper mental models can select people whose true mission is to be a leader - a director of orchestra knowing how to combine different tunes into a symphony of human skills. NLP has the power of erasing resistance. It turns "I must" into "I want".

"If you think you can't, you won't. If however, you think you can, there is every chance that you will"(Steve Shirley)[8]. And if you think you WANT ...

6 Final message

Nevertheless, there is still one more thing...

" Do budding managers have power innately, which makes them into leaders or do they inherit power once they become managers? Whatever the balance of truth is with power you have influence. You can cause change, you can make things happen. But you cannot be a leader unless you act. Thinking is not enough"[8].

Steve Shirley

Life president of the FI System Group

7 Acknowledgements

I would like to thank Dr Walter Baets and all the invited guests and participants of the Summer School on Managerial Complexity for their critical discussion over my paper and numerous precious comments which enriched my knowledge and the contents of the above article.

References

1. Armstrong M., Levers for change. In *HR Management – Strategy & Action*, ed. by Kogan, Page (London, 1992) pp. 42 - 43.
2. Butler J. and Salaman G., Why managers won't learn. In *Managing Learning*, ed. by P. Iles and C. Mabey (Routledge London and New York in association with the Open University, 1994) pp. 34 – 42.
3. Clark L. The essence of change (Prentice Hall International Ltd, 1994).

4. Dale M., Learning organizations. In *Managing Learning,* ed. by P. Iles and C. Mabey (Routledge London and New York in association with the Open University, 1994) pp.22 – 33.
5. Knight S. Neuro Linguistic Programming. NLP w działaniu (Wydawnictwo Studio EMKA, Warszawa, 1997).
6. Masłyk – Musiał E. , Stres i opory wobec zmian. In *Zarządzanie zmianami w firmie* (Wydawnictwo CIM, 1995) pp. 89 – 97.
7. Senge P. M., The leader's new work: building learning organizations. In *Managing Learning,* ed. by P. Iles and C. Mabey (Routledge London and New York in association with the Open University, 1994) pp. 5 – 21.
8. Shirley S., How to give reality to the vision, *Management Today* **4** (1995) pp.5.

COMPLEXITIES OF VIRTUAL ORGANISATION: TOWARDS A FRAMEWORK

SØREN DEJGAARD

Department of Marketing, Copenhagen Business School, Grundtvigsvej 37, 1864 Frederiksberg, Denmark

e-mail: dejgaard@cbs.dk

This paper attempts to build up a framework of virtual organisation on the basis on concepts taken from complexity theory. On basis of the dimensions of interaction and co-ordination, four ideal types of virtual organisation are identified, all illustrating a shift from bureaucratic structure towards network structure. A shift made possible by the introduction of new information and communications technologies. Furthermore, this shift may also be seen as a transformation from a transactional perspective represented by neo-classical economics based on external co-ordination towards a more integrative perspective where networks become institutionalised and where co-ordination takes place internally within the virtual organisation.

1 Introduction

The purpose of this paper is to give an introduction to concepts from complexity theory, to investigate whether they can be an aid in explaining the rising turbulence caused by information technology in the business community, and to give indications as to a framework of different ideal types of virtual organisation. The difficulty in using complexity theory is that it is a new all encompassing, but not very well developed theory. However, modest attempts have been made to apply complexity theory to economics. This has mainly been done by way of explaining stable macro structures with chaotic microelements. This is possible because complexity denotes order resulting from the interactions of many heterogeneous objects. As an example, an organisation may be stable even though it consists of many heterogeneous individuals, each with their own ambitions and goals. Likewise with a market, even though it consists of a number of heterogeneous firms and customers all interacting, co-ordinating, co-operating or competing, the market as such may be said to constitute a stable structure or a recognisable pattern. When this recognisable pattern is found at another – higher – level of description than the interaction of the heterogeneous objects, the pattern is often said to hold emergent properties[1]. Hodgson (1993, p. 246-248) speaks of complex hierarchical systems

[1] To illustrate emergent properties take the relationship between a sentence and the individual words within the sentence. The meaning of the sentence cannot be

where each level has emergent properties, meaning that a higher level of analysis cannot be explained only by knowledge of the components or interaction among these, of an underlying level. As such market behaviour cannot be explained solely by firm behaviour and firm behaviour cannot be explained only in terms of its smaller parts. A consequence is that explanations based on extreme reductionism or methodological individualism[2] must to some degree be questioned, depending on the level of turbulence at the higher degree. This implies that the more turbulent the higher level of description the less accurate it will be to use methodological individualism as a method for analysing economic phenomena.

2 From organisation towards virtual organisation

Due to the accelerated speed of development caused by information and communication technologies (ICT) firms and markets of today are changing, not only in a quantitative or qualitative way, but in a way where the conception or meaning of the terms themselves are under transformation. The evolution of technological progress implies constraint removal (Hirschhorn (1984) that is, the introduction of new technology, here ICT, removes the constraints for what is possible. If firms and markets are seen as clusters of activity (Normann and Ramirez, 1994) then, due to ICT, these activity clusters will be broken up; they become unbundled because a new logic of how value is created is at play. This new logic, which is a consequence of ICT, includes lowered cost of co-ordination, which eventually results in a more co-ordination intensive business and market structure. Furthermore, increased interaction between people both within the single firm and across company boundaries may be anticipated resulting in distinctive patterns of organisation and self-reinforcement (these points will be elaborated below). Eventually, new ways of rebundling activity clusters and hence creating new organisational forms such as the virtual organisation may be the result of more

---

explained by the words alone; it is due to the specific interaction pattern among the words that the sentence gives meaning. Or to qoute Steven Durlauf of the Santa Fe Institute: “When the interactions occur at a level of describtion other than that at which the patterns occur, these patterns are often called “emergent”. Hence magnetism, which is a property of a large number of iron atoms (in this case with spins polarized in one direction) is an emergent property” (1997, p. 1).

[2] Jon Elster (1982, p. 453) defines methodological individualism as: “the doctrine that all social phenomena (their structure and their change) are in principle explicable only in terms of individuals – their properties, goals, and beliefs”. This implies that explanations of social and economic phenomena should be made in individual terms and is as such an important expression of reductionism.

interaction and increased co-ordination. The concept of the virtual organisation should therefore be seen as a working title that attempts to conceptualise an amalgamation of the firm and the market into an activity cluster.

The virtual organisation is furthermore seen as an emergent structure created by the dialectic tension between parts and the whole. The parts may be firms, departments within firms, groups, individuals etc. and the whole constitutes the virtual organisation as an entity. Thus, the virtual organisation is an emergent structure that is created or constructed by the interaction between the two levels that parts and the whole constitute. The parts influence the whole and the whole influences the parts. Therefore, the virtual organisation is never fully developed – it is always in the process of being created. A consequence is that the parts cannot adequately explain the whole; put together, the whole signifies more than its constituent parts because of emergent properties of the virtual organisation.

These emergent properties cannot be explained by economics, as they are not entities that can be accounted for by economic terms, that is, by the method of methodological individualism. On the contrary, it is the emerging "more" that arises from interaction of the constituent parts; as the members of the virtual organisation start to interact they acquire knowledge from one another, they start to share competencies and co-ordinate these in a wider value constellation to mutual benefit; they learn from one another, and it continuously evolves. By "they" and "it" I try to suggest the two emergent levels of explanada, the parts and the whole. The question of how parts become a whole, that is, how the virtual organisation is created, can be answered by introducing two complementary forces that stems from ICT's network effect of lowering co-ordination costs. These are the principles of "internalising the external" and "externalising the internal". Internalising the external occurs when a firm extends its boundary to encompass the whole business network of the virtual organisation. This way, what before was seen as external (suppliers, customers, competitors etc.) has become internalised in the network. A consequence of this is resource interaction (building on resource-based management, Wernerfelt, 1984; Barney, 1991) where value is derived from the interaction of resources (tangible, intangible and knowledge) throughout the network. Resource interaction may be understood as an extension of Porter's (1985) concept of the value chain towards what Normann and Ramirez (1994) has termed value constellation where "*value is co-produced by actors who interface with each other (p. 54)"*. This implies that the emphasis on value creation moves from the single firm towards the entire virtual organisation. Externalising the internal has the opposite effect. The lowered cost of co-ordination makes it attractive for firms to concentrate on their core competencies (Prahalad & Hamel, 1991) and outsource activities that are not core to them to firms to which these activities are core. This way, what before took place internal in the firm is now undertaken by an external relation. The outsourcer and the outsourcee typically form tight relations (Quinn, J. B., 1992) and thus become a part of the larger network that constitutes the virtual organisation.

Combining these thoughts may result in a better understanding of what distinguishes a virtual organisation from a traditional organisation. First concerning constraint removal: the use of advanced ICT may extend the possibilities for organising activities over time, geographic space, and organisational boundaries. The second point was the dialectical relationship between internalising the external and externalising the internal resulting in an ability of the virtual organisation to acquire and maintain critical competencies through its design of complementary value-adding business processes. Third comes the constant tension between parts and the whole network of the virtual organisation. This signifies that virtual organisation is more a state of becoming rather than a stable structure as such. All sums up to what Castells (1998, p. 171) outlines as "*that specific form of enterprise whose system of means is constituted by the intersection of segments of autonomous systems of goals*".

2.1 On interaction and co-ordination

Having described the virtual organisation as a rebundled activity cluster made possible because of constraint removal due to the introduction of new technology, having described the dialectical part-whole process of creation, and having shortly described the complementary forces of internalising the external and externalising the internal, time has now come to go into details with the phenomena released due to the new technology, namely the effects of increased interaction and co-ordination intensive structures.

Interaction and co-ordination may be seen as two sides of the same coin. Interaction implies some degree of co-ordination and vice versa; the difference is not only that interaction is a sociological term and co-ordination an economic term and hence that interaction describes social phenomena among human beings such as talking, buying, selling, etc., and co-ordination is used to describe more economic phenomena such as the price mechanism that co-ordinates supply and demand, or production or logistics that is co-ordinated ex ante. The difference is also that interaction is normally thought of as a process taking place between autonomous wholes whereas co-ordination may be thought of as a structure governing the parts of a whole in a span between autonomy and authority. To say that firms interact with one another is nonsense; it is the people of the firms that interact but by intentions of the firm, however, to obtain the intention co-ordination is needed. This doesn't imply that unintentional interaction is not possible, it merely delimitates the focus of the kind of interaction dealt with in this paper. So when a firm intentionally interacts with another firm, e.g. in order to secure reliable supply of a sparepart to its production process, at the same time logistics, product specifications, payment etc. is co-ordinated.

2.1.1 Interaction

The removal of constraints caused by the introduction of ICT enables an increase in interaction among people both within and across firms, because some of the constraints that will to some degree cease to exist are related to time and geographical space[3]. The technology itself, ICT or its effect, the interconnection of computers within or across firms, is not by itself significant. What is important is that people behind the computers and hence their firms become interconnected in a network. Interconnected in the sense that ICT enables communication, interaction and co-ordination across time and distance and among network members.

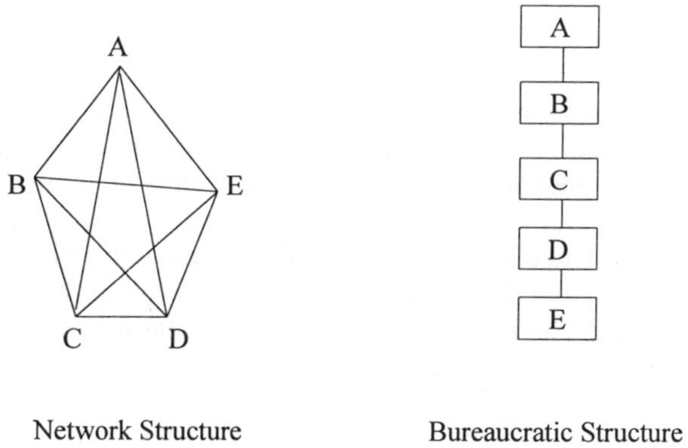

Figure 1 Network and bureaucratic interaction patterns

Figure 1 attempts to illustrate, although simplified, how the interaction patterns taking place in a network structure differs from the interaction patterns in a traditional, bureaucratic structure. Bureaucracy is here understood in a Weberian sense, i.e. characterised by division of labour, a clear authority hierarchy, formal selection procedures, detailed rules and regulations, and impersonal relationships. Where the actors of the network structure is characterised by being connected to all

[3] It takes the same time to e-mail your neighbour and someone in China; both time and geographical constraints to interaction has been eliminated with the result of more interaction.

other in the network, in the bureaucratic structure, the actor doesn't bear this characteristic, and whether interaction takes place here at all or the firms merely transact[4] is an open question.

Another inherent difference within figure one is the parallel and sequential flow of information and communication in respectively the network and bureaucratic structure. Where the flow of information and communication is sequential in the bureaucratic structure, that is, one with one at a time, the flow of the network structure runs parallel, all with all, implying both a more efficient and effective flow. The consequence of the network structure of interaction is self-reinforcement; as the actors exchange information and values, they learn from one another, attain knowledge, either in the tacit or explicit form, and eventually they constitute a totality that is different from and unlike the vertical production structure of the bureaucratic structure.

However, neo-classical economic theory supports the bureaucratic structure as it treats the actors' information and values as given – as irrelevant, it ignores learning and hence the self-reinforceable properties stemming from interaction in networks. Therefore, within neo-classical theory it is possible to say that actors and firms are at the same level of abstraction – firms are not regarded as emergent properties of actors, as an evolving phenomena, but as a from axioms deducible entity. Hence, neo-classical theory cannot "catch" the logic of interaction, that of causal causation where one set of cause and effects leads to another, it captures only causes that leads to steady state equilibrium effects. To quote Veblen (1934, p. 8): "*The question is not how things stabilize themselves in a "static state" but how they endlessly grow and change*". The self-reinforcing aspect of interaction relates to some kind of process of positive feedback. This is in stark contrast to neo-classical economics, where the process towards equilibrium may be characterised by negative feedbacks such as diminishing returns. Furthermore, neo-classical theory regards interaction as taking place indirectly through the price mechanism, here, on the other hand, actors interact directly with one another. This critique of neo-classical interaction is necessary in order to understand the way interaction is conceptualised in this paper and more importantly what the result of interaction might be in regard to self-reinforcement.

Yet another difference between the network and bureaucratic structure of organisation is the relationship between information, learning, and knowledge. Under a bureaucratic structure and within neo-classical theory information and knowledge is regarded as being identical, and learning is ignored as perfect information is assumed. Within the network structure it is necessary to recognise

[4] The difference being the frequence of interaction and recurrence between the same agents.

that data, information, learning and knowledge are different phenomena. Data are bits and bytes, letters and numbers, symbols and signals describing events observed. Information is data presented in a meaningful context. And learning transforms information into knowledge.

According to Argyris & Schön (1978) two types of learning exists. Single-loop learning and double-loop learning. Single-loop learning takes place when errors are corrected in accordance with prevailing norms and values. It is a question of learning a pregiven culture and a set of routines. When forming a virtual organisation, however, often there is neither culture nor norms or routines to determine right from wrong. Within such a setting double-loop learning becomes an option. Double-loop learning is an ability to learn in an ongoing way that mirrors both the interaction of knowledge within a network structure and interaction with the external environment. As such double-loop learning has autopoietic[5] properties. Double-loop learning may be described as a: *"feedback loop which connects the detection of error not only to strategies and assumptions for effective performance but to the very norms of which define effective performance" (Argyris & Schon, 1978, p. 22).* In other words, double-loop learning consists of "learning to learn" and places great emphasis on experimentation, creating new insights and developing a search for effective feedback that is based on continuous redefinition of the organisation. The existence of the double-loop feedback enhances the virtual organisations knowledge and skill base to accommodate a richer and more sophisticated information processing capability that allow for multiple and diverse perspectives to exist simultaneously. Double-loop learning is based on a theory of action outlined by Argyris and Schön in 1974. An important feature of this theory of action is that human beings is both a part of and informed by changes in values, behaviour, norms, leadership, etc. This implies a distinction between an individuals' espoused theory (that is, how a matter is explained; the individual's worldview) and the individuals' theory in use (that is, what the individual does; how he acts). Bringing these two into congruence is a primary concern of double-loop learning. In this respect interaction is an important feature of double-loop learning. It is through interaction with others that this conflict is typically identified, and identification is a prerequisite for solving the conflict.

The above describes the relationship between double-loop learning and interaction in a virtual organisation. But learning was only a mediator between information and knowledge. The traditional definition of knowledge as "justified true belief" is adopted here implying that knowledge, unlike information, concerns beliefs and commitment, is about action to some end, and like information, it is context specific

---

[5] According to Krogh and Ross "the autopoietic perspective reflects the belief that the cognitive activities in organisations are simultaneously open and closed" (1994, p. 33).

and relational – it is about meaning. For interaction to result in self-reinforcement, knowledge creation is imperative.

The view on knowledge creation advocated here is based on Nonaka and Takeuchi (1995) concept of knowledge conversion where "*knowledge is created and expanded through social interaction between tacit and explicit knowledge... between individuals*" (p. 61). Tacit knowledge is knowledge gained through either learning by doing or a process similar to apprenticeship-relationships, where the apprentice follows his master and relies upon him as his trustworthy authority (Polanyi, 1978, p. 53). Examples of tacit knowledge could be swimming, learning to drive a car, a management style etc. It is the kind of knowledge learnt through culture, norms, routines etc. and is as such personal and internalised. It cannot instantly be imitated or copied by others. Due to these subjectivist properties it normally takes longer time to learn tacit knowledge, and the learning, as such, takes place through interaction of actors on a personal or personal-like level. Hence it is often argued within economic theory that the learning of tacit knowledge is both more expensive in money terms, more time consuming, and requires a larger personal investment in the relationship between the present carrier and the receiver of the knowledge, than does explicit knowledge. Explicit knowledge, in opposition to tacit knowledge, is knowledge that has become codified, or made objective in the sense that it can be communicated "*by symbols and language, and thus has the necessary features to be tradable*" (Maskell, 1996, p.4). Explicit knowledge is (former) tacit knowledge that has become codified. An example of explicit knowledge could be a generic production process, a computer programme, or an instructions manual – the main feature is that is has been codified for others too understand. Accordingly, it can easily be imitated and copied; it can be transferred without interaction (at least at a personal level), it can be bought and sold, and as such it is a cost efficient method of knowledge transfer. In regard to the relationship between tacit and explicit knowledge Nonaka and Takeuchi may be cited: "*Knowledge that can be expressed in words and numbers represents only the tip of the iceberg of the entire body of knowledge*" (1995, p. 60). Or to quote Polanyi: "*We can know more than we can tell*" (1966, p. 4).

Hence, through Nonaka and Takeuchi's concept of knowledge conversion indicating interaction between tacit and explicit knowledge it is possible to begin to understand the creative processes underlying knowledge creation as interaction between individuals. Furthermore, with this line of argument self-reinforcement may be understood as an interaction process of tacit and explicit knowledge among individuals within an organisational setting.

2.1.2 Co-ordination

Previously it was argued that an important effect of ICT was reduction of co-ordination costs and further it was stressed that interaction and co-ordination must be complimentary phenomena, without giving a clear notion of what co-ordination is thought to be. Malone and Crowston (1993) discusses a framework for studying co-ordination interdisciplinary including such disciplines as computer science, organisation theory, management science, economics, linguistics, and psychology and reaches a simple general definition: "*Co-ordination is managing dependencies between activities*". The definition focuses around three important topics: "managing; dependencies; and activities". The managing aspect of co-ordination implies some sort of control or governance structure. In this respect, Mintzberg (1983) mentions five co-ordination mechanisms with rising complexity to manage co-ordination. First he refers to mutual adjustment where co-ordination is achieved through simple processes of informal communication. Second comes direct supervision where co-ordination is accomplished by letting a supervisor take responsibility for the work of others, instructing them and monitoring their actions. Then follows three cases of co-ordination through standardisation of either work processes, output or skills and knowledge. Standardisation of work processes equals Taylor's scientific management and standardisation of output implies that co-ordination of tasks is predetermined. Finally follows standardisation of skills and knowledge achieving indirectly what standardisation of work processes or output achieved directly: it controls and co-ordinates the work.

Mutual co-ordination as a co-ordination mechanism may be characterised as autonomous as no hierarchical interference takes place in the interaction of the parts, be it individuals, their intentional activities, and the firms they constitute. This is contrary to direct supervision and the three forms of co-ordination through standardisation, where co-ordination takes place within a hierarchical setting. Here co-ordination may be characterised as being authoritarian. Furthermore, Mintzberg (1983, p. 7) mentions: " *as organizational work become more complicated, the favoured means of co-ordination seems to shift from mutual adjustment to direct supervision to standardisation, preferably of work processes, otherwise of outputs, or else of skills, finally reverting back to mutual adjustment.*" Hence, mutual adjustment or autonomy is the preferred co-ordination mechanism not only with the simplest activities to be co-ordinated, but also with the most complicated. Although there seems to be a circular movement from mutual adjustment over direct supervision and standardisation to mutual adjustment again, this paper will regard it as a continuum with authoritarian co-ordination patterns in the one end and autonomous co-ordination patterns in the other. The reason being that different forms of virtual organisation hold complicated structures thus allowing for the exclusion of the simple version of mutual adjustment.

The second element within the definition of co-ordination offered by Malone & Crowston were dependencies. Dependencies must exist between activities; if no dependency exists there is no need for co-ordination. Dependency takes various forms, to mention but a few, dependency exists whenever two or more parties share resources, assign tasks, engage in relationships, needs to transfer goods etc. For virtual organisations the extension of the value chain of the single firm towards a value constellation network based on reconfiguration of value creating activities in accordance with each firms core competencies and complimentarity within the network, signifies dependencies to be co-ordinated.

The definition of co-ordination also includes "activities". Activities occur by combining, developing, exchanging, or creating resources of a tangible or intangible kind. Mintzberg (1983) regards organisational structure to be created first by dividing activities into distinct tasks and then achieving co-ordination among these tasks. This corresponds to Normann and Ramirez' notion of unbundability of activity clusters currently taking place due to ICT's generation of new ways to co-ordinate activities resulting in a rebundling of activities in accordance with the logic of ICT. Activities then, within this setting, are the parts of which the virtual organisation is created.

Understanding co-ordination, it becomes interesting to know what the result of decreasing co-ordination costs might be. Based on simple concepts from microeconomics, such as substitution and elasticity of demand, it is possible to make some predictions about the potential consequences of declining co-ordination costs on the structure of companies and markets. Historically, an effect of reduced co-ordination costs due to information and communications technology was substitution; human co-ordination was substituted by information technology. This effect took place in the data processing era from around 1960 to about 1980 where mainframe computers were used to automate low level activities like payroll, accounting, and other forms of calculus. As co-ordination costs continued to decline and the usage of computers became more and more widespread the consequence was an increase in the overall use of co-ordination. This began to happen around 1980 with the introduction of the personal computer (PC). The PC was so inexpensive, so easy to use, and so efficient that soon every office had a computer, with co-ordination heavy activities such as production schedules, customer databases, and just in time procurement as a consequence. A third effect of reduced co-ordination costs indicates a shift towards more co-ordination intensive structures – be it companies, markets, or societies - that were previously either not possible or too expensive. Some authors have termed this period the network era (Nolan & Croson, 1995). It is termed "network era" because computers from around 1990 became connected through networks either as local area networks (LANs) or as wide area networks (WANs) constituting the Internet. It is this transformation of firms and markets towards more co-ordination intensive structures resulting in "rebundability" of activity clusters in a new way that gives rise to different forms of virtual organisation.

3 A framework of virtual organisation

Having described interaction and co-ordination in detail the next step will be to develop these concepts into a coherent framework in order to describe different forms of virtual organisation. Such a framework is depicted beneath. However, before going into detail with the framework it is important to acknowledge that it could easily have been designed along other dimensions e.g. potential benefits as a function of ICT's effect of business transformation (see Venkatraman, 1994). Furthermore, interaction and co-ordination patterns cannot promise to give a full picture of virtual organisation, but it is the hope of the author that the framework can be a rough guide in recognising trends concerning possible reconfigurations of activity clusters due to ICT. So the framework does not promise a clear picture, at most it provides rough guidelines as to what form virtual organisations might take.

Short run and long run patterns of interaction indicate the duration of relationships between members of the virtual organisation and not the duration of the virtual organisation as such. As described earlier a virtual organisation is never fully developed – it is always in the process of being created. The co-ordination pattern may be autonomous, that is, mutual adjustment (Mintzberg, 1983) exists between the members, or authoritarian, that is, supervision or standardisation of either work processes, output, or skills and knowledge, vis a vis the virtual organisation and not the individual member companies as such. Hence, as stressed earlier, it is the dialectical tension between parts and the whole that creates the virtual organisation. In addition, all four ideal types depicted within the framework have a network structure.

The four ideal types of virtual organisation are made up of a co-ordination and an interaction dimension. Each dimension is to be regarded as a continuum with two extremes. In between these extremes exist combinations which implies that many forms of virtual organisation may exist. That many different forms of virtual organisation exist can be seen from the confusion that search for a definition of virtual organisations meet within the research community (Führer, 1998). Is a virtual organisation a temporary collaborative network or is it an augmentation of long term industrial relations when the partners become networked? Is it intra-organisational or inter-organisational? Do the partners share resources, concentrate on core competencies and outsource non-core business activities to others? Are

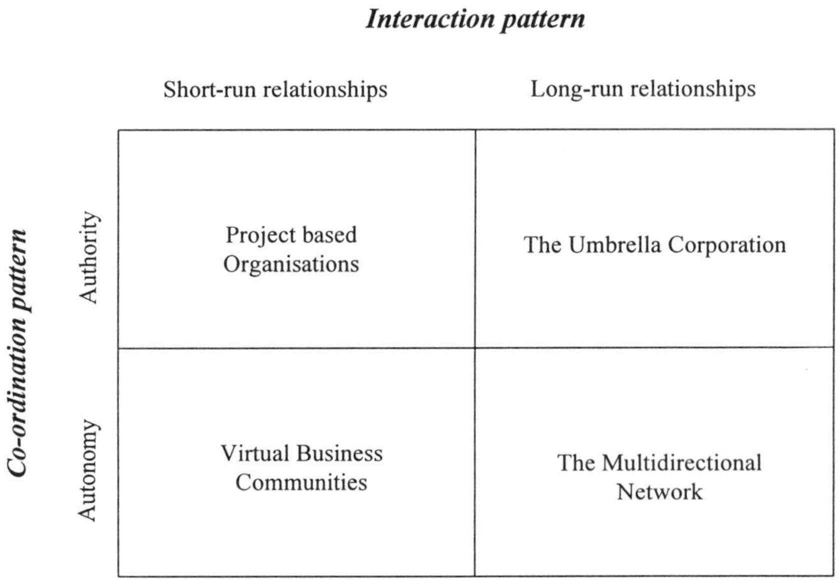

Figure 2 Different forms of virtual organisation

virtual organisations self-organising entities or is it an integrated supply chain of one big company? Many such questions exist and many answers have been given. One possible way out of this search for "one right definition" is to recognise that various forms of virtual organisation may exist simultaneously side by side as different attempts to rebundle activity clusters due to ICT's effect of constraint removal. However, in order to keep the description simple only the four ideal types shown in figure 2 will be scrutinised in detail here.

3.1.1 Project based organisations

Project based virtual organisations are formed around a time based task. The task is often predetermined by a third party (a customer) who needs a complicated assignment done. In order to meet this assignment various firms merge temporarily forming a network in order to utilise each other's core competencies synergistically. When the task is fulfilled the organisation dissolves, therefore the characteristic of the interaction pattern is short run relationships. The co-ordination pattern, on the other hand, is authoritarian in the sense that the aim and time schedule of the project is set by the third party. This is the form of virtual organisation that resembles known organisational structures the most. Project based organisations established around a market opportunity offered by a third party has probably existed since the dawn of industrialism. However, what is new is the network structure's global reach transcending time and space enabling firms to collaborate at a distance.

Project based organisations are often found within the building and construction industry. To illustrate the points above the construction of the Oeresund link between Denmark and Sweden may serve as an example. The Oeresund link consists of two different works: The Oeresund bridge linking Denmark and Sweden, which is planned to open in year 2000 and the land work on both sides of Oeresund, which is planned to be finished by 2008. A/S Oeresundsforbindelsen, a Danish government organisation, and Oeresundskonsortiet, a Swedish government organisation jointly invite tenders for the work to be done according to schedule. These government organisations have divided the land work and the coast-to-coast link into different projects, all of which are contracted out. Each project is run by a meta-organisation consisting of different contractors and their network of suppliers. An example of such a meta-organisation is Sundlink Contractors AB who heads the largest project of the Oeresund Link. It consists of Danish "Højgaard & Schultz" and "Rasmussen og Schiøtz", Swedish "Skanska AB", who are the lead manager of the project, and German "Hochtief AG". These companies have worked together before on similar projects such as the Great Belt Link in Denmark and the High Coast Bridge in Northern Sweden. Sundlink Contractors AB is currently heading the building of the coast-to-coast link between Denmark and Sweden, that is, it co-ordinates the work of the different contractors, also by the usage of information and communications technologies. Sundlink Contractors AB is best described as a virtual organisation as it consists of and connects the different contractors and their suppliers under the roof of a meta-organisation. It holds legal status being a limited company but holds only a time-limited physical presence. Sundlink Contractors AB is simply the companies involved until the project is finished. Then it cedes to exist.

3.1.2 The Umbrella Corporation

This form of virtual organisation is characterised by either a big multinational corporation or a co-ordination centre that reconfigures value creating activities by utilising ICT to bind suppliers and customers into a more or less rigid network constellation with the aim of building value constellation outside its own boundaries. In this respect two themes are relevant to discuss. Boundary crossing activities and the rise of a new form of markets. Traditionally, a "boundary" can be thought of as a division of what goes on inside contra outside the organisation. Hence, within the boundaries of the organisation counts those resources and activities that can be controlled and influenced by the organisation, and left outside are those that cannot be controlled. This control is assumed to be necessary in order to adopt and relate effectively to the environment. Accordingly, *"the organisation ends where its discretion ends and another begins"* is the definition of organisational boundaries proposed by Pfeffer and Salancik (in Ford, 1993, p. 531). To use the metaphor of activity clusters again ICT enables the multinational (or the co-ordination centre) to reconfigure the activity cluster by letting some activities take place outside its boundary. The light version is to create an electronic market that includes all the company' suppliers. This is the option often chosen by multinationals because it permits cost reduction and market transparency. The heavier version is to create a network of value constellation where each member's core competencies and knowledge is used in a complimentary way. The co-ordination centre often describes this option. To describe the electronic market phenomenon General Electric's (GE) may be used as a case study:

The centre of its efforts is its Trading Process Network (TPN), a Web site where GE now does $1 billion-worth of business a year with about 1,400 of its suppliers, singlehandedly exceeding all consumer electronic commerce (on the Internet). It has software that lets GE's purchasers specify whom they want their request for bids to go to and what sort of information, such as drawings, to include; the software then manages the bids as they come back, eliminating low bidders and handling further rounds, finally notifying the bidders of the outcome. The suppliers get a notification and an online form on which to fill out the details of their bid. Pretty simple stuff, by the looks of it; but GE reports some remarkable results. Since the service started early last year, it has cut the length of the bidding process in GE's lighting division to less than half, from 21 days to 10. Because requesting the bids is so easy, purchasing officers approach more suppliers; the increased competition has lowered the cost of goods by 5-20%. Previously, it was often too much trouble to include foreign suppliers. Now 15% of orders go abroad. (The Economist, May 10, 1997)

The co-ordination centre acts as the focal company in the kind of virtual organisation that is built up around a specific idea of how to co-ordinate activities with the aim of creating a network of value constellation. The co-ordination centre

does not have to be a multinational; often it is a rather small company with limited resources possessing a vision of how to co-ordinate activities so as to obtain a global reach. The result is a stretching of company boundaries to encompass suppliers and customers with the aim of creating a network of value constellation. Hence another kind of market may arise. Such a market can be regarded as an internal market of the virtual organisation, or what Arndt (1979) has termed a "domesticated market" where "*transactions ... are usually handled by administrative processes on the basis of negotiated rules of exchange*" (p. 56). The difference between the electronic market and the domesticated market described by Arndt lies in the degree of asset specificity and hence adaptation to a specific constellation of value creating activities and to the frequency and duration of the transaction. Where electronic markets often requires a low degree of asset specification and the transaction is characterised by both a low frequency and short duration, the domesticated market within a network is often characterised by a high degree of asset specificity, adaptation and a high frequency and long duration of transaction. The transaction has turned into interaction enabling the effects of interaction; learning, self-reinforcement, etc., described earlier. However, both kinds of markets may serve as barriers to entry establishing the network structure as the fundamental form of competition in the new global economy (Ernst, 1994). Inside the network structure new possibilities are constantly created as interaction and co-operation focuses on sharing of costs and risks, knowledge creation through joint R&D projects and keeping up with new information. Outside the network structure of virtual organisation survival is becoming increasingly difficult for the neo-classical firm. As Castells (1998, p. 171) puts it: "Under the conditions of fast technological change, networks, not firms, have become the actual operating unit."
The co-ordination centre encompasses a domesticated market and to give an example the Swedish professor Bo Hedberg has given a good account of the Swedish Pyramid Sportswear that is behind the apparent American for clothing brand GANT:

"*Gant may be regarded as a rather clear-cut illustration of how one or more persons with a business concept but with limited resources – or at least a desire to limit the resources committed and concentrate on essentials – can build an enterprise around an idea. The brand name, fashion creators, models, catalogs, stores, factories, customer lists, and inventory systems are co-ordinated into what appears to the customer as the successful GANT company. But behind the scene is Pyramid Sportswear... The brand name, collections, production, and retailers network are all the work of Pyramid. Pyramid Sportswear finds designers, identifies trends, contracts out quality-assured production, and cultivates a retailer network using catalogs, advertising, image-creating activities, and sales support in the form of a customer data base, active inventory tracking systems etc. A systematically designed customer data base, and extremely detailed and up-to-date reporting of shipments and current sales, are two central elements which enable the imaginary*

GANT organization to exist and strive. Remote quality control in interaction with temporary partners is another ingredient. Without modern information technology this supporting structure could not hold... The whole is considerably larger than the sum of its parts. The partner enterprises and the individual actors in the GANT network are linked together in a variety of ways, from commercial contracts to personal contacts and personal chemistry." (Hedberg et al. pp. 6-7)

3.1.3 Virtual business communities

A virtual business community is a spontaneous kind of virtual organisation that breaks with the traditional goal rationality of scientific management. Goal rationality is characterised by a means-end relationship implying that the end is given and the means (for reaching the end) is known. This type of rationality focuses on making the means-end relationship as efficient as possible. Contrary to this view virtual business communities incorporate "randomness" as an organising principle. Randomness of whom the collaborating firms, organisations, institutions, individuals etc. might be. Randomness of task specification and sometimes even randomness as to a specific purpose or goal – a common interest, curiosity, or a guiding vision often suffices. Hence, the kind of rationality characterising this kind of virtual organisation is of an evolving kind; it could be termed network rationality. The reason behind this new logic of business constellation is the network structure of the Internet and hence the Internets ability to create connections and be connected across time and geographical distance. Randomness, therefore, may be thought of as the result of the network structure of the Internet. Furthermore, instead of attempting to limit chance as most business researchers of today are doing, perhaps instead it should be cultivated and incorporated into models of business using a network structure in order to create creativity and innovation.

Currently the computer press is dominated by success stories of how virtual business communities over the Internet has innovated and created new business opportunities. In order to exemplify and highlight the above described theory I will here represent the case of Linux as described in Wired Magazine:

"...Aided by thousands of disparate hackers - a "net.mind" of sorts - Linus Torvalds, a 27-year-old Finn, has also created a new world. That realm, Linux, is an operating system, the force that defines the digital domain of programs just as our own environment determines the characteristics of animal and plant life.
...Linux was started six years ago as a typical programming lark: written to run on a PC with 4 Mbytes of RAM as a free version of the costly commercial Unix operating system. Today, Linux has an installed base conservatively estimated at around 3 million users.

...The saga of Linux has many strands. It is the story of Linus, an arch hacker of unusual wit and charm who single-handedly solved problems that usually require teams of programmers toiling for months. It is also the story of the Internet as a model for distributed collaboration. Indeed, Linux is the Internet's Kalevala, a huge patchwork of code that defines a rapidly growing cybernation, the tightly linked community of those who make and use it. What unites these coders is the drive to create the world's greatest operating system, one more powerful than any commercial Unix, able to run on practically any hardware, and infinitely customizable. An OS, moreover, that is fully the equal of Microsoft's flagship, Windows NT - with true multitasking, virtual memory, shared libraries, TCP/IP networking, and other advanced features. Many see Linux as NT's most serious competitor, the only viable alternative to the Microsoft monoculture - singular proof of the ideal that says we should have a digital choice.
...The growth of the development team mirrored the organic, not to say chaotic, development of Linux itself. Linus began choosing and relying on what early Linux hacker Michael K. Johnson calls "a few trusted lieutenants, from whom he will take larger patches and trust those patches. The lieuts more or less own relatively large pieces of the kernel." The Linux approach is deceptively simple. All hackers are free to work on any additional features or improvements. Even in the earliest stages, new code is freely downloadable for users to try and critique: beta-testing is not a last-minute attempt to catch the worst flaws, but an integral part of the process. When several people work on the same area, they may compete or combine; if they compete, the best code wins through raw Darwinian selection."
(Wired Magazine issue 5.08 – August 1997)

The Linux case of a virtual business community illustrates both the autonomy of co-ordination between the participants not depending on goal rationality but on interest, randomness as an organising principle, and the short run character of interaction. Even though there seems to be a goal of such a community the goal blurs and change character according to circumstances. Hence the rationality implicit evolves along with the community. Where goal rationality is ex ante, the goal and the means to reach it is given – and the task is to find the most efficient way to fulfil the goal, the type of rationality inherent in networks is often ex post, that is, constructed afterwards implying that means-end relationships can first be established after termination of the project. However, a vague goal, like improving a programme, might exist but the how's and who's are unknown or at least undefined from the beginning – instead, the task is often to develop the most effective goal.

3.1.4 The Multidirectional Network

The multidirectional network consists of a number of small and medium sized enterprises (SME's) which cooperate over activities such as finance, training, research and technological development, product design, marketing, export promotion and distribution in order to gain the traditional strengths of large corporations such as economies of scale, scope, and global reach without loosing the flexibility and innovative spirit often characterising SME's. The classical example being the Italian industrial districts and a further case is represented by Taiwanese SME's competitive position within the international market for PC-related products, key components and knowledge intensive services (Ernst, 1998). The multidirectional network is characterised by a high degree of flexibility arising from mainly two sources. One is an extreme degree of specialisation where members engage in single tasks allowing them to avoid heavy fixed capital costs by producing, purchasing, and selling in small lots. The second is the autonomy of each member enabling it to decide its own direction, that is, whether it is to be considered "in" or "out" of a specific assignment. Hence, inside the rather long term arrangement of the multidirectional network temporary patterns of sub-networks arise from assignment to assignment. The interaction pattern is characterised as being long term even though it is up to the individual member to determine whether it is active or passive within an assignment. Because, even though the member is passive at a certain point in time it is not excluded from the network as such.

4 Perspectives on virtual organisation

This section aim at dealing with similarities between different forms of virtual organisation as discussed in section 3, and relate them to the concepts of complexity theory taken up in section 2.

First, the four ideal types of virtual organisation represent a movement away from the traditional bureaucratic structure towards different patterns of network structures. However, it is the interaction of the often dispersed and heterogeneous members that create the distinct pattern of virtual organisation. This interaction, in turn, is allowed by the network structure of the virtual organisation. Furthermore, a characterising phenomenon of the network structure is that it allows co-ordination to take place internally within the network through a domesticated market like arrangement. As such co-ordination within a virtual organisation can be said to have become institutionalised. Co-ordination between the parts of the bureaucratic structure, on the other hand, occurs through transaction like arrangements and is as such an external arrangement.

In connection to this, the rationality inherent within networks becomes an issue. Although network rationality earlier was described exclusively in regard to virtual business communities, put in more general terms it may signify an attribute relevant to all four ideal types. Specifically if it attempts to describe the different preconditions underlying respectively the bureaucratic and network structure. In this respect Castells (1998, p. 171) has made a distinction based on Alain Touraine: "*In a dynamic, evolutionary perspective there is a fundamental difference between two types of organization: organisations for which the reproduction of their system of means becomes their main organizational goal; and organisations in which goals, and the change of goals, shape and endlessly reshape the structure of means.*" The first type of organisation covers what I have termed the bureaucratic structure and the second resembles virtual organisation except that the parts-whole relationship is neglected. Hence, in order to elaborate this distinction to cover network rationality, Castells outline may be used again. He proposes to regard network enterprises as: "*that specific form of enterprise whose system of means is constituted by the intersection of segments of autonomous systems of goals.*" This proposal covers the essence of the network structure; interaction, continual adaptation, and no steady state equilibrium. Some degree of randomness is also inherent in the network structure because the members of the network both depend on and are autonomous in regard to the network. Furthermore, as members of one network simultaneously can be members of other networks, each member continuously bring new goals to the virtual organisation achieving blurring means-end relationships at the aggregate virtual organisation level. The goal of the firm is no longer to secure efficient means-end relations, instead, for virtual organisations with a network structure the goal is to secure effectiveness in regard to processing knowledge into tangible or intangible products.

A given network performance then, will depend on its pattern, that is, on its connectedness and consistence. Connectedness describes how well the bureaucratic structure has been transformed into a network structure, and hence its ability to facilitate interaction between members. Consistency outlines the extent to which there is congruence between member goals and network goals.

5 Conclusion

This paper has attempted to build up a framework of virtual organisation on the basis on concepts borrowed from complexity theory. Intensified interaction and co-ordination were identified as being major new aspects of the structuring of firms and markets. This is mainly a consequence of new information and communications technology's effect of constraint removal, thus allowing for new ways of organisation. On the basis of interaction and co-ordination, four ideal types of virtual organisation have been identified, all illustrating the shift away from bureaucratic structures toward network structures of organisation. Furthermore, this shift may also be seen as a move away from a transactional perspective represented by neo-classical economics based on external co-ordination towards a more integrative perspective, where networks become institutionalised and where co-ordination takes place internally within the virtual organisation.

This paper shows that virtual organisation is a mode of organising more than a specific organisation structure. The virtual organisation is a meta-organisation, an organisation of organisations, at the same time enabled by ICT and utilising ICT to fulfil its purpose.

References

Argyris, C. & Schon, D. (1974). Theory in Practice. San Francisco: Jossey-Bass.

Barney, J.: "Firm Resources and Sustained competitive advantage" Journal of Management, 1991, Vol. 17, No. 1, pp. 99-120.

Castells, M. (1998): The Information Age: Economy, Society and Culture, Volume 1: The Rise of the Network Society. Blackwell Publishers

Durlauf, S.N. (1997): "What should Policymakers Know About Economic Complexity?" Department of Economics, University of Wisconsin and Santa Fe Institute WP 97-10-080

Ernst, Dieter (1994): Inter-Firms Networks and Market Structure: Driving forces, Barriers and Patterns of Control. Berkeley, CA: University of Calefornia, BRIE Research Paper.

Ernst, Dieter (1998): What Permits Small Firms to Compete in High-Tech Industries? Inter-Organisational Knowledge Creation in the Taiwanes Computer Industry. DRUID Working Paper No. 98-3

Ford, D. (1993): Understanding Business Markets – Interaction, Relationships, Networks. Academic Press

Führer, E. & Ashkanasy, N.M. (1998): The virtual organisation: Defining a Weberian ideal type from an inter-organizational perspective. The University of Queensland/Karl Franzens University, Graz.

Hedberg, Bo og Goran Dahlberg og Jorgen Hansson og Nils-Goran Olve (1997): Virtual organizations and beyond - discover imaginary systems. Wiley, Chichester

Hirschhorn, L.: Beyond Mechanisation, Cambridge, MA: MIT Press, 1984

Hodgson, G. M.: Economics and Evolution – Bringing life back into economics, Polity Press, 1993

Krogh, G.v. and Ross, J. (1994): "Organizational Epistemology" St. Martins Press

Malone, T.W. & Crowton, K. (1993) The Interdisciplinary Study of Co-ordination; MIT Center for Co-ordination Science, appeared in ACM Computing Surveys, 1994 (March) 26(1), 87-119. http://ccs.mit.edu/CCSWP157.html

Maskell, Peter (1996): "Knowledge and knowledge creation – some introductory notes to students attending the course IM55" Copenhagen Business School.

Mintzberg, H (1983): Structures in Fives: Designing effective Organisations. Prentice-Hall International Editions

Nolan, R. L. & Croson, D.C. (1995): Creative Destruction, Harvard Business School Press.

Nonaka & Takeuchi (1995): The Knowledge-Creating Company – How Japanese Companies Create the Dynamics of Innovation. Oxford University Press

Normann R. & Ramirez, R.: Designing interactive Strategy: From Value Chain to Value Constellation, John Wiley & Sons, 1994

Porter, M: Competitive advantage - creating and sustaining superior performance, Free Press, New York, 1985

Prahalad, C. K. & Hamel, G.: "The Core Competence of the Corporation" Harvard Business Review, May-June 1990, pp. 79-91.

Quinn, J.B.: "Intelligent Enterprise - a Knowledge and Service Based Paradigm for Industry" The Free Press, 1992.

The Economist, May 10, 1997 http://www.economist.com/editorial/freeforall/14-9-97/index_survey.html

Veblen, T.B. (1934): "Essays on our Changing Order" Ed. L. Ardzrooni (New York: The Viking Press)

Venkatraman, N. (1994): IT-enabled business transformation: From automation to business scope redefinition. Sloan Management Review/Winter 1994, pp. 73-87.

Wernerfelt, B.: "A resource-based View of the Firm" Strategic Management Journal, 1984, Vol. 5, pp. 171-180.

Wired Magazine issue 5.08 – August 1997 http://www.wired.com/wired/5.08/linux.html

THE EFFECT OF NEW ORGANISATIONAL FORMS ON MARKET ORIENTATION: MAXIMISING CUSTOMER VALUE SERVICES

KATY MASON

Warwick Business School, The University of Warwick Coventry, CV4 7AL, England

E-mail: k.masonc@wbs.warwick.ac.uk

The structure of a firm (e.g. Chandler 1962) and market orientation (e.g. Narver & Slater 1990) are said to have a positive effect on business, yet the impact of structure on market orientation has received little attention. With an increasing number of firms pursuing vertical de-integration strategies (VDI's) this paper explores the relationship between market orientation and the level of vertical de-integration of a firm. Two drivers of VDI's are identified: cost and market orientation. In an increasingly competitive global environment it is essential that firms understand how their structure affects market orientation in order to safeguard business performance. The objectives of this research are threefold; 1) to explore the relationship between market orientation and vertical de-integration; 2) to model the components of vertical de-integration as a key mediating variable between market orientation and business performance, and 3) to propose a methodology for testing the relationship between marketing effectiveness and vertical de-integration.

1. Introduction

The past decade has witnessed a fundamental change in direction of both marketing theory and practice. There is a movement towards relationship marketing by building long-term relationships between suppliers and customers (Kotler 1991; Parvatiyar, Sheth & Whittington 1992). The emphasis on these long-term relationships increases with the number of firms pursuing vertical de-integration strategies and in the process, forming web-like networks of independent organisations. These vertically de-integrated structures have been given various titles: *'networks'* (Miles & Snow 1986; Thorelli 1986), *'value-adding partnerships'* (Johnston & Lawrence 1988), *'alliances'* (Ohmae 1989), *'shamrocks'* (Handy 1990) and *'virtual integration'* (Magretta 1998). As Webster (1992) notes, *"...all are characterised by flexibility, specialisation, and an emphasis on relationship management instead of market transactions."* This network paradigm recognises that global competition occurs increasingly between networks of firms (Thorelli 1986). In the future, the effective competitor will be required to deliver customer value within some kind of network (Webster 1992, Doyle 1995). Understanding how these networks impact on a firm's ability to align itself with the market (i.e. become more market oriented) is crucial to improving business performance. There is a need for strategic and applied research that identifies and provides an understanding of networks and which anticipates the effects that these new organisational forms will have on marketing strategy.

This paper explores the relationship between the level of vertical de-integration of a firm and market orientation. We explore the nature of vertical de-integration and the characteristics of market orientation required for a strong business performance. First we examine the nature of vertical de-integration and suggest how this construct should be conceptualised. Second we theorise that successful vertical de-integration requires a high level of market orientation. Finally we propose a methodology for testing the relationship between vertical de-integration and market orientation.

2. Theoretical Background

2.1 Customer Value: The Supply Chain & The Value Chain

The essence of marketing is to maximise value of a product offering for the consumer. To enable us to do this we need to understand *how* and *where* added value is created. Taking a supply chain perspective and using Porter's (1985) value chain as an analysis tool, we set out to examine these two key questions.

2.2 The Supply Chain

The primary objective of the firm is to maximise business performance. Business performance is defined in terms of return on capital employed, profitability and market share (Kaplan & Norton, 1992). Firms achieve such objectives by *'satisfying customer needs profitably'*, (Chartered Institute of Marketing, 1989). Customer needs are satisfied when they [the customer] attach sufficient *value* to the commodity being offered to make the purchase. A firm therefore, strives to add value in the process of transforming a product from its raw state to a *'customer solution'*. Kay (1993) convincingly argues that added value and corporate performance are closely related. Therefore we assume that firms attempt to add value to products to increase business performance.

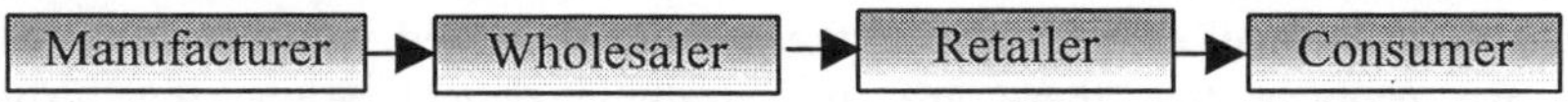

Figure 1. The Supply Chain

Getting the product[1] from its raw state to the customer involves a supply chain whereby the product is created and then distributed to the end consumer (see Figure 1). The supply chain comprises of a number of functions that must be performed such as designing and marketing products. These functions when vertically integrated, are carried out within a single firm. Vertical integration has been defined as "... *the combination of technologically distinct production, distribution, selling, and/or other economic processes within the confines of a single firm.*" (Porter, 1980, p.300). Alternatively the functions can be outsourced and carried out by individual firms. We identify this structure as *'vertical de-integration'*. Understanding vertical de-integration requires distinguishing between a range of marketing relationships dependent on the type of contract.

Taking a supply chain perspective, we examine the strategic decisions of vertical de-integration strategies and their impact on a firm's ability to achieve market orientation. We suggest that the greater the level of vertical de-integration a firm has, the higher the level of added value created for the customer and by implication, the more market oriented a firm becomes. What interests us is which functions create significant added value for the customer and where these functions are positioned within the supply chain. If we can understand this, we can identify which functions of the firm should remain in-house (i.e. vertically integrated) and which functions would be best outsourced (i.e. vertically de-integrated). As Quinn (1992) points out, firms must consider outsourcing all those activities, which do not add high value or offer some compelling means of achieving synergy. There appears to be a consensus that companies should *"stick to their knitting"* by specialising in certain areas of excellence and leaving the rest to other companies (Peters & Waterman 1982; Hamel & Prahalad, 1994).

It should be noted that this research is concerned with the integration of processes that directly affects the end product offering, i.e. the level of integration of the supply chain functions as opposed to the integration of organisational functions such as payroll, firm canteen, cleaning service.

2.3 The Value Chain

In competitive terms, value is the amount the buyer is willing to pay for what a firm provides. As noted at the beginning of this section, it becomes imperative to understand how and where this value is created along the supply chain. We have suggested that the level of VDI effects the amount of added value that is created. Further, we propose that there are two strategic considerations that drive a company's decision to adopt a vertically de-integrated structure; cost and market

---

[1] In order to avoid repetition the term *'product'* has been used where *'service'* or *'product'* may apply.

orientation. As Porter (1985) notes there is a need "... *to understand the behaviour of costs and the existing and potential sources of differentiation."* (p.33).

Cost has long been recognised as a primary driver of outsourcing decisions (Ford *et al.*1993; Venkatesan, 1992). Differentiation is understood to be the way in which a firm adapts its product to create unique and added value over and above that which a firm's competitors offer. Market orientation is the strategy applied to achieve closer alignment to the market, i.e., adding value for the customer (Narver & Slater, 1990; Kohli & Jaworski, 1990).

Hayes & Wheelwright (1984, Ch.9.) refer to the cost driver as *'process positioning'* (dealing with the ways or means chosen to conduct activities e.g. for production, logistics, R&D and distribution) and the marketing orientation driver as *'product positioning'* (dealing with the strategic portfolio with regard to the markets in which the firm sells). Two distinct categories of strategic decision making emerge from the literature; those that apply to financial objectives (cost behaviour, cost saving objectives, asset risk, Return On Investment) and those that apply to marketing strategy objectives (customer retention, market share, market position).

By adopting Porter's (1985) value chain as a an analysis tool we can systematically examine all activities a firm performs and how they interact in order to investigate the sources of competitive advantage created by each part of the supply chain. Porter's (1985) generic value chain is composed of nine generic categories of activities which are linked to each other thereby demonstrating the specific activities a particular firm performs such as designing, producing, marketing, delivering and supporting its product (see Figure 2). The value chain divides into three distinct areas; margins, primary activities and service support. These value activities represent the building blocks of competitive advantage. How each activity is performed combined with its economics will determine whether a firm is high or low cost relative to competitors. These links extend to the firms' suppliers, channels and buyers.

The value chain is embedded in a larger stream of activities termed the *'value system'*. Suppliers are recognised as being in a position to influence a firm's performance in many ways i.e. through the quality they deliver, timeliness, cost efficiency, reliability, design expertise etc. (Venkatesan, 1992). *Downstream* of the value chain are the *Channels.* Many products pass through the value chains of channels on their way to the buyer. Activities that affect the buyer also influence how the firm's own activities take place. The value chains of each part of the supply chain ultimately adds value to the buyers value chain. *'Gaining and sustaining competitive advantage depends on understanding not only a firm's value chain but how the firm fits in the overall value system.'* (Porter, 1985, p.34)

The value chain is embedded in a larger stream of activities termed the *'value system'*. Suppliers are recognised as being in a position to influence a firm's performance in many ways i.e. through the quality they deliver, timeliness, cost efficiency, reliability, design expertise etc. (Venkatesan, 1992). *Downstream* of the value chain are the *Channels.* Many products pass through the value chains of channels on their way to the buyer. Activities that affect the buyer also influence how the firm's own activities take place. The value chains of each part of the supply chain ultimately adds value to the buyers value chain. *'Gaining and sustaining competitive advantage depends on understanding not only a firm's value chain but how the firm fits in the overall value system.'* (Porter, 1985, p.34)

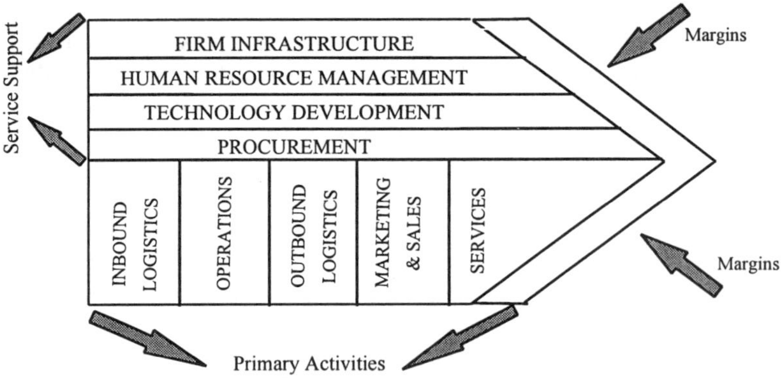

Figure 2. The Generic Value Chain (Source: Porter, 1985)

The importance of cost and economics in this value equation becomes increasingly clear. We suggest that the more cost oriented a firm is, the more likely they are to pursue vertical de-integration strategies. However, it should be noted that value instead of cost must be used in analysing competitive position since firms often deliberately raise their costs in order to command a premium price via differentiation. It is also perhaps worth clarifying here the distinction between *value chain* and *value added.* The value chain as we have seen is an analytical tool used to examine both cost behaviour and differentiation opportunities. Value added is the selling price less the cost of purchased raw materials. This has often wrongly been used (in Porters view), as the focal point for cost analysis because it has been viewed as the area in which a firm can control costs. Value added incorrectly distinguishes raw materials from the other purchased inputs used in a firm's activities. Neither does it identify the tie between a firm and its suppliers that may reduce cost or enhance differentiation. As Venkatesan (1992) comments it is very often firms inappropriate costing systems that lead to them making unsuitable *'make or buy'* decisions. Ford *et al.* (1993, p.207) note that a decision to buy a product or service from an outside supplier rather than use its own facilities to produce it can positively affect the company's cost structure and competitiveness.

We have so far looked at how these new organisational forms might be created within the supply chain and what is likely to drive there development. These organisational forms are now explored in more depth to help gain insights into how they may impact on marketing strategy.

2.4 The Nature of Vertical De-integration

Understanding VDI requires distinguishing between the range of marketing relationships dependent on the level of vertical de-integration. Thorelli (1986) describes vertical de-integration as a flat structure, dependent on administrative processes but not on hierarchies. This lack of hierarchy suggests a need for stronger relationships, as organisations must operate in a network structure of independent firms. The question arises as to whether a firm that pursues VDI strategies is engaged in transactions that are sporadic or that are within a relationship which involves negotiation; whilst they are influenced by market forces this market-based process is not the sole foundation of the business. These new organisational forms are a response to a range of pressures that promote closer alliances between customers and suppliers. *'Contracting out'* enables the firm to respond quickly and flexibly to accelerating change in marketplace dynamics, technology, competition, shorter product life-cycles, the revolution in quality and JIT[2] and customer preferences (Burnett 1992; Webster 1992; Macbeth, 1994; Macbeth *et al.* 1994; Doyle 1995).

A common theme describing these organisational forms emerges from the literature. Each network has a stable core with flexibility coming from the web of other companies (suppliers, distributor, marketing services) surrounding that core. Morgan (1989, p.67) refers to such a model as,

"... a loosely-coupled organic network.....[where] the firm is really a system of firms – an open ended system of ideas and activities, rather than an entity with a clear structure and definable boundary".

In other words, the purpose of these new organisational forms is to increase added value by creating a more flexible structure and capturing the core competencies of outside firms for activities that cannot be matched in-house.

Porter (1980) recognises the importance of the *'vertical de-integration'* decision in his identification of buyers and suppliers as part of his *'five competitive forces'* framework. This framework determines the intensity of competition in an industry and the total value of profits created in a particular industry (Reimann 1989). Achieving and maintaining a competitive advantage largely depends on a firm increasing its overall value while pursuing a cost leadership, differentiation and/or focus strategy. Therefore if long-term relationships can be formed with suppliers and buyers of the core firm (reducing the threat from the bargaining power of suppliers and buyers), then developing a network structure appears an appropriate strategy to increase added value and compete in the global environment. This underlies our argument that added value can be increased through VDI strategies.

---

[2] Just In Time

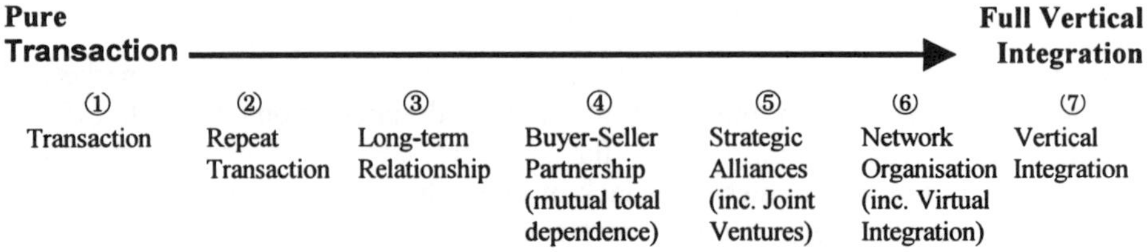

Figure 3. The Range of Marketing Relationships (Source: Webster, 1992)

Further, the range of market relationships can be compared to the degree and type of vertical de-integration. We suggest that the more long term a relationship between a supplier and their customer, the higher the level of market orientation the firm can achieve. There are numerous models that comment on these relationships (see Porter's Five Competitive Forces, 1980; Burnett, 1992; Webster 1992; Macbeth *et al.*, 1994). As Webster (1992) observes the structure and relationship model is a continuum involving more bureaucratic control and less market control as they pursue economic efficiency (see Figure 3.).

2.5 The History of Vertical Integration

In the two decades after the Second World War, organisations were vertically integrated into large, bureaucratic, hierarchical organisations, in which the logic of economies of scale equated efficiency with size (Levitt, 1983). With the introduction of Strategic Business Units (SBU's) and the parallel growth in sophistication of the consumer during the 1970s and 1980s, organisations began to delegate operating and profit responsibilities to the SBU's. Purchasing became recognised as being of strategic importance and as a contributor to business success (Gadde & Hakansson, 1994). With the further globalisation of markets and the increasing search for advantage through niche marketing, organisations had to downsize and cut costs in order to compete as the high administrative costs of integrated business structures became more apparent (Quinn, Doorley & Parquette, 1990; Webster, 1992; Harrison & Kelly, 1993). Many industries are now reversing their historical pursuit of vertical integration. Ford, Chrysler and General Electric buy over half of their component parts from external suppliers. Vertical de-integration seems to make the most sense in industries that have global sources of supply. We suggest that this added flexibility allows companies to better serve their target markets and thus become more market oriented. This underlies our argument that the more market oriented a firm is, the greater the impact vertical de-integration strategies will have on that firms business performance.

2.6 *The Make or Buy Decision*

As we have seen vertical de-integration is a strategic decision and has implications for the overall corporate strategy of the firm. The purchasing literature has traditionally referred to these decisions as *'make or buy'* decisions. In recent years, many companies have been moving significantly away from *'making'* towards *'buying'* (Dobler, Burt & Lee, 1990). However, research carried out by Ford *et al.* (1993) has revealed that VDI decisions are rarely taken within a thoroughly strategic perspective. As McIvor, Humphreys & McAleer (1997) note many firms adopt a short-term perspective and are motivated primarily by the search for short-term cost reductions. They argue that VDI decisions are made most frequently by default with little consideration for the long-run competitiveness of the organisation. Porter (1985) recognises the importance of procurement in the value chain. There is sufficient evidence to suggest that vertical de-integration decisions taken on the basis of cost cutting exercises do not increase and may even damage business performance. Cost reduction have often involved outsourcing, resulting in a sacrifice in quality and added value (Venkatesan 1992). Therefore we suggest that the higher the level of cost orientation a firm has, the more likely the there is to be a negative impact of vertical de-integration on business performance.

Moreover, we suggest that VDI is motivated by either a desire to reduce cost and asset risk, or a desire to improve market competitiveness, or both. Vertical de-integration has been facilitated by technological advances provided by computer and telecommunications technologies. This presents new opportunities for work to be carried out wherever it can be most effectively and efficiently done and offers organisations the opportunity to network and form inter-organisational, knowledge-based relationships, which may transcend former physical and intellectual boundaries (Reich 1991). However as Sherman (1992) observes, one third of VDI ventures end in failure. It is recognised that there may be many contextual factors that effect the success or failure of a network, however we propose that the level of market orientation is central to successful vertical de-integration.

Understanding cost and market orientation as drivers of VDI is important. VDI based on cost orientation alone may lead to short-term cost savings and higher asset turnover but at the expense of lost opportunities for long-term market competitiveness (Ford *et al.* 1993). VDI based on market orientation alone may pose similar problems of inflexibility as vertical integration as long-term relationships become difficult and costly to terminate (Webster 1992). VDI may facilitate market orientation through the generation and dissemination of market intelligence within this more flexible structure.

For an organisation to achieve consistently above-normal market performance, it must create a sustainable competitive advantage (SCA) (Porter 1985). The desire to create superior value for customers and attain SCA drives a business to foster and maintain a culture that will produce the necessary behaviours. Market orientation is an organisation culture (cf. Deshpande & Webster 1989) that effectively and efficiently formulates the necessary behaviour for the creation of superior value for buyers and thus continuous superior performance for the business (Shapiro 1988; Webster 1988; Kohli & Jaworski 1990).

2.7 Market Orientation

There are two broad approaches by which organisations seek to maximise returns for shareholder; 1) cost reduction and 2) increasing value. We have already looked at some of the issues of cost. The delivery of customer value will now be examined in more detail. The process of delivering customer value through the alignment of a firm to its markets is referred to as *'market orientation'*. Market orientation (MO) has been defined as *"Organisation wide generation of market intelligence pertaining to current and future customer needs, dissemination of intelligence across departments and organisation wide responsiveness to it"* (Kohli & Jaworski 1990, p.6). There is growing evidence that market orientation has a positive effect on business performance. Absent from the literature is specific recognition that the level of VDI a firm has impacts on the level of market orientation a firm may achieve. We suggest that there is a positive relationship between the extent of market orientation and the level of vertical de-integration.

To maximise its long-run performance, the business must build and maintain a long run, mutually beneficial relationship with its buyers. Accordingly, a market-oriented seller decides how best to share with its buyer the superior value it creates for them (Jackson 1985). As outlined above, the level of vertical de-integration will influence the marketing strategies adopted by a firm because vertical de-integration determines the level of *'value chain'* or *'channel'* integration. If a firm structure is one of vertical de-integration, opportunities still exist for vertical channel integration where companies within a network co-ordinate their efforts to reach a desired target market. This approach allows the *'core'* member of the value chain to regard other members as an extension of their own operation. The core company thus co-ordinates and develops the marketing strategy for the value chain. Marks & Spencer and IKEA act in precisely this fashion with many of the products that they sell.

Only recently have researchers constructed a theory of the antecedents and consequences of market orientation, developed a valid measure of the construct and tested its effects on business performance (Narver & Slater 1990). Market orientation is recognised as having three behavioural components – customer orientation, competitor orientation, and inter-functional co-ordination; and two decision criteria – long-term focus and profitability, (Narver & Slater 1990). By using this framework as measure of market orientation, we will explore the relationship between market orientation and vertical de-integration.

3 Modelling VDI as a Mediating Variable

Our theory implies what we label the *Vertical De-integration Mediating Variable* (VDIMV) model of market orientation (Figure 4), which focuses on one party in the supply chain structure and that party's level of vertical de-integration and market orientation. Because we hypothesise that vertical de-integration and market orientation are key constructs in business performance outcomes, we position them as mediating variables between five important antecedents and six outcomes.

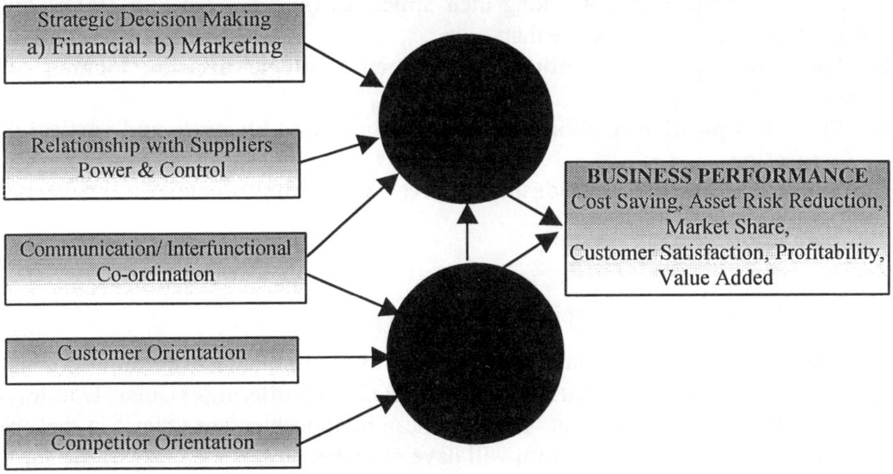

Figure 4. The VDIMV Model of Marketing Orientation

3.1 Precursors of Vertical De-integration and Market Orientation

Drawing on two decades of theory and empirical research on outsourcing and vertical de-integration literature and the recently developing network and new organisational forms literature in marketing, we identify six major precursors of the vertical de-integration and market orientation. Specifically as shown on the left-hand side of Figure 4, we posit that (1) strategic decision making (2) relationship with suppliers, (3) communication, (4) competitor orientation and (5) customer orientation directly influence market orientation and through market orientation, indirectly influence VDI.

3.1.1 Strategic Decision Making

A common assumption in the literature is that there are two categories of strategic decision making, a) financial objectives and b) marketing objectives (Ruekert 1992; Ford *et al.* 1993). On the financial side Harrison & Kelley (1993) identify that it is today's dynamic business environment and the global competition that has increased cost consciousness. Outsourcing provides opportunities for reducing fixed and quasi-fixed costs. On the Marketing side Webster (1992) argues that it is the need of organisations to react quickly to the changing needs and preferences of customer that has lead to organisations re-thinking their structures (also see Ruekert 1992; Doyle 1995). Therefore we hypothesise that:

H_1: There is a positive relationship between strategic decision making and vertical de-integration.

H_{1a}: There is a positive relationship between financial objectives and vertical de-integration.

H_{1b}: There is a positive relationship between marketing objectives and vertical de-integration.

3.1.2 Relationship with Suppliers

A dynamic competitive environment requires firms to continuously seek out products, processes and technologies that add value to their own offerings (Quinn Doorley & Paquette, 1990; Morgan & Hunt, 1994). These relationships are crucial in that they determine the level of control a firm will have over the various elements of the supply chain and thus the product offering ultimately made to the customer (Johnston & Lawrence 1988; Webster 1992; Doyle 1995). Therefore we hypothesise that:

H_{2a}: There is a positive relationship between the type of relationship with suppliers and vertical de-integration

H_{2b}: There is a positive relationship between power/control and vertical de-integration

3.1.3 Communication

Narver & Slater (1990) identify *'inter-functional co-ordination'* as one of the three factors driving market orientation. They define this term as *'the co-ordinated utilisation of company resources in creating superior value for target customers'*, (p.22). They also observe that any point in the buyer's value chain affords an opportunity for a seller to create value for the buyer firm. Kohli & Jaworski (1990) consider the importance of co-ordination and communication in their model of market orientation. For the purpose of this research we use the term *'communication'* to represent the importance of both internal and external communication and co-ordination between the functions of the firm along the supply chain. Anderson &

Narus (1990, p.44) note that communication *'can be defined broadly as the formal as well as informal sharing of meaningful and timely information between firms'*. It is this emphasis on the importance of using communication along the supply channel that leads us to believe the communication is of fundamental importance to the VDI decision. *'Communication can be described as the glue that holds together channels of distribution'* (Mohr & Nevin 1990, p.36). Therefore we hypothesise that:

H_3: There is a positive relationship between communication and vertical de-integration.

H_4: There is a positive relationship between communication and market orientation.

3.1.4 Customer Orientation

The concept of customer orientation has its roots in the development of the marketing concept within both marketing and management literatures. Narver & Slater (1990) define customer orientation as *"the sufficient understanding of one's target buyers to be able to create superior value for them continuously"*(p.21). It is argued that to do this a seller must also understand a buyer's entire value chain (Day & Wensley 1988). Jaworski & Kohli (1993) identify the importance of a firms ability to respond to such information and in being able to do so a firm becomes market oriented (also see Ruekert, 1992; Webster 1992). It follows that in a search to listen to and then respond quickly and effectively to customer needs, a firm must seek out expertise from outside the firm. As we want to examine the indirect relationship that cost has VDI, we therefore hypothesise that:

H_5: There is a positive relationship between customer orientation and market orientation.

3.1.5 Competitor Orientation

The seller has a need to understand the short-term strengths and weaknesses and long-term capabilities and strategies of current and potential competitors (Porter 1980, 1985; Aaker 1988; Day & Wensley 1988; Narver & Slater, 1990). Whilst neither Ruekert (1992) or Jaworski & Kohli (1993) specifically identify the importance of understanding principle competitors, they do emphasis the importance of gathering and using information from the marketplace in the development of marketing strategy. Therefore we hypothesise that:

H_6: There is a positive relationship between competitor orientation and market orientation.

3.2 *Outcomes of Vertical De-integration and Market Orientation*

Although as components of business performance improvement, market orientation and vertical de-integration are desirable outcomes, we posit six outcomes. First, cost saving and asset risk reduction flow directly from vertical de-integration. Second, market share and customer satisfaction are direct results of market orientation. Third, and most importantly we propose that value added and profitability arise directly from both market orientation and vertical de-integration. We theorise that these outcomes, especially value added, promote business success.

3.2.1 Cost Saving

Cost has long been recognised as a primary driver of contracting decisions (Porter, 1980; Venkatesan; 1992; Ford *et al.* 1993). McIvor, Humphrey & McAleer (1997) note that many firms are motivated primarily by the search for short-term cost reduction. Harrison & Kelley (1993) argue that contracting out allows greater numerical flexibility by reducing fixed and quasi-fixed costs. Contracting out avoids investment on what are considered to be non-core activities (Bettis, Bradley & Hamel, 1992). Therefore we hypothesise that:

H_7: There is a positive relationship between cost saving and vertical de-integration

3.2.2 Asset Risk Reduction

It has been argued that new organisational forms have been created in order to react quickly to customer demands (Webster 1992). Firms wishing to achieve such objectives risk exposing themselves to over investment in vertically integrated systems. New organisational forms potentially solve these problems by sourcing external suppliers to deliver the added value required by the customer (Magretta 1998). Contracting out allows fixed costs to become variable costs. Lonsdale & Cox (1998) cite the example of a firm manufacturing components for machinery in-house. It is not necessary to produce these components consistently throughout the year. Despite sporadic usage, the cost of maintaining this function exists all year round. A solution is to contract out this function. We hypothesise that:

H_8: There is a positive relationship between asset risk and vertical de-integration.

3.2.3 Market Share

Previous work measuring the impact of market orientation on business performance has measured the financial performance of the firm (Narver & Slater 1990; Pitt, Cruana & Berthon, 1996). There is recognition in the marketing literature that business performance needs to be measured on market performance criteria, i.e. market share, customer satisfaction, customer retention levels etc. (Green *et al.*, 1995). We theorise that if a firm is better able to achieve market orientation through pursuing vertical de-integration strategies, it will be in a position to increase its market share. Therefore we hypothesise that:

H_9: There is a positive relationship between market share and vertical de-integration.

3.2.4 Customer Satisfaction

Successful companies have satisfied customers. There is a strong association between customer satisfaction and profit (Ambler & Kokkinaki, 1997). Customer satisfaction requires close monitoring of customers needs and wants. Buyers' satisfaction is influenced by the product service delivery system, performance of the product, the image of the company, customers' perceived price-value relationship, employees' level of performance and competitors' advantages and weaknesses (Gulledge, 1988). Considering this in the context of the level of vertical integration a firm has, we investigate the impact of cost as a driver of VDI. It is suggested that transaction cost economics[3] largely ignores the learning effects which might be associated with the establishment tapping into a network of highly technically qualified suppliers facilitating an increase customer satisfaction. By vertically integrating, to avoid potential opportunism, the organisation may forego access to what might be vital information in the subcontracting network. It is this learning and opportunity for adaptability to customer demands that interests marketeers in understanding the new organisational forms (Ruekert, 1992; Webster, 1992; Doyle 1995). Therefore we hypothesise that:

H_{10}: There is a positive relationship between customer satisfaction and vertical de-integration

3.2.5 Profitability

The relationship between firm structure and business performance has been intensively researched over the past decade (e.g. Hay & Morris, 1990), as indeed has the relationship between market orientation and firm performance (see Jaworski, 1996). Ruekert (1992) comments, *"A dominant view of organisational functioning suggests that organisational actions such as the degree of market orientation are inextricably linked to the organisational structures, systems and processes created*

---

[3] The basic concept of transaction cost economics is that a variety of transactions each have a cost associated with each customer - supplier link in the value chain. These costs are not purely financial include a number of limiting factors (see Williamson , 1975)

to sustain them." (p.230). Therefore we theorise that both vertical de-integration and market orientation impact on profitability and further, that vertical de-integration mediates the market orientation/business performance relationship. We hypothesise that:

H_{11a}: There is a positive relationship between profitability and vertical de-integration.

H_{11b}: There is a positive relationship between profitability and market orientation.

3.2.6 Value Added

Quinn, Doorley & Paquette (1990) comment that if firms don't take advantage of new technologies to provide opportunity for lowering costs, restructuring the organisation and redefine strategic focus to more closely match customer needs and create higher added value, they will loose competitive advantage. Suppliers are recognised as being in a position to influence a firms performance in many ways i.e. through the quality they deliver, timeliness, cost efficiency, reliability, design expertise etc. (Venkatesan, 1992; Lonsdale & Cox, 1998). Quinn & Hilmer (1994) comment, *"Perhaps the greatest [way in which managers can leverage their firm's resources when outsourcing is through] the full utilisation of external suppliers' investments, innovation, and specialised professional capabilities that would be prohibitively expensive or even impossible to duplicate internally... ".* Therefore we hypothesise that:

H_{12a}: There is a positive relationship between value added and vertical de-integration.

H_{12b}: There is a positive relationship between value added and market orientation.

3.2.7 Market Orientation & Vertical De-Integration

Finally we suggest that there is a *'two way'* relationship between market orientation and vertical de-integration whereby market orientation drives the level of vertical de-integration and vertical de-integration drives the level of market orientation. We have seen evidence to support the need for understanding customers needs, and additionally the needs of the customers' customer though focusing attention of the value chain and the supply chain (Day & Wensley 1988). We theorise that if each member of the supply chain is market oriented i.e. focused on its customer, and its customers' customer, market orientation will increase. Therefore we hypothesise that:

H_{14}: There is a positive two-way relationship between market orientation and vertical de-integration.

4 Methodology

This study seeks to contribute to an understanding of the characteristics of successful VDI strategies from the perspective of the marketer through a more empirically based approach than currently exists. Empirical evidence existing to support the hypotheses:

- *that there is a positive relationship between the level of vertical integration and business performance* (e.g. Galbraith, Stiles, 1983) and
- *that there is a positive relationship between market orientation and business performance* (e.g. Narver & Slater, 1990),

However, the evidence that firm structure has a mediating effect on the market orientation and business performance relationship is wholly anecdotal. Also, the market orientation relationship has been tested previously against financial measures of performance (i.e. Return On Assets) and not against market measures of performance (i.e. market share, customer satisfaction) which as we suggest, would give a more balance view of business performance.

This study will utilise a cross-sectional survey design to assess the effect of firm structure on market orientation of a *'matched pairs'* sample of British companies competing in the UK. The purpose of the design is to correlate the scores of vertical de-integration tests with the scores on levels of market orientation in order to measure the interrelationship of market orientation and vertical de-integration.

4.1 Pilot Study

Using multiple interviews with senior managers within firms, the pilot study will measure the relative success of the business, the extent of market orientation and vertical de-integration. These semi-structured interviews will take place at ten leading UK companies in various industry sectors. Managers will be encouraged to comment on the questionnaire and amendments to the questionnaire will be made as appropriate. The Chartered Institute of Purchasing & Supply (CIPS) has agreed to make available their list of 3000 members for the main survey and have given their backing to this research project. The ten companies to be studied at the pilot stage will be randomly selected from this list. The Chartered Institute of Purchasing & Supply's list includes firms predisposed to outsourcing issues. A concern here is whether their understanding of market orientation will be sufficient. The multiple interviews within each organisation are designed to address this issue.

After the questionnaire has been tested in face to face interviews with managers to check reliability and validity a pilot mailing will be executed. This will be distributed to a random sample of 40 managers from the CIPS list. To create a matched pair's sample of high VDI firms with low VDI firms, secondary data analysis will use an input-output matrix tool developed by Maddigan (1981). It is anticipated that published company accounts will provide the necessary information for this analysis.

The mailing will include:

- A cover letter declaring support from the Chartered Institute of Purchasing & Supply and (hopefully) the Marketing Council, detailing the importance of their participation and response deadlines.
- The questionnaire
- A guide to completing the questionnaire
- A pre-paid and addressed reply envelope
- Contact details to help with any questions arising

This mailing will be followed up with phone calls and faxes to insure a maximum response. The response rate will be noted, as this will influence the size of the final mailing. The preliminary results will be analysed to check the compatibility of data with the research objectives. Final amendments will be made to the questionnaire as thought necessary.

4.2 Main Survey

The main survey will then be mailed to a matched sample of 2000 leading UK companies in various industries. Often value chains are complex. By executing the survey at the SBU level, the value chain will be simpler and more easily measured (Porter, 1985; Narver & Slater 1990). To control extraneous variables even though all the firms in the sample operate internationally the study will be limited to one country. The minimum response required from the mailed survey is a matched sample of 30 high VDI and 30 low VDI companies competing in similar markets. The primary objective is to collect an additional 30 high MO firms and 30 low MO firms (30 being the minimum response required for statistic analysis of each stratum). The expected response rate is anything between 8% and 10%.

4.3 Measurement

All measures are to be analysed for validity and reliability following guidelines offered by Anderson & Gerbing (1988). It is expected that the following measures will be used. Here, we briefly discuss the origin of the measures used.

4.3.1 Market Orientation

Narver & Slater (1990) identify three behavioural components in their tested model of market orientation. These inferences about the behavioural contents of market orientation (i.e. customer orientation, competitor orientation and inter-functional co-ordination) are consistent with the findings of Ruekert (1992), Kohli & Jaworski (1990), Shapiro, (1988) and Webster (1988). Because the Market Orientation Scale of Narver & Slater (1990) taps the major facets of market orientation its 15 items will be adopted to measure the construct. It is also relevant to note that this measure has been verified and validated at the SBU level and in several countries including the UK (see Pitt, Cruana & Berthon, 1996).

4.3.2 Vertical De-integration

Given our conceptualisation of vertical de-integration, it is essential that its measure should capture both the importance of the VDI decisions to managers (the respondents) and their beliefs about working towards a particular firm structure (i.e. highly vertically integrated). Maddigan (1981) offers a measurement of vertical integration. She notes that today different industries can often serve as input suppliers to each other. Also, one industry's' output is often sold as an intermediate product to several other industries. The network characteristics of these transactions can be highlighted using the input-output matrix that she develops. Vertical integration is often defined as the linking of successive stages of production. However, it is much more than the joining of production processes. Vertical integration describes the firm's strategy of exercising ownership control in the production of products that are used as inputs to each other. Maddigan (1981) comments, *"The firm's management may use vertical integration to increase profitability or decrease the firm' s risk. VIC can be employed as an independent variable in models designed to specifically test the impact of vertical integration on a firm's market power and risk*[4]*. It adds to the arsenal of tools used to investigate the impact of structure on performance."* (p.335).

4.3.3 Business Performance

In the marketing literature has historically used both financial and marketing measures for business performance. Green *et al.* (1995) maintain that market share and profitability are the most common performance dimensions. They define performance as the degree of market success attained by a product at market maturity or the point at which product boundary changes. Other measures of performance they identify include customer satisfaction, ROI, ROS, unit sale,

---

[4] For a summary of the theoretical models of vertical integration, see Kaserman (1978)

customer loyalty, market acceptance and market share growth. These indicators will be adopted for our measurement of business performance thereby allowing us to compare the outcomes of both financial and marketing objectives. The authors suggest that the selection criterion for performance should be the importance of the measures for the players in the product market.

4.4 Data Analysis

Analysis of the data will follow the line of previous research in this field and will test for instrument reliability and validity. It is intended that the data will be randomly split into two samples before assessing reliability and validity (Churchill 1979). The reliability analyses will be conducted on the first sample and replicated on the second, then tests for construct validity on the combined samples will be conducted. Evidence of construct validity will be sort by comparing the pattern of correlations among variables with the predicted theory. Where they conform we have construct validity (Cronbach 1970, p.143; Kerling 1973, p.463).

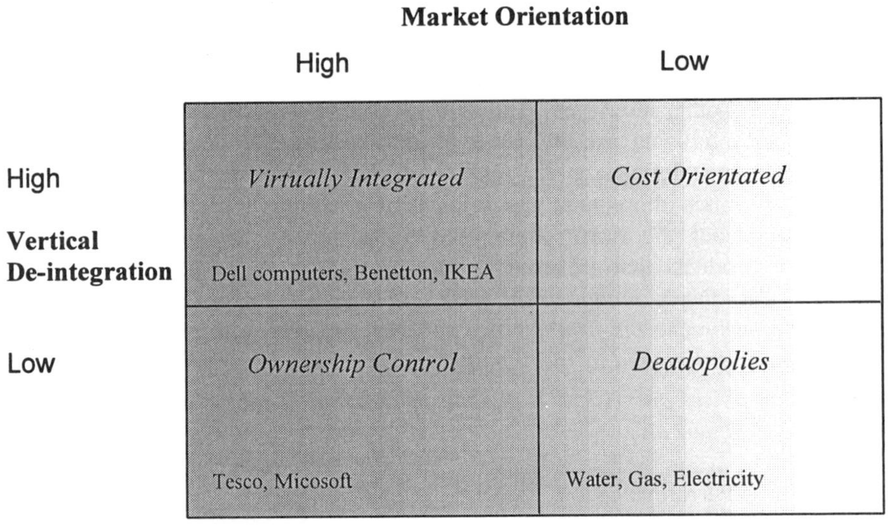

Figure 5 Matrix of The Marketing Orientation & Vertical De-Integration Relationship

The three areas of measurement are market orientation, and vertical de-integration and business performance. Company demographics will act as the control variables, (i.e. Industry sector, firm size, turnover). As noted previously, the survey sample will include leading UK companies in various industries. The relative success of the business is to be measured and strategies of market orientation identified using multiple interviews with senior managers within the businesses. It is expected that the level of market orientation achieved will have a positive relationship with the degree and control over vertical de-integration and business performance. It is expected that the optimal level of vertical de-integration in order to achieve high market orientation and high business performance will be dependent on industry sector, maturity of the industry and firm size.

SPSS will be used for preliminary analysis. Various forms of multivariant analysis will be adopted (factor analysis, MANOVA, cluster analysis). The expectation is that vertically de-integrated firms will score differently from vertically integrated firms in their measure of market orientation. The matrix below demonstrates the expected finding where the type of integration is less important generically than the level of market orientation. It is thought that the level of integration may help improve market orientation dependent on industry sector.

4.5 Limitations of Research

The cross-sectional design does not have a time dimension and thus we cannot use this to look causal influences. Instead it attempts to simulate this by the statistical elimination of differences between the two groups (High VDI and low VDI). The assumption is that if we can match the two groups in all ways except on the experimental variable (the level of vertical de-integration) then any differences in the dependent variable (market orientation) must be due to the effect of the experimental variable.

In the end we cannot be sure that we have completely matched the two groups or controlled for all the factors that might possibly lead to the observed differences between groups. We can only do our best and control for factors about which we have information and which, on the basis of previous research, have been shown to be relevant.

As Venkratraman & Ramanujam (1987) note,

> *"The use of a single informant per unit of analysis to collect data on organisational-level performance constructs limits our ability to rule out the possibility of functional or response bias (Huber & Power 1985, Philips 1981). However, since in our study respondents are organisational mangers, we can argue that they are representative of the organisations. Nevertheless, as noted by Venkatraman & Grant (1986) the need to employ multiple respondents to measure organisational-level constructs such as Business Economic Performance cannot be dismissed by researchers."* (pp117,118).

This argument underlies the approach to be taken in the pilot study for this research but the limitations of not using multi respondents in the main survey are duly noted.

6 Contribution to Knowledge

Research into how new organisational forms are affecting firm performance is at the centre of theoretical concerns in marketing and strategic management. This research will make an original theoretical contribution to the debate surrounding networks, market orientation and firm performance in the context of the global business environment in which UK firms have to compete.

The relevance to practitioners will come from guidelines developed from an understanding of what networks are doing to achieve market orientation and why they are doing it. Gaining an understanding of why certain strategies are appropriate will help practitioners create solutions to their own unique set of problems. If managers can be provided with a framework explaining the relationship between cost, market orientation and firm structure decisions, they will be in a better position to make appropriate strategic decisions and measure their outcomes.

References

Aaker D. 1988. *Strategic Market Management, 2nd Ed.* New York: John Wiley & Sons Inc.

Ambler T and Kokkinaki F. 1997. "Measures of Marketing Success." *Centre for Marketing Working Paper* 97-902(April):1-29.

Anderson J C and Gerbing D W. 1988. "Structural Modelling in Practice: A Review and Recommended Two Step Approach." *Psychological Bulletin* 103(3):411-23.

Anderson J C and Narus J A. 1990. "A Model Of Distributor Firm And Manufacturer Firm Working Partnerships." *Journal of Marketing* 54 (January):42-58.

Bettis R, Bradley S, and Hamel G. 1994. "Outsourcing and Industrial Decline." *Academy of Management Executive* 6(1):19.

Burnett K. 1992. "Strategic Customer Alliances." *Financial Times* (London).

Chandler A D. 1962. *Strategy & Structure: Chapters In The History Of American Industrial Enterprise.* Cambridge: MIT Press.

Day G S and Wensley R. 1988. "Assessing Advantage: A Framework for Diagnosing Competitive Superiority." *Journal of Marketing* (April):1-20.

Deshpande R, Farley J U, and Webster F E Jr. 1993. " Corporate Culture, Customer Orientatio, and Innovativeness in Japanese Firms: A Quadrad Analysis." *Journal of Marketing* 57(1):23-37.

Dobler D W, Burt D N, and Lee L. 1990. *Purchasing and Materials Management.* New York, NY: McGraw-Hill.

Doyle P. 1995. "Marketing In The New Millennium." *European Journal Of Marketing* Vol.29 (13):23-41.

Ford D, Cotton B, Farmer D, Gross A, and Wilkinson I. 1993. "Make-or-Buy Decisions And Their Implications." *Industrial Marketing Management* 22:207-14.

Gadde L and Hakansson H. 1994. "The Changing Role Of Purchasing: Reconsidering Three Strategic Issues." *European Journal of Purchasing and Supply Management* 1(1):27-35.

Green D H, Barclay D. W. and Ryans A. 1995. "Entry Strategy and Long-Term Performance." *Journal of Marketing* 59(4):1,16.

Gulledge L. 1988. "Measure Satisfaction, Performance to Meet Customers' Expectations." *Marketing News* (March 14):34-35.

Hamel G and Prahalad C K. 1994. *Competing For The Future*. Boston, MA: Harvard Business Press.

Handy C. 1990. *The Age of Unreason*. Boston: Harvard Business School Press.

Harrison B and Kelley M R. 1990. "Outsourcing and the Search for Flexibility." *Work Employment and Society* 7(2):213-35.

Hayes R H and Wheelwright. 1984. *Restoringour Competitive Edge*. New York: Wiley.

Jackson B B. 1985. *Winning and Keeping Industrial Customers*. Lexington KY: Lexington Books.

Jaworski B and Kohli A K. 1993. "Market Orientation: Antecedents and Consequences." *Journal of Marketing* 57(3):53-70 .

Johnston R and Lawrence P R. 1988. "Beyond Vertical Integration - the Rise of Value-Adding Partnership." *Harvard Business Review* (July-August):95-101.

Kay J. 1993. *Foundations of Corporate Success: How Business Strategies Add Value*. New York: Oxford University Press.

Kohli A K and Jaworski B J. 1990. "Market Orientation: The Construct, Research Propositions and Managerial Implications." *Journal of Marketing* 45(2):1-18.

Kotler P. 1991. *Marketing Management, Analysis, Planning, Implementation and Control*. Prentice-Hall.

Levitt T. 1983. "The Globalisation of Markets." *Harvard Business Review* (May-June):92.

Lonsdale C and Cox A. 1998. *[Out]Sourcing: A Business Guide to Risk Management Tools and Techniques*. Bath, UK: Earlsgate Press.

Macbeth D K. 1994. "Partnering Strategy: Some Theoretical Arguments in Favour." *Paper Delivered at 3rd IPSERA International Conference* (28-30th March).

Macbeth D K, Ferguson N, and Neil G C . 1994. *Partnership Sourcing- An Integrated Supply Chain Management Approach*. London: Pitman Publishing.

Maddigan R J. 1981. "The Measurement Of Vertical Integration." *Economics and Statistics* 63(3):328-35.

Magretta J. 1998. "An Interview With Dell Computer's Micheal Dell." *Harvard Business Review* 76(March-April):86-100.

Miles R F and Snow C C. 1986. "Organisations: New Concepts for New Forms." *California Management Review* (3):62-73.

Mohr J and Nevin J R. 1990. "Communication Strategies in Marketing Channels: A Theoritical Perspective." *Journal of Marketing* 54(October):36-51.

Morgan G. 1989. *Creative Organisation Theory* . California: Sage.

Morgan R M and Hunt S D. 1994. "The Commitment-Trust Theory Of Relationship Marketing." *Journal Of Marketing* 58(July):20-38.

Naver J C and Slater S F. 1990. "The Effect Of A Market Orientation On Business Profitability." *Journal of Marketing* (October):20-35.

Ohmae K. (1989). "Managing a Borderless World." *Harvard Business Review* (May-June).

Parvatiyar A, Sheth J N, and Whittington F B Jr. 1992. "Paradigm Shift in Interfirm Marketing Relationships: Emerging Research Issues." *Working Paper* (Emory University).

Peters T J and Waterman R H. 1982. *In Search of Excellence*. New York: Harper & Row.

Pitt L, Cruana A, and Berthon P R. 1996. "Market Orientation And Business Performance: Some European Evidence." *International Marketing Review* 13(1).

Porter M E. 1990. *The Competitive Advantage of Nations*. New York : Macmillan.

———. 1980. *Competitive Strategy: Techniques for Analysing Industries and Competitors*. New York, NY: The Free Press.

Quinn J B . 1992. *Intellegent Enterprise* . New York: New York Free Press.

Quinn J B, Doorley T L, and Paquette P C. 1990. "Beyond Products:Services Based Strategy." *Harvard Business Review* (March-April):58-67.

Quinn J B and Hilmer F G. 1994. "Strategic Outsouring." *Sloan Management Review* (Summer):43-55.

Reich R B. 1991. "Who Is Them?" *Harvard Business Review* (March-April):77-88.

Reimann B. 1989. "Sustaining The Competitive Advantage." *Planning Review* 17(2):31.

Ruekert R W. 1992. "Developing A Market Orientation: An Organizational Strategy Perspective." *International Journal Of Research In Marketing* 9(3):225-45.

Shapiro B. 1988. "What the Hell Is 'Market Oriented'?" *Harvard Business Review* (November/December):119-25.

Thorelli H. (1986). "Networks: Between Markets and Hierarchies." *Strategic Management Journal* 7:37-51.

Venkatesan R. 1992. "Strategic Sourcing: To Make Or Not To Make." *Harvard Business Review* 70(6):98-107.

Venkatraman N and Ramanujam V. 1987. "Measurement of Business Economic Performance: An Examination of Method Convergence." *Journal of Management* 13:109-22.

Webster F E Jr. (1992). "The Changing Role of Marketing in the Corporation." *Journal of Marketing* 56(4):pp.1-17.

———. 1988. "The Rediscovery of the Marketing Concept." *Business Horizons* 31(May-June):29-39.

DATA MINING IN DIRECT MARKETING DATABASES

PETER VAN DER PUTTEN
Sentient Machine Research, Amsterdam
and Dept. of Computer Science, Leiden University
E-mail: putten@cs.leidenuniv.nl

In direct marketing large amounts of customer data are collected that might have some complex, non linear relation to customer behavior. Data mining techniques can offer insight in these relations. In this paper we give a basic introduction in the application of data mining to direct marketing. Best practices for data selection, algorithm selection and evaluation of results are described and illustrated with a number of real world cases. We suggest two lines of research which we consider important to put data mining in the hands of the marketeer: automating data mining techniques and integration of data mining in an open knowledge management framework.

1 Introduction

In marketing, there are two opposed approaches to communication: mass marketing and direct marketing. In mass marketing, a single communication message is broadcast to all potential customers through media such as print, radio or television. Such an approach always implies a high waste: only a small proportion of the customers will actually buy the product. Now that competition increases and markets get more fragmented the problem of waste worsens. Moreover, in spite of huge investments in market research and media planning, it is still hard to quantify the benefits of mass marketing.

These developments have led to an increased popularity of direct marketing, especially in the sectors of finance, insurance and telecommunication. The ultimate goal of direct marketing is cost-effective, two-way, one-to-one communication with individual customers. For this it is essential to learn present and predict future customer preferences. In today's chaotic business environment, customer preferences change dynamically and are too complex to derive straightforwardly.

Data mining, the continuous analysis of customer behavior patterns, may offer a flexible solution to this problem.[1,2,3] In this paper we will give a practical introduction to data mining for direct marketing purposes.

We will not discuss any theoretical algorithmic issues, nor will we describe experiments in detail. We only aim to offer managers a tutorial style introduction to current data mining best practices: how is data mining commonly applied and evaluated, and which data and algorithms are most appropriate, given common direct marketing tasks.

In the first part we will describe the data mining process in a direct marketing context. A case from insurance is added to give an impression of the practical issues related to data mining projects, including the evaluation of data mining results. In the second part we will focus on lessons learned with respect to the selection of data and algorithms, based on eight data mining projects carried out in co-operation with the Dutch association for direct marketing, sales promotion and distance selling. We conclude with suggesting directions for research.

2 Data mining

Data mining can be defined as the extraction of valuable patterns that are hidden in large amounts of customer data.[4] The whole process of steps involved in data mining is often referred to as the knowledge discovery cycle (see figure 1). This includes definition of the objectives, selection and preparation of the data and evaluation of the results with technical and business criteria.

Within the loop of a single project, it is not uncommon to go through the knowledge discovery cycle a number of times. For instance, by doing data mining analysis one might discover that some important data was not selected or was not prepared in the appropriate format. By performing different data mining projects repeatedly, an organization starts to learn more and more about customers. This can be considered to be a second loop of learning.

The current success of data mining in businesses is facilitated by a number of technical factors. Growing amounts of customer data are collected and made accessible in corporate data warehouses. Powerful new data analysis algorithms are discovered by researchers from statistical pattern recognition and artificial intelligence fields such as machine learning, neural networks and evolutionary computation. Today, ordinary office computers are powerful enough to run these advanced data mining algorithms.

In a direct marketing context, two prototypical data mining objectives can be distinguished: *prediction* and *description*. Prediction involves predicting unknown or future customer behavior from known customer attributes. Description aims at discovering human interpretable patterns in the data. Best practice application of prediction and description in direct marketing are described below. For a detailed example we refer to the insurance case described in the next section.

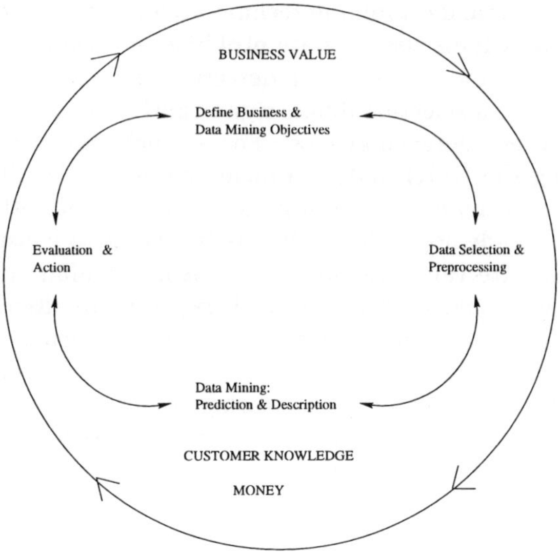

Figure 1. The Knowledge Discovery Cycle.

2.1 Prediction

The classical case in direct marketing for prediction is response modeling. Usually, the relative number of customers that responds to direct mail is very low (5% or less). Predictive models can be built to identify the prospects most likely to respond. Historical data about previous mailings or product ownership are used to construct the model. If such information is unavailable, for instance when selecting prospects for a new product, a test mailing is performed to collect information. The resulting model can be applied to filter prospects from the existing customer base or from address lists acquired from list brokers. Although response analysis is by far the most common type of predictive modeling, other applications, such as predicting customer retention or estimating turnover potential, are promising for direct marketing as well.

2.2 Description

A shortcoming of prediction is that it produces models that are black boxes. A response prediction model is perfect for selecting interesting prospects. But it does not give the marketeer more insight into the reasons why these customers respond.

There are a several data mining techniques for descriptive modeling. The main subtasks we will discuss here are profiling and segmentation.

Profiling, also known as deviation detection, is used to discover for which attribute-value pairs a selection differs significantly compared to the database as a whole or other reference groups. For example, we considered profiling useful for mining the Dutch national media survey (SUMMO). Every questionnaire contained hundreds of questions on media interests, product consumption and socio-demography, so it was infeasible to construct the profile of deviating attribute values manually. By using profiling for an analysis of vodka drinkers we found that vodka drinkers are more often students, drink more Bacardi rum and are more frequent visitors of cinemas.

In segmentation, the goal is to discover subgroups in data. Customers within a cluster should resemble each other as much as possible, where as the clusters themselves should differ as much as possible. For example, in the vodka case we found out that the average vodka drinker does not exist. Instead, subgroups were found that could be described as "cocktail drinking teenagers", "young student couples" and "travelling salesmen".

3 Insurance case

The business objective in the insurance case was to expand the market for an existing consumer product, a caravan insurance, with only moderate cost investment.[a] We identified two data mining objectives: selecting individual prospects and describing existing customers.

3.1 Data selection and preprocessing

Each customer was characterized by a selection of 95 attributes. The attributes could be divided in two groups. The product usage attributes defined the product portfolio of an individual customer, so these attributes can be considered to be internal (company owned), behavioral attributes. We also purchased socio-demographic survey data that had been collected on zip code level. All customers belonging to the same zip code area have the same value for these attributes. This included information on education, religion, marital status, profession, social class, house ownership and income. The selection of attributes to be used was made based on expert domain knowledge and exploratory data analysis (correlation with attributes to be predicted).

[a] Some details in the insurance case have been changed for reasons of confidentiality, without loss of generality.

A number of preprocessing steps were taken, some of which are provided directly by the data mining environment we used.[b] Most numerical attributes were transformed to categorical values. For each attribute, normalization factors were computed so that all attributes had the same standard deviation. Missing values were identified so that the algorithms that were going to be used could handle these values correctly.

3.2 Construction and evaluation of the response prediction model

To select prospects we constructed a model to predict the attribute "owns a caravan policy" given all other attributes. A random sample, the training set, was drawn from the customer base. The training set was used to construct a nearest neighbor prediction model. In a nearest neighbor model, the prediction for a given customer is computed by averaging over the response of the most similar customers from the training set.[5] Similarity between customers is computed by combining the distance per attribute, taking several influencing factors into account, such as attribute types (categorical or numerical), normalization factors and missing values. Missing values for the attributes of the customer for which prediction was to be made were simply ignored (don't care values); missing values for customers from the training set resulted in some constant distance for this attribute (don't know values). The relative importance of different attributes was computed by running a profile analysis of the policy owners compared to all customers in the training set. A number of attributes were assigned very low importance, so the actual number of attributes taken into account to compute the resemblance was reduced.

The prediction model was applied to the testing set, a random sample of customers, disjoint with the training set. For each test customer, the response score was computed to indicate the potential to buy a caravan insurance.

Given a low response rate of 6.5%, the naive prediction model which scores all records as non-respondents, already achieves 93.5% classification performance. So this standard data mining measure, which counts the relative numbers of cases for which the classes were predicted correctly, did not suffice. This suggests that other evaluation criteria were needed to evaluate the accuracy of the prediction model.

For this kind of analysis, often a cumulative response chart is used (see figure 2). All records are ordered from left to right on the x-axis with respect to their predicted response rate. If only the top 25% is mailed, the cumulative response $RespCum_m$ (relative number of respondents in mail selection) is

[b]The DataDetective^{TM} data mining environment.

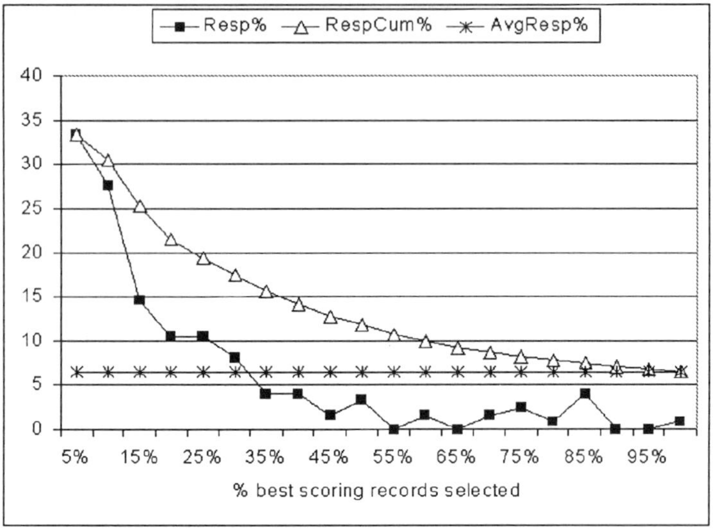

Figure 2. Cumulative Response Chart Insurance Data.

21.5%, which is more than 3 times higher than the response level $RespCum_r$ achieved when records are selected randomly. Another way to evaluate the model is shown in figure 3. Here the relative part of all respondents that is found is plotted:

$$TotResp_m = \frac{RespCum_m * s}{RespCum_r}$$

with s selection size). If the top 25% customers are selected, nearly 75% of all respondents are found (!). The Gain measure:

$$Gain_m = \frac{(RespCum_m - RespCum_r) * s}{RespCum_r}$$

indicates when the gains of using a prediction model start to decrease.

The optimal mail selection size depends on the gain, cost and profit per mail piece, marketing strategy, mining costs etc. Assume that in the insurance case the strategy was just to mail the top 25% of the customers with the highest response score and we wanted to maintain the same absolute number of respondents. So the customer base we had to select from had to contain 100/75 % customers. We only had to mail $25 * 100/75$ % $= 33.33$% of the amount of customers we would have had to mail without a prediction model.

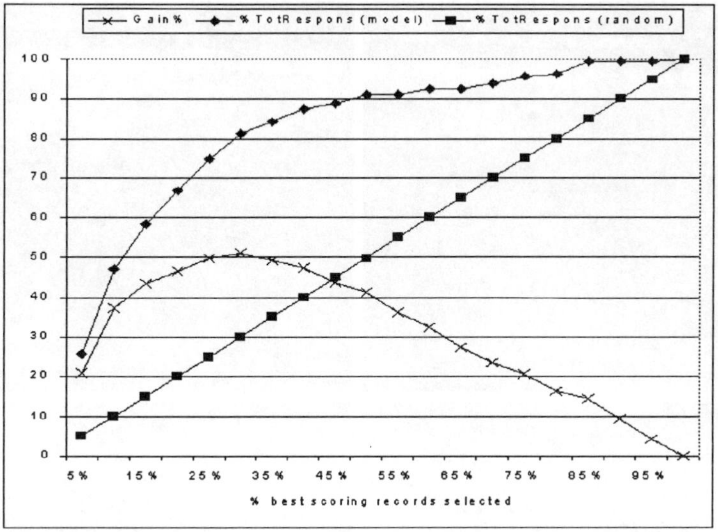

Figure 3. Gain Chart Insurance Data.

Considering that mass mailings to millions of customers at 1 US$ per mail piece are common,[6] reductions in cost can be substantial.

3.3 Descriptive segmentation of respondents

Contrary to prediction, description does not aim to offer results that can be measured and interpreted in a single, objective manner. There exist few standard algorithm-independent error measures for segmentation techniques, and even less for profile or rule discovery. The quality of the description rather depends on the way the marketeer interprets the descriptions, the conclusions that are drawn and the actions that are taken, which is, of course, subjective and hard to measure.

In the insurance example, we wanted to find segments of caravan insurance owners which should be marketed with a different tone of voice, different offers etc. For this we used an iterative projection algorithm, which projected all owners from high (95) dimensional space on to 2 dimensions in a number of steps (see figure 4). In each step, every customer performs a nearest neighbor search to find resembling customers and moves a little bit in that direction. When the projection process had converged, we built customer profiles of the

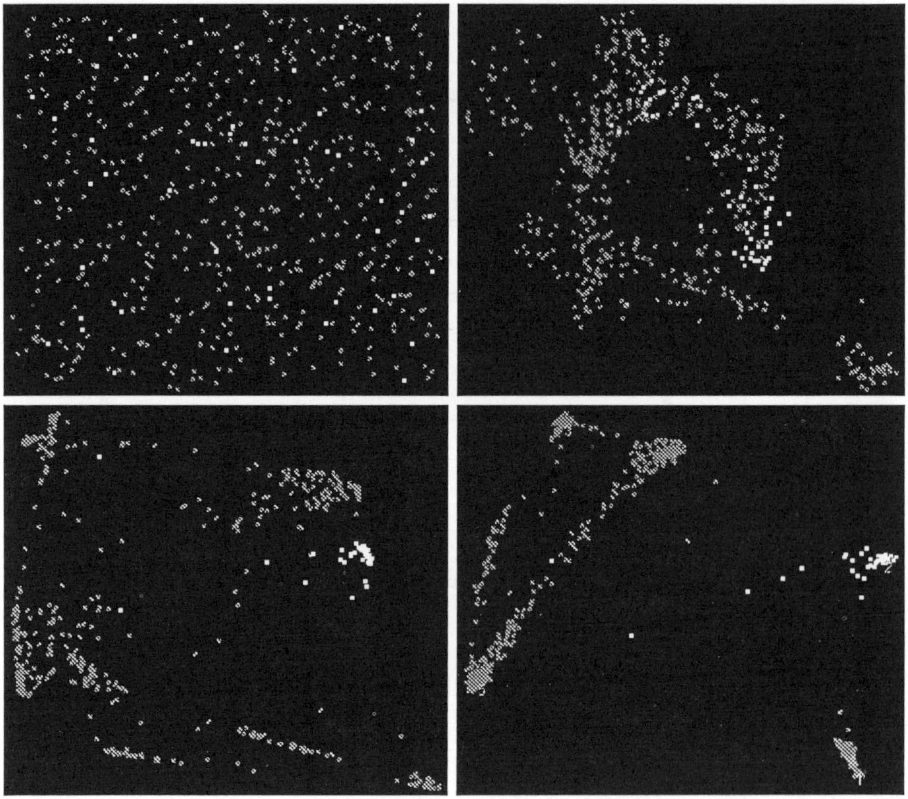

Figure 4. Iterative nearest neighbor projection reveals clusters in data. The figure at the right hand bottom side shows the end result. The user assigns the points to clusters depending on the final state of the projection and the resulting customer profiles. The white points belong to cluster 2.

resulting clusters. For instance, if we compared cluster 2 to all customers and the other clusters, we found out that the cluster contained relatively loyal customers (two car insurance policies, more often in a high turnover category), who where living in residential areas with a high rate of privately owned houses and who were more often belonging to a religious, senior citizen family structure.

Organization	Sector	Business Objective
Fund 1	charity	segmentation fund members
Fund 2	charity	upgrading, estimation donation potential
NS	transport	identifying suspects for prolongation of railways discount card
ABN AMRO	banking	selecting prospects for a new service (balance check by telephone)
VSB	banking	cross selling saving programs to current account holders
Readers Digest	publishing	selecting prospects from a list broker database
Centraal Beheer	insurance	selecting potential converters from a list broker database
NV Databank	list broker	selecting prospects for marketing of a new product

Figure 5. Cases in the DMSA project.

4 Data Mining in Marketing Practice: the DMSA cases

In this section we will propose some best practices for data selection and data mining algorithm selection, based on data mining experiences from a variety of direct marketing projects.

Developing a single data mining algorithm that offers the best peak performance (e.g. predictive accuracy) on all possible data mining problems might seem a good approach for research. However, a recent theorem, known as the "No Free Lunch Theorem", seems to suggest that such an approach will probably not be succesful. This theorem states that measured over all possible data sets and cost functions, there can be no single algorithm that performs best .[7] So it makes more sense to identify the requirements for which a certain class of algorithms performs better.

Furthermore, researchers sometimes assume that the data set is given. However, the choice of data is probably even more important than the algorithm used. In the specific context of marketing, it might be possible to develop best practices for collecting appropriate data.

These issues have been the focus of a research project Sentient Machine Research has performed in co-operation with the Dutch Organization for Direct Marketing, Sales Promotion and Distance Selling (DMSA). To be more exact, the objective of the research was to identify under which circumstances and for which data the relatively "new" data mining algorithms such as neural networks, rule induction and evolutionary algorithms performed "better". We compared these algorithms to classical statistical techniques such as linear regression and discriminant analysis. Experiments were performed in eight real world data mining projects for several organizations, including banks, insurance companies, publishers, a railway company and charities[6] (see figure 5). In the next subsection, we would like to share some of the lessons learned from this project.

4.1 Collecting the right data

A general result was that the data used was the most important factor for the success of the data mining projects. Data collection was always a critical constraint in project planning. Data preparation amounted to up to 80% of total work invested. The requirement for what kind of data to use depends primarily on the nature of the data mining task at hand.

For prediction tasks, the data should possess as much predictive power as possible. Firstly, the number of attributes plays an important role. The more attributes are used, the higher the probability becomes that intricate non-linearities between attributes occur which are important for predicting customer behavior. On the other hand, the so called "Curse of Dimensionality"[8] limits the amount of attributes that can be used. If the number of attributes increases, the density of the data set in pattern space drops exponentially and complexity of models can grow linearly or worse. Complex models (i.e. a large number of parameters) have a higher chance of overfitting to the training data and will not perform well on new data (low generalization).

Secondly, the type of attributes to be used is of importance. The best data to use is internal behavioral customer data which relates directly to the products to be marketed or the customer behavior to be predicted. Examples are product usage, information requests, response, account balances etc. Traditional marketing attributes such as social class, education, age and gender do not suffice to predict modern customer behavior. Empirical support for these claims is shown in figure 6. Highest gains were achieved in projects based on data as described above.

For description tasks, however, raw summaries of product usage etc. do not suffice to inspire marketeers. Descriptive attributes, such as socio-demographic attributes that are collected at zip code level, should be added. These attributes typically possess much less predictive power, but offer more insight to marketeers.

4.2 Choosing the right algorithm

In the DMSA project, we roughly distinguished between adaptive pattern recognition techniques such as neural networks, rule induction and genetic

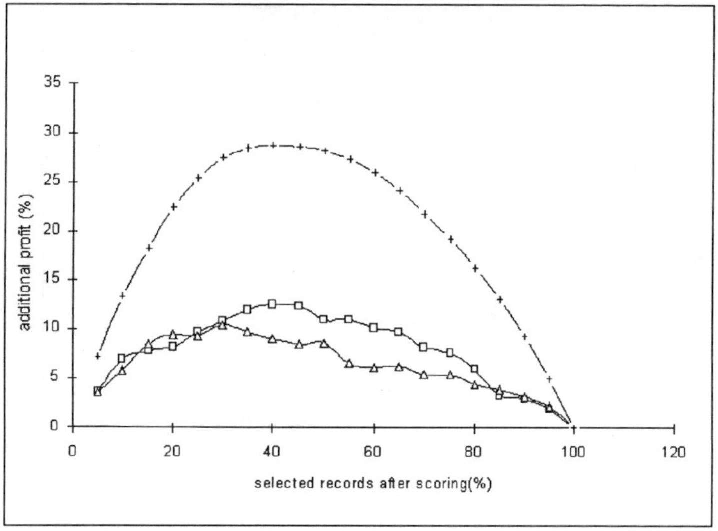

Figure 6. Chart with gains of several projects. Including internal data results in higher gains for response prediction tasks.

algorithms and classical linear statistical techniques such as regression and discriminant analysis. The main advantage of adaptive techniques in general is that these techniques are able to model highly non linear relationships. Furthermore, it is often claimed that these algorithms are less parametric, i.e. make less assumptions about the relation between customer attributes and predicted behavior. For linear regression for instance, this relation is assumed to be linear, which is a pretty tough assumption.

However, what is often not mentioned is that adaptive techniques require implicit assumptions about the data as well. Also, it is sometimes claimed that a high noise level might linearize the non-linear relations that are present in real world marketing data. Practical aspects such as speed, amount of data preparation needed and ease of use are also important. For example, linear regression is still one of the fastest algorithms around when it comes to scoring large data sets.

Overall we found that when the number of attributes was large enough for non-linearities to occur, neural networks performed best in "creaming the crop": selecting the top customers in a prediction task (figure 7). We achieved cumulative response rates that were up to twice as high as the runner up al-

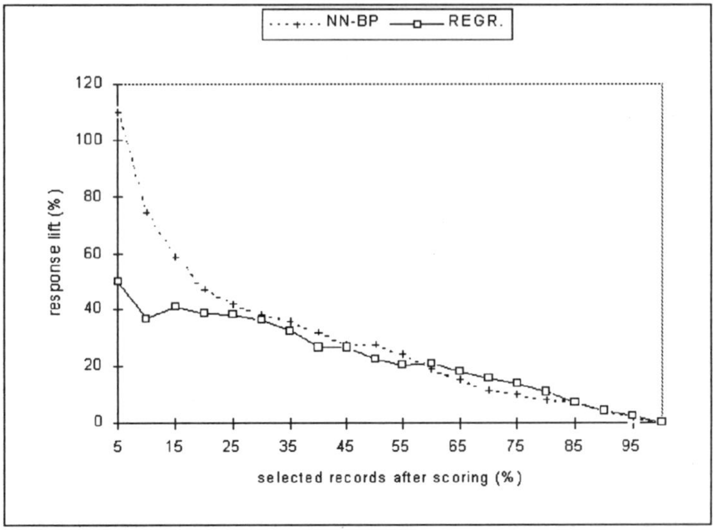

Figure 7. Creaming the crop with backpropagation neural networks. This graph displays the lift $RespCum_m/RespCum_r$ for linear regression versus backpropagation neural networks for one of the DMSA projects.

gorithms, which can correspond to considerable savings. For larger selections, the advantage of neural networks diminishes. This is reasonable to expect, because the relation between customer attributes and response for customers that have an average score is very weak. So if the marketeer is mainly interested in two-way, high quality relations with top customers, which are likely to be interested in the offer, application of adaptive techniques for data mining makes sense. If a marketeer is more interested in performing large bulk mailings with a high response in absolute numbers, wants to mail a relatively large proportion of the customer base, and is willing to accept a high rate of non response, classical techniques such as linear regression will suffice and will be faster to apply.

5 From Data Mining to Knowledge Discovery

The goal of data mining should be the transition to a learning, customer oriented organization. Customer analysis should be performed on a regular basis, or customers should even be monitored on line. Prediction should not be limited to response analysis, but should be a business activity aimed at modeling higher level customer attributes, needs and attitudes, such as "willing to take risk", "high probability of churn" or "early adopter".

Reality is still different. A general problem of most data mining projects is that data collection is a major bottleneck, data mining algorithms still require a lot of parameterization and data preparation and reusability of data mining results is poor. To shorten the knowledge discovery cycle we suggest two important directions for research: automating data mining techniques and integration of data mining in an open knowledge management framework.

5.1 Automating data mining algorithms

Although current data mining algorithms generally require less parameters to be set and less data preparation to be performed than classical statistical algorithms such as linear regression, users still need to have a low level understanding of how a specific algorithm works.

We identify several possible directions to solve this problem. First, in a practical approach, one could identify heuristics for making reasonable choices for specific applications, data mining techniques and steps in the knowledge discovery cycle. These best practices could be properly documented in some kind of data mining methodological framework, or ideally, these best practices are incorporated in intelligent assistants which guide the user through data preparation and prediction processes.

There are also less heuristic and more general algorithmic approaches, which are sometimes referred to as meta learning methods. These algorithms learn to make choices which were normally made by the data mining analyst. For example in ensemble learning, a set of models with different parameter settings is constructed, and the predictions are combined linearly. Recently, various theoretically founded methods were developed to set the coefficients of this linear combination.[9] Another algorithmic approach is Bayesian learning. In Bayesian learning, there is no single parameter setting, but rather a probability distribution over the possible parameter values, which is updated if new data arrives.[8]

5.2 Combining knowledge management and data mining

Whereas it might be fruitful to aim at automating the data mining algorithms, research into the direction of a more open and integrated framework for storing and reusing data mining processes and results would be useful as well. In cognitive psychology there is an important lesson known as the Learning Paradox: "He who knows nothing can learn nothing". There have only been a few attempts to introduce knowledge management to data mining.[10,11] In our view research must not focus on a a single analysis or algorithm, but on a number of analyses in a learning organization. For this it is required that analyst and domain knowledge is represented, a library of data mining results is maintained and intelligent assistance is offered based on this knowledge base.

6 Conclusion

Data mining can be a helpful tool for managers and organizations to cope with a dynamically changing and complex business environment. We identified best practices for application of data mining for direct marketing, selection of data and algorithms and evaluation of results. The key to successful application of data mining will be integration in information infrastructure and business processes.

References

1. R.P. ter Heide. Neural Networks applied to Direct Marketing. In *Proceedings SNN Symposium Amsterdam*, 1995.
2. Charles X. Ling and Chenghui Li. Data mining for direct marketing: Problems and solutions. In *Proceedings 4th International Conference on Knowledge Discovery in Databases (KDD-98)*, New York, 1998.
3. Michael J.A. Berry and Gordon Linoff. *Data mining techniques for marketing, sales, and customer support.* John Wiley & Sons, New York, NY, 1997.
4. Usama M. Fayyad, Gregory Piatetsky-Shapiro, Padhraic Smyth, and Ramasamy Uthurusamy. *Advances in Knowlege Discovery and Data Mining.* MIT Press, Cambridge, Mass., 1996.
5. Belur V. Dasarathy. *Nearest Neighbor (NN) Norms: NN Pattern Classification Techniques.* IEEE Computer Society Press, Los Alamitos, CA, 1991.
6. E. Wagenaar. *Data Mining in Marketing Databases.* DMSA, Amsterdam, NL, 1997. In Dutch.
7. D.H. Wolpert and W.G. MacReady. No free lunch theorems for search. Technical Report SFI-TR-95-02-010, The Santa Fe Institute, 1996. ftp://ftp.santafe.edu/pub/wgm/nfl.ps.
8. Christopher M. Bishop. *Neural Networks for Pattern Recognition.* Oxford University Press, Oxford, UK, 1995.
9. Michiel C. van Wezel, Walter A. Kosters, and Joost N. Kok. Maximum

likelihood weights for a linear ensemble of regression neural networks. In *Proceedings ICONIP'98*, Kitakyushu, Japan, 1998.

10. R. Wirth, C. Shearer, U. Grimmer, Th. Reinartz, J. Schlösser, Chr. Breitner, R. Engels, and G. Lindner. Process-oriented tool support for kdd. In *Proceedings of the 1st European Symposium on Principles of Data Mining and Knowledge Discovery*, Trondheim, Norway, 1997.
11. Robert Engels, Guido Lindner, and Rudi Studer. A guided tour through the data mining jungle. In *Proceedings of the 3rd International Conference on Knowledge Discovery in Databases (KDD-97)*, Newport Beach, CA, 1997.

RESEARCH DIRECTIONS ON COMPUTER INTEGRATED ENTERPRISE

IRINA NEAGA
"Gh. Asachi" Technical University, Department of Manufacturing Engineering,
Blvd. Copou, no. 11, 6600 Iasi, Romania
E-mail: ineaga@sigma.tuiasi.ro

Nothing is permanent in the business, manufacturing industry, and organizational structure or information technology infrastructure of an enterprise. Therefore, the concept of classical enterprise is changed. Within a complex framework that changes, the enterprises become computer integrated organizations based on ***information technology (IT)*** and they are defined as dynamic, and chaordic systems. The researches of ***Computer Integrated Enterprise (CIE)*** are directed to develop the ***virtual enterprise*** that will be able to rapidly respond to a changing market place The engineering functions of the integrated enterprise will be accomplished by using multi-agent systems based on intelligent computational tools and methods such as expert systems, fuzzy logic, artificial neural networks and genetic algorithms.

The paper deals with the application of ***Distributed Artificial Intelligence (DAI)*** and of the ***Object-oriented programming method*** in order to improves the various aspects of the production and management processes within an integrated enterprise, especially in the area of manufacturing engineering. However, the conclusions of the paper could be applied for other types of enterprises.

1 Introduction

The increasing role of computers in all aspects of manufacturing activities including design, production planning, and marketing has been an important first step in the development of Computer Integrated Manufacturing (CIM). The following approaches have been determined by the evolution of the computer science together with the communication industries that allows working more closely each other over massive divides of space and time and brings a new potential to the companies. Therefore, the background of Computer Integrated Enterprise (CIE) have been created.

The main outcomes of the area of CIE are the definition and the development of the virtual enterprise based on IT. Generally, a virtual organization uses computers to access information, data and knowledge, and it can be defined by following characteristics:

- an integrated standard organization structure;
- IT provides access to information, data and knowledge;
- IT supports decision-making and aids in action execution.

Research directions of Computer Integrated Enterprise include:

- analysis of complexity of computer integrated systems, and if it is possible supervising it;
- development of concurrent engineering systems for design and manufacture;
- development of an integrated supply chain management system;
- definition of information systems for business process re-engineering, and decision-making;

The development of concurrent design and manufacturing environments implies the following items:

- use of knowledge processing techniques for knowledge exchange and sharing;
- design of the multi-agent systems formed by intelligent agents for concurrent design, process planning and scheduling;
- using the World Wide Web as communication medium for sharing information and knowledge;
- incorporate powerful software tools based on the intelligent computation methods.

Based on the theory of open system thinking (OST), the designing of chaordic enterprise defined by F.M. van Eijnatten and L.A. Fitzgerald [6] could be accomplished as a computer integrated complex, dynamic, and non-linear system. At the same time, the virtual enterprise will be a networking enterprise defined as an open medium. Combining a paradigm or a theory with intelligent computational methods, and IT we will obtain a generic standard model that will support the enterprise engineering functions.

The enterprise integration has to incorporate human resources. To achieve a high level of integration it is necessary that units of the enterprise be they human or machine based, be able to communicate and understand each other. Therefore, the language of communication is a very important direction of research, but at the same time a distinct topic.

2 The Current State-of-Art of CIE

2.1 TOVE – Toronto Virtual Enterprise

Within the Enterprise Integration Laboratory (Department of Mechanical and Industrial Engineering) at the University of Toronto, Canada, researches on enterprise integration have been accomplished.

This project is called ***TOVE*** (Toronto Virtual Enterprise) and it is described within [10] and at http: //www.ie.utoronto.ca/EIL. The virtual enterprise researches are focused in the following directions:

- Ontologies for enterprise modelling – it is not possible to operate virtually unless their is a shared set of concepts embodied in the language of communication. That is what ontologies provide.
- Agent architectures - virtual enterprise will have a significant amount of autonomous operation. Agents play a major role in operating these types of enterprises. The agent research provides the services necessary for autonomous operating agents.
- Constraint-directed reasoning/coordination - at the heart of an agent lies a constraint-directed problems solver that forms the basis of its own reasoning and the basis of inter-agent negotiation. Both are necessary components of virtual enterprise.

These researches include enterprise modelling, and design, concurrent engineering, and integrated supply chain management. The goal of the project is to conduct research leading to the creation of an information system to support enterprise design (also known as business process re-engineering) and execution [9].

2.2 Enterprise modelling

Enterprise modelling defines the actions and the constraints for plans and schedules which are constructed to satisfy the goals of the enterprise. For enterprise modelling the ***theory of complex actions*** is used [7]. The goal of enterprise modelling is to create a generic representation of knowledge related to activities, resources, products, organization structure, cost, and quality. The theory of complex actions is developed using an extension to **the situation calculus** that is a sorted second order language with equality.

Generally, situation calculus is the name for a particular way of describing change in first-order logic. It conceives of the world as consisting of a sequence of situations, each of which is a "snapshot" of state of the world. Situations are generated from previous situations by actions [13].

2.3 Generic Reference Architecture of an Enterprise

In order to create the integrated enterprise some generic reference architectures and methodology have been used.

Based on the results of TOVE project, and CIM-OSA model developed within the ESPRIT program of the European Union, GERAM (generic enterprise reference architecture and methodology) has been developed. This architecture covers all types of enterprises and it is a framework which incorporates the previously systems. Within [3], the requirements of generic enterprise architecture are presented. The generic reference architecture and methodology contain the following items:

- The architecture should cover all activities involved directly and/or indirectly in designing of the organization structure of the enterprise;
- The development of a modelling environment;
- The existence of a detailed methodology according to the enterprise engineering process;
- The description of a unifying perspective for designing and manufacturing activities that implies an effective communication mechanism between these activities.

3 The Background of Intelligent Agents Area

Conventional programming requires a-priori closed-world assumptions and thus lacks sufficient flexibility for complex tasks such as process planning, scheduling, decision-making, and operations management. For this reason the concept of ***agent*** has become important and a hot topic in the area of Artificial Intelligence. ***Intelligent agents*** are programs that can adapt themselves to changing environment and operate autonomously even on complex tasks.

The development of applications has migrated to Distributed Artificial Intelligence (DAI) in which multiple problem solving agents cooperate to achieve a common objective. The DAI systems are a rapidly and emerging technology that offers the possibility to handle complexity and changes within dynamic environment. However, yet, there have been relatively few reported cases of such systems being employed to tackle real-world problems in realistic domains. The main reason is that the solutions of incorporating pre-existing systems such as expert systems into a community of cooperating agents are under research and in the development phase.

Intelligent agents are basic computational entities that form distributed systems called ***multi-agent systems*** (MAS). According to [14] the intelligent agents area is divided in ***agent theory***, ***agent architecture*** and ***agent languages***.

- Agent theory is related to the intelligent agent's definition and the mathematical formalism for knowledge representation and reasoning mechanisms;
- Agent architecture is related to the designing of systems based on agent theory;
- Agent languages are software systems for programming of the agent.

The intelligent agents described as hard-software systems have the following main properties [14]:

- autonomy: agents operate without the direct intervention of humans and have some kind of control over their actions and internal state;
- social ability: agents interact with other agents by some agent-communication languages;

- reactivity: agents perceive their environment (which may be the physical world, a user interface, a collection of other agents, the Internet, or all of these combined) and respond in a timely fashion to changes that occur in it;
- pro-activeness: agents do not simply act in response to their environment, they are able to exhibit goal-directed behavior by taking the initiative.

For implementing intelligent agents the following methods and techniques are usually used:

- search with bounded resources;
- constraint satisfaction;
- abstraction and problem decomposition in constraint satisfaction;
- temporal reasoning;
- reasoning with uncertainty;
- case-based reasoning.

For the development of a multi-agent system the application is divided into agents and then the negotiation and cooperation between agents is carried out.

4 Applying Intelligent Agents

Using the intelligent agents for enterprise integration it is not a new idea because the TOVE project is also based on DAI, but we intend to include complexity, that by default exists, in order to reduce it, if it will be possible.

Within a Computer Integrated Enterprise a multi-agent system should have the following functions:

- allow the decision maker to access relevant information wherever it is situated within the enterprise;
- allow requesting and obtaining information from other departments and other enterprises;
- identify and deliver the information;
- inform the decision maker of changes which modify the decision framework;
- identify the components that may be interested by the outcomes of the decision making activity.

Applying the intelligent agent's theory within the framework of the enterprise is directed to define some architectures with the intelligent agent's properties that facilitate the ***concurrent engineering***. A concurrent manufacturing engineering system aims to address the issues of a product simultaneously. Within a concurrent system, the design and manufacturing information is distributed among the constituent intelligent agents. The agents will possess the necessary knowledge for dynamically accessing, transferring, and evaluating much of the concurrent engineering information. Using the intelligent agents, we will obtain an advanced integrated system that will be able to communicate with other intelligent software applications using the Internet as communication environment.

A multi-agent system for concurrent engineering in mechanical equipment design and manufacture comprises a design agent, a geometric interface agent, feature agents, the part agent, machine agents, and an environment manager that is defined as a co-ordinator agent.
The design agent uses a CAD system for geometric part representation.

The geometric agent acts as an intermediary environment for translating information between geometric part representation and the manufacturing features of the part.

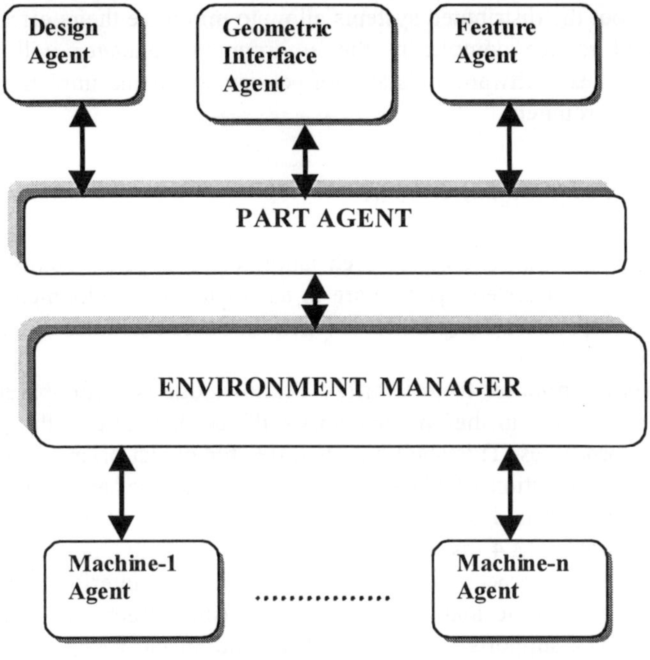

Figure 1. Architecture of a multi-agent system for manufacturing engineering

Feature agent describes the geometric and technological constraints. The feature-based modelling is based on feature extraction or/and design by features.
The Part Agent dynamically updates and maintains the information and data converted into knowledge related to the part description. The knowledge representation is dependently by the class of part processed (rotational, prismatic, etc.) and technological constraints. Within an intelligent manufacturing system the knowledge representation methods are the objects or/and the frames which provide an integration of Computer Aided Design (CAD) and Computer Aided Process

Planning (CAPP) systems. This is directed to obtain a concurrent system for design and process planning.
The machine agents select, sequence and cluster the operations in order to obtain the process-plan of the part. The architecture of a multi-agent system for manufacturing engineering is shown in figure 1.
Within the system every agent has an independent function but at the same time communicates with other agents concurently. The communication between intelligent agents is implemented using communication languages [14].

Generrally, a high level of integration increases the complexity of the system, and the chaos, but the distributed systems allow to minimize theirs effects. In order to control, and reduce complexity the environment manager will incorporate software tools. These software tools are not yet developed and implemented as part of a complex environment.

5 Expert and Multi-agent Systems

Within many enterprises and/or research laboratories, expert systems (ES) called also knowledge-based systems (KBS) are in use or under development. ES can be developed from scratch or using ES tools (shells). Some examples of expert system tools are as follows:

M.4 software product from Cimflex Teknowledge is a rule-based shell that captures the constraints in the and/or decisional tree that reflects the product and manufacturing processes. This shell has facilities for object-oriented programming which implies the creation of classes, multiple inheritance, and message-passing methods. The background of object technology will be briefly presented into the following paragraph. M.4 as powerful software tool for building KBS, and embedding intelligent processing in applications has an interface, which allows accessing data in database management systems using structured query language such as Oracle. M.4 supports both Visual Basic and Visual C++, and has facilities for the development of client-server applications.

The ***software system Exsys Professional*** (Multilogic Exsys) has been used for building an expert system for process planning with the representation of knowledge by frames, rules and decision nets as it is described in [5]. This shell also supports fuzzy logic. The shells, which support multiple knowledge representation methods, are defined as ***hybrid tools***. A hybrid system combines multiple representation paradigms into a single integrated programming environment.

The ***software system XpertRule*** (Attar Software Ltd.) integrates KBS, learning from data and genetic algorithm optimization into one environment, which could be used in decision-making.

The knowledge representation method is based on decision trees and the XpertRule application is constructed graphically as a hierarchy of chained tasks that contain decision-making knowledge and/or control knowledge.
The stand-alone expert systems can be transformed into a multi-agent system using generic development tools. A general framework for the development of cooperating agents for industrial applications is described in [8]. The architecture of a ***Generic Environment*** is shown in figure 2.

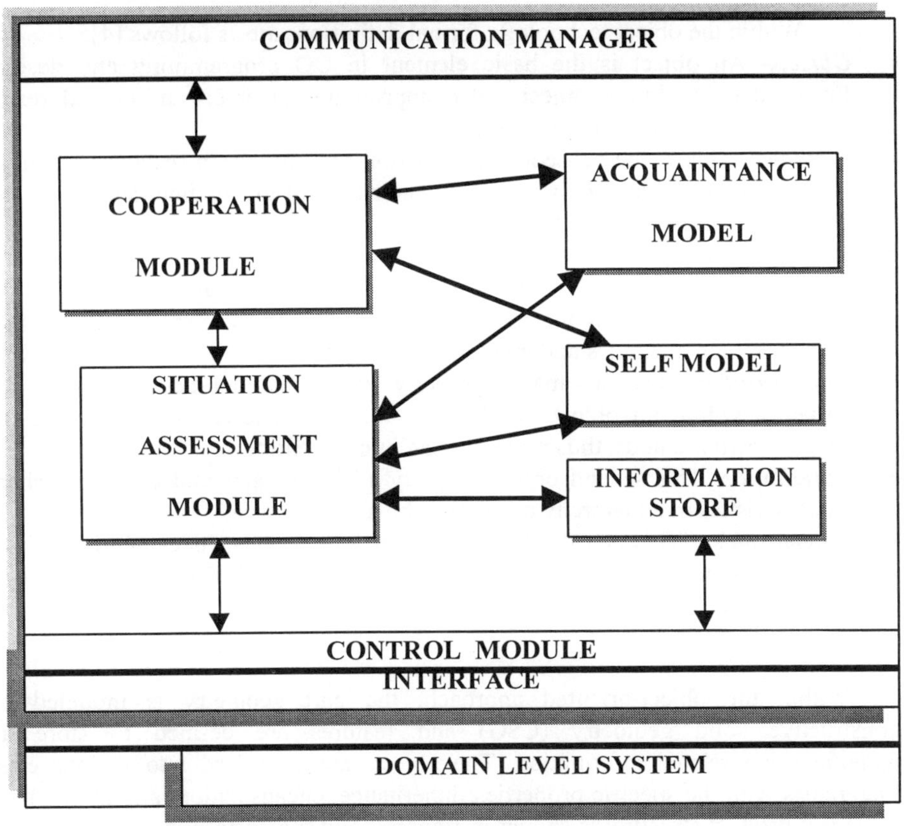

Figure 2. Architecture of Generic Environment for transforming expert systems

6 Using Object Technology

In order to develop computer integrated systems it is also possible using Computer-Assisted System Engineering (CASE) tools based on the ***object-oriented (OO)***

programming method. It is generally accepted that conventional software tools and procedures do not accomplished the needs of contemporary system developers because software systems and furthermore, the multi-agent systems become large and complex, difficult to conceptualize, design, implement, and manage. Object technology offers some solutions, but at the same time this technology is itself complex. Therefore, the conclusion is that the complexity can be supervised by complex system.

Within the object technology formal definitions are as follows [4]:

- ***Objects-*** An object is the basic element in OO programming and design. Encapsulated within an object is data appropriate for object and procedures or methods, which interact with that data.
- ***Method*** – This is the procedure or function that initiates a response when an object receives an incoming message. The nature of the method determines the object's response to such messages.
- ***Message*** – Interaction among objects is enabled by message. The object's methodology then manages the specific response to a message.
- ***Class*** – This represents the aggregation of a set of harmonious objects that share common attributes and methods.
- ***Encapsulation-***This firs major property of objects bundless methods and variables within an object or class so access is allowed only through the object's own methods, thus producing an integral entity.
- ***Inheritance-*** This second object property allows objects and classes to share characteristics in a hierarchy of relationships.
- ***Polymorphism-***This is a third major property of objects. Its capabilities allow different objects to interpret the same message in accordance with the nature of the object receiving it.

Within the object-oriented approach, the part geometry is modeled by constructive solid geometry (CSG) and features are defined to store all manufacturing related information. The objects are organised into various class hierarchies with the specific properties: inheritance, encapsulation, and abstraction. A typical product structure is shown in figure 3, and the corresponding knowledge representation by means of objects is shown in figure 4.

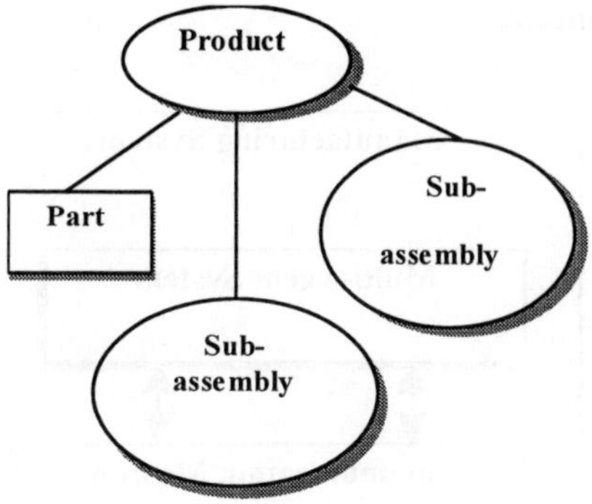

Figure 3. A product structure

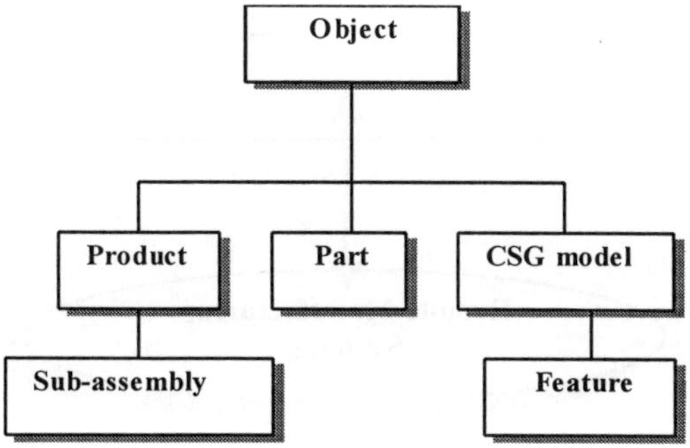

Figure 4. Object-oriented model

The Rational Rose products use specific object-oriented analysis and design techniques such as Booch-93 and OMT. These software products are based on industry open-system standards and also support the C++, Visual Basic, and Java™. The Rational Rose family product improves the productivity of software developers and provides an integrated standard solution for software changes, and management. They run on Windows or Unix systems.

7 Further Developments

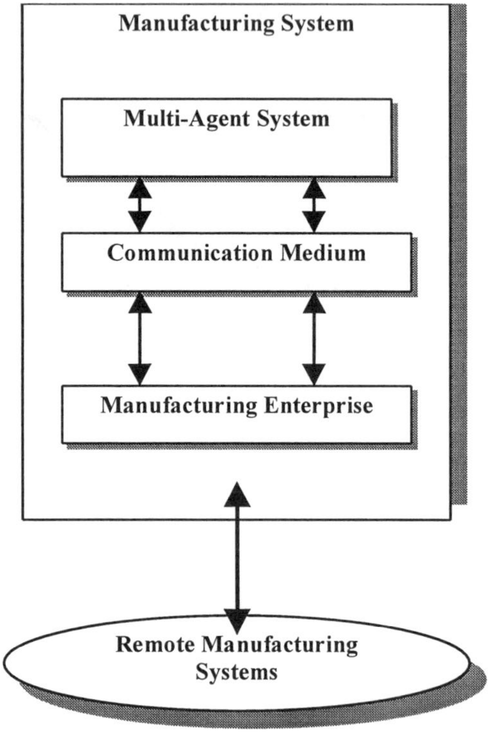

Figure 5. General organization of a multi-agent networking system

Enterprise integration focuses on the communications, the coordonation, and optimization of enterprise decisions and processes in order to achieve higher level of productivity, flexibility, and quality. Therefore, the main problem remains to integrate the software systems into an environment which could share information or/and knowledge. This would be possible by using software interfaces for World Wide Web developed in ***Java***. The general organization of a multi-agent networking system is shown in figure 5. The organization of manufacturing systems could be extended to other types of enterprises.

Furthermore, the approach of CIE based on DAI creates the basis of ***agile manufacturing*** that describes a new paradigm of the manufacturing enterprises. Within an enterprise organized on three levels (strategic, operational, and execution) the agility can be defined at various levels such as marketing, production, design, organization, and management. This concept represents the

ability to quickly respond to change and opportunities, and the decision-making structure is accomplished by agents.
In order to control and to minimize complexity into a practical framework it would be necessary tò develop software tools based on the complexity theory.
The backgroud of applying the chaos theory is created by the following decision support tools developed by Physics Academic Software:

- ***Chaos Data Analyser*** is a research tool consisting of 14 programs that detect hidden deterministic in seemingly random data;
- ***Chaos Simulations*** are interactive tools that simulates the non-linear systems demonstrating deterministic chaos.

7.1 Conclusions

The contributions of the paper are related to:

- the definition of the virtual enterprise in manufacturing engineering;
- the definition of a multi-agent oriented networking system based on IT as an open medium;
- dealing with some practical solutions that could be a basis for the further realization of a manufacturing system as virtual environment;
- analysis of the solution based on DAI by comparing with OO technology;
- the definition of a complex system as a collection of interacting agents that reduce a high problem complexity by functionally decomposing the problem solving;
- the possibility of controlling and reducing complexity of virtual environments by software tools.

The researches on enterprise integration are currently in progress, and this paper presents a comprehensive state-of-art of CIE, and draws some future directions. The further approaches will include:

- artificial neural networks, and genetic algorithms applied in production planning, and management of enterprise;
- information technology and future dimension of the enterprise;
- development of Java software products and implementing in the industry.

8 Acknowledgements

The author would like to thank Dr. Walter Baets, scientific co-ordinator of the first Summer School on Managerial Complexity, a meeting supported by European Union programmes: “Human Capital & Mobility” (HCM), and “Training & Mobility of Researchers” (TMR).

References

1. Balasubramanian, S., Norrie, D.H., A Multiagent Architecture for Concurrent Design, Process Planning, Routing, and Scheduling. Concurrent Engineering: Research and Applications, Vol. 4, No. 1 (1996), pp. 7-16.
2. Baets, W., Organizational Learning and Knowledge Technologies in a Dynamic Environment (Kluwer Academic Publishers, 1998)
3. Bernus, P., Nemes, L., A framework to define a generic enterprise reference architecture and methodology. Computer Integrated Manufacturing Systems, Vol. 9, No. 3, (Elsevier Science, UK, 1996) pp. 179-191.
4. Cashin, J, Object Technology: The New Approach to Application Development (Computer Technology Research Corp., 1996)
5. Duda, J., Habel, J., Pobozniak, J., Control mechanism of expert system with representation of knowledge in the form of hierarchical decision nets. Proceedings Applications of Artificial Intelligence in Engineering XII, (Computational Mechanics Publications, UK, 1997) pp.141-144.
6. Eijnatten, F.M., Fitzgerald, L.A., Designing the Chaordic Enterprise, Paper presented at the 14th Colloquium, Maastricht (1998), and at the First Summer School on Managerial Complexity, Granada (1998).
7. Gruninger, M, Pinto, J., A theory of Complex Actions for Enterprise Modelling. Working Notes AAAI Spring Symposium (Series 1995) Extending Theory of Action: Formal Theory and Practical Applications, Stanford, http://www.ie.utoronto.ca/EIL/public/complex.ps.
8. Jennings, N.R., Varga, L.Z., Transforming standalone expert systems into a community of cooperating agents. Intl. J. of Engineering Applications of Artificial Intelligence, Vol. 6, No. 4 (1993) pp. 317-331.
9. Fox, S.M., Research in Enterprise Integration. Proceedings Artificial Intelligence and Manufacturing Research Planning Workshop, (AAAI Press, USA, 1996) pp. 40-52.
10. Fox, S.M., The Tove Project: A Common-sense Model of Enterprise. Proceedings Artificial Intelligence and Manufacturing Research Planning Workshop, Industrial and Engineering Applications of Artificial Intelligence and Expert Systems, Lectures Notes in AI (Springer-Verlag, 1992) pp. 25-34.
11. Luger, G.F. (editor), Computation & Intelligence Collected Reading (AAAI Press, 1995)
12. Luger, G.F., Stubblefield, W.A., Artificial Intelligence Structures and Strategies for Complex Problem Solving (Benjamin Cummings Inc., USA, 1993)
13. Russell, S., Norvig, P., Artificial Intelligence-A Modern Approach (Prentice Hall International, USA, 1995)
14. Wooldridge, M., Jennings, N.R. , Intelligent Agents: Theory and Practice, The Knowledge Engineering Review, Vol. 10, No. 2 (1995) pp. 115-152

A FRAMEWORK TO EVALUATE THE FEASIBILITY OF NEURAL NETWORK APPLICATIONS IN BUSINESS AND FINANCE

B. KAMP
Tilburg University / BDO CampsObers, P.O. Box 90153, 5000 LE Tilburg, The Netherlands
E-mail: b.kamp@kub.nl

This paper discusses several conditions and constraints that determine whether neural networks are likely to be successful for real-life (business) organizations. These conditions and constraints are linked together in a coherent framework. This framework points to the critical conditions when deciding whether to develop a neural network solution for a particular (management) decision process. These conditions will be illustrated by potential business applications of neural networks as found in the literature.

1 Introduction

Research on neural networks can be divided into fundamental research and applied research. Within fundamental research, new network structures and learning algorithms are developed. This research often involves extensive mathematics. Applied research on neural networks explores the feasibility of neural network solutions for real-life problems [9]. Although in a laboratory setting many of those models perform well, the implementation in a real-life organization is not straightforward.

This paper discusses several conditions and constraints that determine whether neural networks are likely to be successful for real-life (business) organizations. Although most of these constraints are well-known [18], they are not described in a formalized or structured manner. In this paper, these constraints are set in a coherent framework.

This framework points to the critical conditions when deciding whether to develop a neural network solution for a particular (management) decision process. These conditions will be illustrated by potential business applications of neural networks as found in the literature.

Although neural networks are meant to support complex decision making, the decision to invest in neural network technology itself is also complex. The framework can give somedirection to this decision.

2 The role of artificial intelligence for organizations

Artificial intelligence may be helpful in complex management decisions. Especially if the decision involves many concurrent evaluations or complex interactions between factors, computer modeling may capture the information better, and render better or more consistent solutions. The use of artificial intelligence in organizations may have different purposes:

- Finding better solutions than the human expert: by advanced searching or reasoning techniques, AI may capture knowledge that was unknown to the human expert. For example, the computer may find unknown relationships on the capital market, which can be used to develop a stock trading strategy that is superior to the existing strategies based on human analysis and judgment. (Note that if consistency is the main element of the quality of a decision process, expert systems may also be suited in such environments.)

- Acting as a reference for the human expert: For decisions that involve professional judgment, the decisionmaker may want to compare his initial views to another source. For example, when granting a loan, the loan officer may want to compare his view on the client's financial position to the outcome of a formal model. Any divergence between both views may direct to matters for further analysis by the loan officer.

- Acting as an educational tool for new employees: Like an expert may use an AI system to validate his own professional judgment, for new employees the system can be a guidance to become familiar with the applicable expertise. Some large accountancy firms use expert systems that teach new employees how to set an audit program [10].

- Replacing the human expert: from a cost perspective, it may be attractive to replace the human judgment or decision by a less expensive computer program. For example, complex scheduling problems in production plants could be more efficiently solved by computer modeling than by a qualified engineer. Also zip-code reading systems may save personnel costs at the postal sorting process in post offices.

3 Expert systems and neural networks

Within AI, a distinction can be made between expert systems and neural networks. Expert systems emerged in the seventies, and use logical inferences like IF-THEN rules. Neural networks search for relationships between input and output patterns within large data sets. Therefore, the main difference between expert systems and neural networks is that expert systems are based on known rules, while neural networks derive an internal representation of the problem based on data pattern, independently from the knowledge of the expert. This points to two criteria when choosing between expert systems or neural network: the availability of rules and/or data.

Since the internal representation of the problem within a neural network is derived independently from the expert, its structure is often very complex. Therefore, the neural network remains a 'black box'. Several techniques are proposed to extract rules from trained neural networks, but their effectiveness has not been proved yet [1]. The outcome of the model cannot be explained easily by the input values. In contrast, expert systems often have explanation utilities: it can show which rules has been applied in which manner for a particular case. This way, the user may understand better why the expert system reached that particular outcome. The lack of explanation limits the application of neural networks for some generic purposes. For example, if the system is meant to educate new employees, an explanation of the decision making will be necessary, since the employee needs to acquire an understanding of the decision process. But also for other applications, the lack of explanation can be a serious problem.

4 The Framework

Although Rules availability, Data availability and Need for explanation are well-known, they are not described in a formalized or structured manner. Wong makes several classifications of neural network applications in business and finance as found in the literature, among them:

- classification by level of management (strategic, management- and operational control);
- classification by level of task interdependence (single user, group- and organizational decision).

Although Wong notes that neural networks in strategic decision making are rare due to the nonroutine nature and the lack of sufficient data, it does not give much direction for the evaluation of new opportunities of neural network applications.

The criteria Rules availability, Data availability and Need for explanation are combined into a framework, which is depicted in figure 1. The aim of the framework is to envisage the economical/organizational constraints on AI modeling with respect to the typical characteristics of neural networks. If one or more of the described constraints is not met, successful implementation of neural networks becomes unlikely. Note that in this paper the distinction between expert systems and neural networks is more strict than as found in practice. Often, applied AI models have a hybrid nature by combining rule-based, statistical and neural network components.

Table 1. The framework

<table>
<tr><td colspan="4">Are Rules available?</td></tr>
<tr><td rowspan="10">Yes</td><td colspan="3">No
Have single decisions far-reaching consequences?</td></tr>
<tr><td rowspan="3">No</td><td colspan="2">Yes</td></tr>
<tr><td colspan="2">Can the problem be decomposed?</td></tr>
<tr><td>Yes</td><td>No</td></tr>
<tr><td colspan="2">Data availability

-Problem-related data collection difficulties
-Cost-related data collection difficulties</td><td>NN not suited</td></tr>
<tr><td colspan="2">Many or few data patterns obtainable?</td><td rowspan="2"></td></tr>
<tr><td rowspan="4">Many</td><td>Few</td></tr>
<tr><td colspan="2">Assumed nonlinearity?</td></tr>
<tr><td>Yes</td><td>No</td></tr>
<tr><td>Emphasize validity/ reliability tests</td><td>Compare to alternative statistical techniques</td></tr>
<tr><td>Costs of development Expert System

-Costs of knowledge acquisition (rules)
-Costs of modeling/ programming
-Maintenance costs</td><td colspan="2">Costs of development NN

-Costs of knowledge acquisition (data)

-Costs of modeling/programming

-Maintenance costs</td><td>Costs of development linear model</td></tr>
<tr><td colspan="3">↖ cost comparison Exp.Sys - NN ↗</td></tr>
<tr><td colspan="3">Cost Benchmark</td></tr>
</table>

This strict distinction is made to emphasize the essence of both approaches, although typical characteristic of neural networks can also be found within expert systems. For example, an expert system could lack a clear explanation function (either being absent or having a too formal computer oriented structure). In such cases, the considerations regarding the need for explanation are also valid for expert systems.

The framework has the structure of a decision tree, starting at the top, and descending along the line of met constraints. Note that the framework does not follow the sequence of the system development process, but represents an instant evaluation before starting any investments. Before making extensive analyses on one constraint (for example, the costs of data collection), first all other constraints should be checked: if at first glance it can be said that explanation of the decision will be necessary (and the problem cannot be decomposed), neural networks will not be suitable for that problem, regardless the costs of data collection. In the remainder of this paragraph, the distinctive constraints will be discussed.

4.1 Are rules available?

The first question to consider is whether there are rules known to solve the studied problem. In general, deductive rules as used in expert systems will be preferable over inductively derived relationships as derived by neural networks. However, for many complex problems rules are not available. Especially when the aim of the system is to improve the decision making, the system has to acquire superior knowledge compared to the currently available knowledge. To some extent, the computer model may improve the decision making by simply applying the known rules in a more consistent way than a human experts is able to. For example, if the decision requires extensive reasoning with many considerations, like complex scheduling problems, the computer may capture the entireproblem domain better than the human expert.

To acquire superior knowledge, existing knowledge will not be sufficient. To improve the decision making, the computer has to find knowledge that the human expert has overseen so far. Examples are technical analysis applications in stock market prediction [14]. For decisions that are mainly based upon professional judgment, the acquisition of explicit rules is often difficult. Although the human expert can perform his decision task well, he might not be able to explain his own reasoning on a detail level that is necessary for developing an expert system.

Therefore, if the system requires yet unknown knowledge, or rules are hard to derive since the decision is highly judgmental, neural networks may be more effective than expert systems. On the other hand, by choosing neural networks, a priori known rules are hard to implement in the neural network. A possible method could be to preset some of the weights between input and output, and leave these weights unchanged during the training of the neural network. However, no empirical studies has been found on this matter. Consequently, the neural network has to find these known relationships by its own. This may be rather inefficient compared to rule-based expert systems.

4.2 Are single decisions far-reaching?

This question refers to the need for explanation. If individual decisions made by the system have large consequences, the user is likely to demand a solid explanation of the decision by the system. For example, if the system is applied for the evaluation of loan applications, a rejected client will demand an explanation from the bank loan officer. Even if the loan acceptance decision can be modeled well [6] a naive reply that "the computer says so..." will not be acceptable. When using an expert system, the rejection may be traced back to the combination of a lack of collateral compared to the bank's guidelines and poor income prospects. Such breakdowns of the decision is not possible with neural networks.

Another example are stock market prediction models. To take the full advantage of such models, the investor has to follow the recommendations (hausse/baisse positions) by the system closely with large amounts. Since the decision has a large effect on the investment performance, the investor does not want to follow the model blindly. Especially if the investor is an expert, he will be very reluctant to follow the system's recommendations if these differ from his own view. The system can only convince the investor by showing the soundness of its considerations. This requires an explanation of the decision, which is not (yet) possible with neural networks.

In some cases, it may be possible to decompose a decision with far-reaching consequences into smaller components. For example, if a bank loan acceptance decision is mainly based on the customer's income, collateral, and lifestyle, separate systems may be developed to model each of these components. This way, a rejection can be more easily addressed to an underperformance of one or more of these components. Such indications can be considered as partial explanations.

On the other hand, there are many applications where single decisions have a low impact on the overall performance on the modeled task, since the task is routineous. For example, when using neural networks for clustering customers into homogeneous groups for the purpose of target/direct marketing, a single incorrect classified customer only leads to a potential unsuccessful attempt to acquire this client [8]. Such incidental losses will be compensated by -on average- a more suited selection of potential customers. For stock investment applications, neural networks may be used to determine the volatility exposure of a total portfolio by estimating the volatility of the many different stocks within the portfolio.

Also for complex scheduling problems, an explanation will normally not be necessary. As long as a suggested solution fits into the problem constraints and the solution is likely to be better than alternative solutions suggested by human experts, the user will not be very interested to know how the system reached this solution. On the other hand, if the aim of the model is to improve judgmental decisions, explanation can become more important. A small Dutch investment fund developed a stock trading system that outperformed human judgment when applied to out-of-sample data. To achieve this outperformance in practice, the recommendations of the system had to be followed strictly. However, since trading transactions had to be approved by the same human expert that developed the model, his own judgment still affected the ultimate trading policy: if the model produced a recommendation that was counterintuitive to the human expert, the proposed transaction was canceled or changed, since the apparent better solution by the system could not be documented by explanation. Such a problem of interference of human judgment to analytical methods that have been proved to be superior, could be solved by a strict segregation between the development of the model and the actual execution of the recommended trading strategy.

4.3 Availability of data

Since neural networks have to be trained on data patterns, the availability of such data can be a constraint. This constraint can be inherent to the problem, or relate to the incurred costs for data collection.

4.3.1 Problem-related data collection difficulties

The developer has to have a general notion of the variables that affect the modeled problem. Although some suppliers of commercial neural network software claim that neural networks can retrieve relevant information from vast databases of partly irrelevant data, no scientific empirical proves have been found on that yet. Therefore, some assistance from a human expert will remain necessary.

Second, the required data must be existent. For example, neural networks may be employed to improve the acceptance decision of a bank loan officer. Former loan application data could be used as training data. However, in former periods, the bank loan officer may have rejected some potential good applications. Improvement of the loan acceptance task could be realized by a neural network that recognizes good applications better than the human expert. However, since these applications were rejected, no data is available to test which of the rejected applications would have been good. Another example is stock market prediction where the objective is to predict crashes. Since the number of (severe) crashes is relatively small, extensive databases on crashes are non-existent.

A special case of data existence problems applies to the most popular architecture of neural networks, the error-back propagation neural network. Backpropagation can be considered as an advanced nonlinear regression technique: based on an input pattern of several input variables, the neural network produces on or more output values. This structure of the model requires that for each input pattern, for all input variables a value is applicable and available. However, some problems have such a structure that for some cases, not all input variables are meaningful. For example, the credit rating of a firm in the oil & gas industry depends to a certain extent on the capacity to replace used oil & gas reserves.

Therefore, the characteristics of these reserves, like size and exploration costs, will be relevant input variables. However, if the neural network is designed to provide a prediction of the credit rating for firms in different industries, these oil & gas reserve input variables will be meaningless for -for example- firms in the food industry. In other words, input factors, the relevance of certain inputs depends on the value of other input variables.

Without any further adjustments, the neural network cannot distinguish whether a value of nihil for the size of the reserves indicates that the firm encounters replacement problems or that reserves are not applicable at all. If the credit rating would be modeled by an expert system, a tree-like structure would be used: if the firm belongs to the oil & gas industry, additional information on the reserves is requested from the user, otherwise, these questions are skipped. The structure of backprop neural networks are less suited for such decision-tree-structured problems. One solution may be to develop industry specific models, but this inherently limits the number of available data patterns per model. Another solution may be to add dummy variables that indicate that a particular condition (i.e., being an oil & gas firm) is applicable. However, every added input variable requires an extra number of data patterns to estimate the corresponding parameters reliably.

Third, the required data can be classified. For example, many attempts have been made to model the bond rating decisions by Moody's and Standard & Poor's [3,12,15]. It is well-known that these bond rating agencies have access to inside information of the rated firms, and that they have rather secret internal guidelines to interpret financial data [4]. Therefore, it is very difficult for an external developer to gather sufficient data to build a neural network on. In some cases, this problem can be circumvented. Although sales in retailing shops like supermarkets are normally anonymous, some retailers experiment with personal discount cards, which offer the possibility to gather information on individual purchase behavior. By combining the identification of the customer via his or her discount card and the sales as registered by the POS-terminal, individual purchases can be identified. Customers are encouraged to apply for the card by offering special discounts on purchases.

4.3.2 Cost-related data collection difficulties

In general, the more data patterns are used, the more reliable the neural network model will be. However, the costs per data pattern vary strongly between different problem areas. In some cases, the development of neural networks is typically data driven: companies possess large databases, and the management wants to extract as much value from these assets as possible. Such added value could be achieved by data mining with neural networks. For example, credit card companies and large retailers have enormous databases with information of customers' behavior. A well-known application by VISA is a fraud detecting system [5]. Since the data is already recorded for the ordinary business processes, little additional costs have to be incurred to make the data suitable for neural network modeling. Other examples of existing large databases are accounting systems of large banks and commercial stock market databases like Compustat and Datastream. Note that for these large databases still problem-related data collection difficulties can occur: the databases are primarily designed for the ordinary business process. Some kinds of information that is required for neural modeling may be absent in the database since it is not relevant for its original purpose.

On the other hand, if data has to be recorded solely for the neural network, the costs of data will be much higher. For example, Oestreicher & Piotrowsky assumed that verbal information in companies' annual reports can be valuable input data for a financial default prediction model. Since this kind of information has to be collected manually, the affordable number of data patterns is relatively small. In 't Veld et al. [16] developed a neural network to estimate the costs of information system development processes, based on the complexity of the project. Since measuring complexity (function point analysis) of such processes is rather time consuming, only a small number of data patterns could be collected. Gathering data from judgmental decisions from an expert may require expensive protocol analysis or the computer coding of paper forms. In some cases like scheduling optimization, data can be generated by (computer) simulation techniques.

4.4 Neural networks based on few data

Since neural networks are highly nonlinear and flexible, special considerations should be paid to validating the model and estimating its reliability. If data patterns are superfluous, the reliability can be tested by applying the neural network to out-of-sample data. These data are not used during the training process, but are used to check whether the performance of the model is generalizable. In cases where data is relative costly, the number of available data patterns will be limited. For the estimating of the weights, as many data patterns as possible should be used during training. However, this way no data patterns are left for estimating the reliability of the model. Consequently, a trade-off has to be made between estimating the model and testing its reliability. Some statistical methods are discussed for neural networks to cope with this problem, like jackknifing, cross validation and weight decay [17].

Any limits on the available number of data patterns also puts a constraint on the number of explanatory variables that can be incorporated in the model. Like (other) statistical techniques, a minimum number of data patterns is necessary to estimate each parameter reliably. Although an exact minimum is hard to define, this minimum will be higher for neural networks than for linear modeling techniques, since any degree of nonlinearity requires additional parameters to be set. Therefore, if only few data patterns are available, it should be considered whether the problem has a nonlinear nature. If such nonlinearity is not a priori assumed, it may be worthwhile to evaluate linear techniques first. If neural networks do not outperform linear models significantly, linear models are preferable since they have a less complex internal structure. Comparison between neural networks and alternative methods can be found in Kim et al. [7] (bond rating), Refenes et al. [14] (stock performance), Bell [2] (bank failure) and Mauser & Magazine [11] (scheduling).

4.5 Comparing the costs of neural network applications

The costs of development of a knowledge based system can be divided into:

- costs of knowledge acquisition (rules and data);
- costs of modeling/programming;
- maintenance costs.

4.5.1 Costs of knowledge acquisition

The framework illustrates that the choice between expert systems and neural networks depends on the incurred costs to obtain either rules or data. In the former paragraphs, it has been discussed that the costs of data collection depends on whether the data already has been recorded for other purposes. If data has to be gathered, the costs will vary widely for different cases. If commercial databases like Compustat or Datastream are used, the costs can be estimated from the offerings by their suppliers. If data is gathered from customers by issuing discount cards, the costs are determined by the implementation costs of the cards, and the opportunity costs of the offered discounts. If data has to be extracted from documents, the costs depend on the necessary time and equipment of data typists.

The cost of gathering rules from an expert when building an expert system depends on how time-consuming this process is, and the wage or salary of the expert, since the expert will not be productive in the ordinary business process when taking part in the expert system development process. VISA chose to use neural networks for its fraud detecting system since too many rules had to be modeled if they would deploy expert system techniques.

4.5.2 Costs of modeling/programming

When the data or the rules are collected, the model has to be constructed. Although is might seem that a neural network "does all the work by itself", in practice many trainings have to be performed, using different network architectures and different combinations of inputs and outputs, and different preprocessings of the data. This task will be performed by specialized knowledge engineers and programmers. On this level, there are no fundamental differences between expert systems and neural networks. Although VISA claimed that it saved on expensive statisticians by choosing for a neural network approach, a sound (statistical) test on validity is also still required for neural network applications. Therefore, the costs of model building and -programming will be comparable for both techniques. Other costs of implementation relate to instruction and education of the users of the models. Since expert systems can be more interactive than neural networks, these costs could be higher for expert systems, since the user has to know more about the structure of the program. However, it has been discussed that applications where both expert systems and neural networks are applyable, the interactive explanation function plays only a minor role. If explanation is important, neural networks are not likely to be feasible.

4.5.3 Maintenance costs

Since most of the problem domains in a business environment are dynamic, intelligent systems have to be flexible to adjust to changes. The flexibility towards adjusting to changes in the environment is different for expert systems and neural networks, and depends on the nature of the problem. For example, if the modeled decision is primarily driven by fuzzy rules, a structural change in the rules is more easily adjusted for in an expert system than in a neural network (assuming the human expert is still available to provide the necessary knowledge). For example, a stock market index predicting system should take the introduction of the Euro into account. The neural network has to retrain on a sufficient number of data patterns from the new situation to derive the new state of the relationships. On the other hand, if the environment changes slowly and gradually -for example, economic indicators during economic cycles- neural networks may be more efficient to pick up trends. Note however, that for neural networks it is not straightforward to determine when the environment has changed enough to urge a revision of the parameters. Within expert systems, it may be more easy to check whether certain changes are captured by the present internal rules. Since the internal representation of a neural network is too complex to analyze, it will be less clear whether the model is still valid. For example, during 1993, the Dutch stock market was primarily driven by the changes in interest. In the next years, the interest seemed to decrease in relevance. Even if a stock market predicting neural network employs interest variables, it remains unclear whether it takes the changing relative importance into account. If the availability of data is not a bottleneck, the neural network may be retrained on a regular basis. If data is scarce, each new training requires a full reestimation of the model's reliability, which may be costly.

4.6 Benchmarking of the costs

By comparing the above mentioned costs for expert systems and neural networks, a choice may be made between both techniques. However, even the most cost efficient technique should also render sufficient savings or additional profits to justify its application. Therefore, a cost/benefit comparison has to be made. As often in cost/benefit analyses, the benefits may be hard to estimate. However, from a prudent perspective, the estimated costs should never exceed the maximum expected benefits. Therefore, starting from the annual costs of the system (depreciation and maintenance costs), the minimum level of benefits like cost saving or additional earnings to offset the costs can be derived (break-even). Often, common sense is sufficient to assess whether this minimum performance of the system is achievable. For example, if already advanced methods are used for a particular scheduling problem, a cost saving of at least 50% to recover the investment in a neural network is unlikely to succeed.

On the other hand, note that knowledge systems can have large economies of scale. VISA invested $ 2 mio. in its fraud detecting neural network. Although this could be regarded as a large amount to invest, when applied to the massive number of transactions of VISA, the cost savings are a multiple of the invested amount. Similarly, a bank saved 25% on direct marketing mail, reducing response numbers only by 4% [5]. Alternatively, reasoning from the maximum expected benefits, a zip-code reading system should not cost more than the employees it replaces.

The benchmark of corresponding benefits for a particular problem may differ between expert systems and neural networks. As described before, if a neural network is designed to improve the decision making process, a higher level of savings may be accomplished compared to an expert system that only mimics the current state of human decision making.

5 Summary and Conclusions

Although in the literature many studies can be found where neural networks turn out to have good prediction, classification, or optimization capabilities, such a well performance is not a sufficient condition to implement neural network applications successfully in a real-life situations. To implement a neural network successfully in a real-life business and finance environment, some limitations of neural networks have to be taken into account. The main constraints relate to the need of explanation of the system's outcome, the costs of collecting either rules or data, and the costs of maintenance.

The need of explanation typically occurs when a single decision has far-reaching consequences. The responsible manager will only be willing to accept the proposed solution if he can verify its validity. Also if the system is meant to educate novices, explanation of the outcome is essential. In such an environments, neural networks are less suited because of their black-box nature. If the modeled decision can be broken down in several components, some degree of explanation may be achieved by developing separate neural network models for each component. On the other hand, if neural networks are applied for routinous decision making processes, explanation is often less important. Occasional wrong outcomes will be compensated by a larger number of correct outcomes.

If the objective of the system is to produce better solutions than as performed by the human expert, neural networks may outperform expert systems, since rule-based expert systems can only perform as well as the human expert, while neural networks may find yet unknown relationships in data.

The costs of data selection depend largely on whether the data has already been recorded for the regular business process. In some industries like banking, large databases are already present. However, even when large databases are available, it has to be ensured that the database contains a sufficient number of relevant variables. A well-known example of not-available data is the potential quality of rejected bank loan applications.

If data has to be collected solely for developing a neural network, costs can be relatively high, resulting in only small databases. Special precautions have to be made to ensure statistical reliability of the model, like cross validation and weight decay. If nonlinearity is not assumed a priori, it is worthwhile to evaluate also conventional statistical techniques that are better suited for advanced statistical tests.

To be economical successful, the costs of a neural network should not exceed the costs of alternative techniques like expert systems, and the costs should not exceed the corresponding cost savings or additional earnings. The annual costs of a neural network application consist mainly of the depreciation costs of the initial investment, and recurring costs of maintenance. The need of maintenance depends on the nature of the problem. If the problem environment is rather stable, adjustments only have to be made after sudden changes. If the problem environment is dynamic and changes along continuous trends, revisions of the model have to be made periodically. In the former case, expert systems may be more easily adjusted, while in the latter case neural networks are more easily updated based on recent data. If data is scarce, a trade-off has to be made between accuracy of the model, and the relative high costs of ensuring unchanged reliability of the model. Due to the black-box nature, the determining whether the neural network still fits the current state of reality is rather arbitrary.

Based on the estimated costs, a minimum degree of improvement (cost savings) can be determined for the neural network to be cost-effective. Based on prior knowledge of the problem area, the management has to evaluate whether this minimum required improvement is likely to be accomplished.

The framework as discussed in this paper places the above mentioned criteria in their logical coherence, thereby guiding the evaluation whether a proposed neural network application is not at forehand unlikely to succeed.

References

1. Andrews R., Diederich J. and Tickle A.B., Survey and critique of techniques for extracting rules from trained artificial neural networks. *Knowledge-Based Systems* **8** (1995) pp. 373-383
2. Bell T.B., Neural Nets or the Logit Model? A Comparison of Each Model's Ability to Predict Commercial Bank Failures. *International journal of intelligent systems in accounting finance & management* **6** (1997) pp. 249-264
3. Dutta S. and Shekhar S., Bond rating: A non-conservative Application of neural networks. *Proceedings of the IEEE Conference of San Diego* (1988)
4. Ederington L.H. and Yawitz J.B., The bond rating process, in: *Corporate financial distress*, by E.I. ALTMAN, (Wiley, New York, 1983)
5. Gandy T., Brainwaves in fraud busting. *Banking Technology* **15 (**1996) pp. 23-25
6. Glorfeld L.W. and Hardgrave B.C., An improved method for developing neural networks: the case of evaluating commercial loan creditworthiness. *Computers & operations research and their application to problems of world concern* **23** (1996) pp. 933-944
7. Kim J.W., Weistroffer H.R. and Redmond R.T., Expert systems for bond rating: a comparative analysis of statistical, rule-based and neural network systems. *Expert Systems* **10** (1993) pp. 167-188
8. Kvaal K. and Djupvik H., Prediction of customer segments with neural nets. *Marketing and research today : the journal of the European Society for Opinion and Marketing Research* **24** (1996) pp. 247-253
9. Li E.I., Artificial neural networks and their business applications. *Information & Management* **27** (1994) pp. 303-313
10. Davis J.T., Massey A.P. and Lovell R.E.R., Supporting a complex audit judgment task: An expert network approach, *European journal of operational research* **103** (1997) pp. 350-372
11. Mausser H.E. and Magazine M.J., Comparison of neural and heuristic methods for a timetabling problem. *European journal of operational research* **93** (1996) pp. 271-287
12. Moody J. and Utans J. Architecture selection strategies for neural networks: application to corporate bond rating prediction. *Neural networks in the capital market* (John Wiley, New York, 1994)
13. Oestreicher A. and Piotrowski-Allert S., Klassiflkation und Beurteilung von Unternehmen mit Hilfe von Selbst- organisierenden Neuronalen Netzen. Ergebnisse einer prototypischen Untersuchung zur Verwendung von Kohonen-

Feature Maps im Rahmen der Bilanzanalyse. *Schmalenbachs Zeitschrift fuer betriebswirtschaftliche Forschung* **48** (1996) pp. 335-371

14. Refenes A.N., Zapranis A., and Francis G., Stock Performance Modeling Using Neural Networks: a Comparative Study With Regression Models. *Neural Networks* **7** (1994) pp. 375-388
15. Singleton J.C. and Surkan A.J., Modeling the Judgment of Bond Rating Agencies: Artificial Intelligence Applied to Finance. *Journal of the Midwest Finance Association* **25** (1991) pp. 72-80
16. Veld L.J. in 't, Koppelaar H. and Pellikaan R., Neurale begrotingsmodellen. *Informatie* **34** (1992) pp. 349-355
17. Verkooijen W.J.H., *Neural Networks in Economic Modelling, An Empirical Study.* CentER dissertation (Tilburg University Press, 1996)
18. Wong B.K., Bodnovich T.A. and Selvi, Y., Neural network applications in business: A review and analysis of the literature. *Decision Support Systems* **19** (1997) pp. 301-320

THE CHANGING NATURE OF WORK AND ITS IMPLICATIONS FOR KNOWLEDGE MANAGEMENT AND INFORMATION SYSTEMS

FRANCK TÉTARD

Institute for Advanced Management Systems Research

Åbo Akademi University

Datacity - FIN-20520 TURKU - FINLAND

Fax: +358-2-2654809

E-mail: frank.tetard@abo.fi

Abstract: In this paper, we expose the main characteristics of work transformation taking place in organisations and the properties of the emerging nature of work. This transformation of work has several implications for management practices, which can be articulated around three themes: (i) how work should be structured, (ii) how knowledge should be managed so that it serves the activities of work and (iii) how to apprehend information systems to support the requirements of work. Futhermore, evidence from the field is brought along the development of this paper.

Keywords: work transformation, knowledge work, knowledge management, information systems

Introduction

It has been fairly mentioned in the business literature that the coming of the knowledge society and information technology (IT) innovations will change the way people work and bring obvious benefits to the organisation and its employees [7]. Especially, Business Process Reengineering (BPR) initiatives were claiming for more substantive, knowledge-intensive, satisfying and value-adding work [12]; still these benefits have not yet been reaped. The coming of the knowledge-based organisation requires from us to think differently about how work is done in these organisations. As markets are becoming global and competition is increasing, organisations need to respond adequately and quickly to changes: this calls for a new reorganisation of work. As we will show it, the nature of work is clearly changing from divided, specialised and monitored work to integrated, multifunctional and self-coordinated work.

These transformations of work have several implications:
- First, the new nature of work is more knowledge-intensive. Knowledge-intensive work is not as easily managed as specialised work: it resists indeed to structured approaches and it has led managers to let knowledge workers organise themselves their own work. Therefore, knowledge management approaches should be structured accordingly.
- Second, the tools (information systems) that we use to perform work have to be adapted. People have for long used many ways to exchange and process information, even before IT was widely available, and it seems that little attention has been paid to thinking about how people use information (and not how people use machines).

In this paper, we will first begin by defining important notions (such as knowledge), then we will show how the nature of work has been changing over time, which state it has reached today and what the future direction is. Then we will look at the implications for management practice related to this work transformation: (i) how to manage knowledge work at an individual level, (ii) what knowledge management approaches should we consider at an organisational level and (iii) how knowledge work will influence the way we apprehend and use information systems (IS).

1 Concepts

In this section, we will expose some important concepts: first, we will distinguish between two models of thought and approaches to understand reality. These concepts will be relevant when looking into the nature of work. Then we will present what we understand with knowledge and propose a working definition of knowledge for our work.

1.1 Models of thought: Rational Thinking versus Systems Thinking

It is argued that two models of thought, that is how thought or reflection works, exist: *separation* and *integration.*

- Separation means the decomposition of entities, through the process of analysis, in order to identify smaller entities.
- Integration is the assembly of distinct entities, through the process of synthesis, in order to create a larger entity.

When trying to understand their world, humans typically divide and integrate entities (data, information) to analyse internal mechanisms of their world or to synthesise information in a larger context (Zeleny, 1989) [22]. These principles of separation and integration will be useful when we will look at the nature of work: for many years, we have been influenced by a rational way of thinking, that is we understand the world by observing it as a passive observer, decomposing the world into small entities and describing it as precisely as possible.

Now the trend seems towards a systemic approach, that is we reassemble entities to understand a problem situation as a whole. There are still sub-systems (sub-problems) which we are aware of but the overall objective is to understand the interdependence of factors in the system and its dynamics. To do this, we need to integrate our knowledge of the smallest entities to create global knowledge of the system (Checkland, 1981) [2]. Still, we are not excluding one or the other, analysis and synthesis are two necessary elements of human thinking (one can not synthesise a concept which has not be analysed).

1.2 *Knowledge*

Knowledge is a subject loaded with different meanings, it explains why there is no agreed definition of what knowledge is. Here, we will present a working definition of knowledge which is appropriate for the purposes of this paper. Many authors in the management literature have tried to define knowledge. First, some have made the distinction between data, information and knowledge as a means to identify specific properties of knowledge (Davis and Botkin, 1994, Earl, 1994) [6, 8]. According to these authors, data is understood as the primary entity of knowledge, which is unorganised and unmeaningful. Then, information is the particular organisation of data in order to create meaningful and valuable information. Knowledge is defined as the application and productive use of information. However, there are shortcomings to this definition: the distinction between information and knowledge is not clear and it leads people to use both terms interchangeably.

Still, in the management literature, we find a new approach to define knowledge: the best known representatives of this approach are Nonaka, Takeuchi, Hedlund [13, 15]. Their definition of knowledge is rooted in the comparison between Japanese and Western culture and a number of studies of Japanese and Western firms. The differences emerging from these studies provide a base to understand why some firms are better than others at managing knowledge. They identify different dimensions of knowledge: explicit-tacit and individual-group-organisational. Effective knowledge management happens at the organisational level when humans create some forms of dynamics to convert and transfer knowledge between the tacit and the explicit dimensions (Spiral of Organisational Knowledge Creation) and also between the human, group and organisational dimensions. The tacit and explicit dimensions of knowledge mentioned have been defined by Polanyi [17]: tacit knowledge can be formulated as "an individual knows more than he knows". The studies made by Nonaka, Takeuchi and Hedlund reveal that Japanese would be better at managing knowledge because of distinctive cultural traits: in Japan, the group is more important than the individual and people work for the success of the group; therefore Japanese are more inclined to share knowledge than individiuals in Western countries.

The definition provided by Nonaka brings an improvement in the sense that it englobes many dimensions and it is a dynamic definition. However, we could criticise that this definition assumes that knowledge is an object which is easy to identify, to assess and therefore to transfer and manage. Here, we introduce a definition worked out by Zeleny, this definition is as dynamic as Nonaka's but also has a temporal dimension (Zeleny, 1989). His definition is rooted in in Maturana's and Varela's process view of knowledge [14]. For Zeleny, knowledge is the **reflective** and **active** interpretation of data (which can be tacit or explicit: for example, description of objects) in a temporal process. In other words, knowledge gives the grounds for action: *Knowing is doing and doing is knowing*. An individual will be qualified as knowledgeable if he/she can build and interpret relationships of data, and put them into action (action is the ultimate test validating knowledge: knowledge without action is just information). To summarise, we define knowledge as a purposeful co-ordination of actions based on reflective and active interpretation of tacit and explicit data.

2 The Nature of Work

2.1 Taylorism: Division of Work and Division of Knowledge

The nature of work, as we have understood it for a number of years, has been influenced by the work of Frederick W. Taylor's research [20]: work is divided into smaller, standardised and measurable tasks. According to Taylor, there is a single best way to accomplish any task. This single best way is put in practice by letting management generate instructions which are performed by workers. Workers are considered as appendages to machinery. As a result of the search for technical efficiency, division of work has emerged: division of work means that work is divided into tasks which are executed by different workers (labour force). Work is also divided between planning and performing activities. The most illustrating example of the tayloristic perspective is provided by one of the most tayloristic industry: the automobile industry.

This division of labour is accompanied by division, dispersion and disaggregation of knowledge (Zeleny, 1989): knowledge about processes is dispersed among so many people that, at the end, no one knows anymore how to perform the work process from the beginning to the end. An other explanation, related to managerial reasons, is provided to division of knowledge (Osterman, 1990) [16]: knowledge, indeed could be retained on the managerial level in order to keep control over the work force, and hence power. Division of work and knowledge has its cost: the more work and knowledge will be divided, the more difficult and costly it will be to co-ordinate processes. These coordination costs cannot be negligated: they will grow as the organisation will specialise the tasks its workers perform (it is argued that a human should not supervise more than five or six other individuals). It is not unusual to see large organisations that have 10-15 managerial levels (Zeleny): ultimately, the cost engendered by co-ordination makes division of work less attractive.

If we relate our argument with our definition of knowledge as a purposeful co-ordination of action, we can affirm that separation of knowledge (represented by managers) and labour (represented by line workers) is a nonsense and a shortcoming of today's management. Still, removing levels of managerial control is likely to have a destructing effect if this is not implemented for other reasons than possible cost savings: these managerial levels are created by the division of work and knowledge, therefore we have to reduce this division, in other words we have to reintegrate work and knowledge.

2.2 Reintegration of Work

As Zeleny underlines it, reintegration of work will naturally take place as division of labour and knowledge will reach an extreme level and become less attractive. Now, it seems that we are approaching the start of "the reintegration wave": in today's business vocabulary, this has been called Business Process Innovation (Davenport, 1993) or Business Process Reengineering - BPR - (Hammer and Champy, 1990) or Business Process Change (Grover, 1995) [5, 10, 11,]. Behind these buzzwords, we find the term "business process" which is defined as a "set of logically related tasks performed to achieve a defined business outcome" or also as "a collection of activities that, taken together, create value for a customer". Still, reengineering is only one of the first steps towards the powerful wave of work reintegration. Ultimately, work reintegration will reach such a level that we could name as "self-service" or "do-it-yourself": division of work and knowledge is minimised so that one person can perform one process entirely by himself.

Reintegration of work means that tasks, labour and knowledge are being reintegrated by reducing: (i) the number of tasks (and therefore integrating work processes, simplifying work, diminishing production costs and increasing quality), (ii) labour division (which means that a process is performed by as few persons as possible), (iii) knowledge division (achieved by delegation of control and decision-making to the level where work is done). These are the ingredients that we typically have in the literature on BPR. As a result of reintegration, work acquires the following characteristics: integrated, multifunctional and self-coordinated. Globally, reintegration has the effect of reducing co-ordination levels and costs of the organisation. In figure 1, the left hand-side represents an organisation where work is extremely divided: managers M1 and M2 receive instructions from the top level and send instruction to workers which have one specific task to perform; the right hand-side represents an organisation where work has been reintegrated: instructions are sent by the manager M directly to workers which perform many tasks (processes).

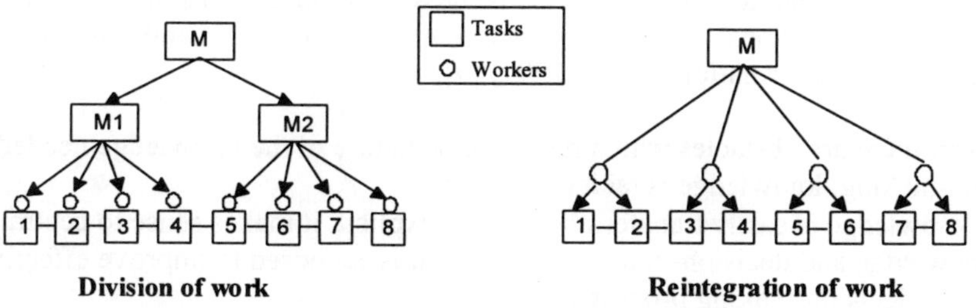

Figure 1: Division and Integration of Work

Furthermore, reintegration of work is supposed to eliminate buffers which emerge between tasks (for example, when some material stocked before it is processed). Eliminating process buffers creates smoother work processes which are less costly and add value to customers.

This idea of reintegration of work is supported by litterature on work design and transformation. Burchanan exposes five generation of job design theories, they are classified as follows [1]:

- 0 generation theory: it corresponds to the Tayloristic model, work is specialised (very few tasks) and own responsibility is limited.
- 1st and 2nd generation theories: they correspond to work enlargement, work englobes more tasks and specialisation needed to perform tasks is reduced.
- 3rd and 4th generation theories correspond to job enrichment in the sense that workers have more responsibility for their work (process ownership), tasks can be performed independently.

Reintegration of work has implications from the knowledge point of view: first, requirements from workers are different in the sense that we expect from them to have a better and larger understanding of their working environment (they receive knowledge from different organisational functions and have to cope with it); second, knowledge should be made available where it will be used (let us recall that from the Tayloristic perspective, workers are merely putting into actions instructions given by management and that there is little decision-making authority delegated to workers).

Figure 2 attempts to explain in which situation knowledge should be available where decisions are made. The starting point is that knowledge and decision-making are typically separated from the tayloristic perspective: for example, knowledge is owned by line-workers, managers take decisions upon the knowledge they receive from the bottom line. We suggest that decision-making and knowledge should be co-localised in order to reinforce effectiveness of organisations in terms of the quality of the decisions, speed of decision-making, accuracy and timeliness of decision-making information (Grant, 1997) [9].

However, there are obstacles which reside in the nature of the knowledge needed for decision-making; knowledge is tacit or explicit:

- if knowledge is easily transferrable, that is explicit, it makes sense to centralise knowledge and decision-making authority (it is supposed to improve effectivity of the decision-making process);
- if knowledge is tacit, which is hard to extract and transfer, then the decision-making power should be decentralised to the point where knowledge resides.

	Case 1	Case 2
Nature of knowledge	Explicit knowledge (easily transferrable) ↓	Tacit knowledge (hardly or non-transferrable) ↓
Location of decision-making	Transfer knowledge to a centralised decision-making point	Allow local decision-making where knowledge resides

Figure 2: co-location of knowledge and decision-making authority

Co-location and decentralisation of knowledge and decision-making authority is justified since hierarchy is inefficient to deal with knowledge integration as managers (decision-makers) do not have the capability to fully integrate the tacit knowledge of their subordinates.

2.3 Knowledge work and workers

As work is changing dramatically (integrated, multifunctional and self-coordinated), we can propose that work is becoming knowledge work which is performed by knowledge workers. Knowledge work is defined as the acquisition, creation, application and packaging of knowledge, performed by professional or technical workers with a high level of skill and expertise (Davenport, 1996) [4].

Knowledge workers are typically professional or technical workers with high expertise. Prahalad [18] identifies 4 necessary characteristics for workers of next generations, which will make of them knowledge workers in many respects:

- *Systems thinking*: the ability to conceptualise and synthesise knowledge (tacit and explicit) and to see connections between elements of the world (analysis capabilities).
- *Intercultural competence*: because firms are increasing their dependence on other cultures.
- *Continuous training*: workers should be exposed to new ideas, technologies, business practices and cultures.
- *Personal standards and standards of behavior*.

A lot of attention has been paid to knowledge work and workers, especially, one has tried to understand knowledge work in order to make it more effective. The fact is that knowledge work has been considered as a "black box" for some years: that is we can identify specific inputs and outputs of knowledge work but we do not understand its internal mechanisms. This work needs flexibility, autonomy and resists heavily structured approaches that we have used in job design.

3 Implications for Knowledge Management

In this section, we will review the implications of the nature of work which is emerging. We will particularly put the emphasis on knowledge management since the knowledge content of work is increasing dramatically.

3.1 Improving Knowledge Work

This is what we identify as one of the most challenging points in the agenda of organisations: we mentioned already that knowledge work resists structured approaches and that knowledge workers are not willing to give up easily their autonomy. Autonomy of workers is reinforced by the increasing number of virtual offices - work can be performed at all hours and anywhere.

Table 1 summarizes two extreme approaches to understand and structure knowledge work (Davenport, 1996): (i) first, the *laissez-faire* approach considers knowledge work as a black box where each individual applies its own work rules and structures, (ii) second, the *reengineering* approach is applied when management attempts to understand how work is done and proposes a way of performing work common to all the persons involved.

	Laissez-faire approach	**Reengineering approach**
Strategy	Let people do their work	Chane their work
Focus	Input and outcomes	Activities
Evaluation	Long-term	Short-term
Level	Individual	Group
Analytic emphasis	Understand environment	Design new environment

Table 1: Approaches to Improve Knowledge Work

In order to improve knowledge work, one objective to aim at is to free workers to do their work by reducing their administrative tasks, and to relieve knowledge workers from the burden of routine work. We have also learned in the SMARTER research project that, when put together, organisation of work and Information Technology (IT) are to blame if we think about knowledge workers' productivity: this problem has been identified as Fragmentation of Working Time. (The research project involves 6 small and medium-sized companies in the region of Turku, Finland and is coordinated by the Institute for Advanced Management Systems Research, in Turku. The project aims at analysing and tackling the phenomenon of Fragmentation of Working Time in these companies).

Fragmentation of working time can be explained as a result of a paradox: For years, workers' efficiency (front-line workers, managers and even top management) has been limited by lack of information (quality and up-to-date information); then, thanks to IT developments, information barriers have been removed and organisation's information flows have grown: nowadays, we are in such a position that we receive more information and knowledge than we can actually process. Especially, key people are getting involved in so many activities that they are not able to focus long enough on any task. We have built tools for improving productivity but we are becoming less productive.

Furthermore, small and medium-sized firms are subject to strong competitive pressures which lead them to look for cost-effectiveness and trim their organisation. Combined with the emergence of new organisational models (network and lean organisations), it has the effect of involving people in too many activities. Especially the working time of key people of the organisation is consumed by daily routines, and they hardly find the necessary time to focus on their core activities, at which they are good at, and to use their skills at their full potential. Also, involving people in many teams (which is good, in a sense, since it promotes teamwork) has the vicious effect to increase the need for information and knowledge. Fragmentation of working time is in itself a symptom which reveals that the allocation of competences, resources and capabilities in the organisation is wrong.

Other observed effects, linked to organisation of work, are:

- restricted freedom in organising working time
- work processes badly organised (related routines badly organised)
- increased availability to others without priorities - deficient teamwork planning
- fragmented communication
- unclear individual work plan

Upon these symptons, we can suggest an agenda for management of knowledge work. This agenda will be best put in practice if management understands and acknowledges that knowledge work can not be managed the same way as routine work. We can articulate the agenda around 4 points:

- *carefully define work assignment and priorities*: we have observed that workers are quickly submerged by routines, redundant work and unexpected tasks; management should make an effort to reduce this routine work and to establish priorities for the availablity of human resources.
- *team-building*: people are involved in too many teams which are often self-coordinated and badly managed; here, management is expected to provide guidelines for building and managing teams. For example, frequency of meetings and reports, communication means,
- *availability of knowledge*: one frustration of workers is that knowledge exists but is so much dispersed that it is not immediately usable; instead, knowledge sources should be integrated and made available (but not pushed!) to workers.
- *use IT*: IT is applied in concert with the previous points to support their implementation. We will typically find IT implementation with reengineering approaches to improve knowledge work by influencing work flows, creating knowledge management tools (knowledge bases, ...) and telecommunication applications.

3.2 Knowledge management

We have seen that learning is the form of labour that is becoming dominant, and knowledge-intensive work requires appropriate approaches from the knowledge management point of view. In our argument, we argue that knowledge management should recognise the demand for intellective skills and develop a learning environment in which such skills can develop. We will see that today's knowledge management practices are facing three dilemmas: *knowledge*, *authority* and *technology* (Zuboff) [23]. We will review the two first dilemmas now and look at the implications for technology in the next section.

We should first acknowledge that many attempts to manage knowledge have been carried out more or less successfully. However, one of our most important criticism is that knowledge has been too much considered as an object which is easy to identify, code, transfer and manage: this is not the case, knowledge is often split between tacit and explicit dimensions (the explicit dimension is the one which is easy to code into a knowledge base). The tacit dimension is owned by people, and actually *people add value to knowledge* by interpreting it and putting it into action. People interpret knowledge differently, depending on their experience. In the organisatons we visit, it is not rare to observe that knowledge bases (document management systems, databases, ...) are implemented and widely distributed to everyone for use (which is a laudable initiative), however little attenton is paid to the quality of knowledge from the user's point of view: the level of detail can be too high or too low, the content is not designed with enough flexibility so that it can be organised, summarised or analysed. For example, our experience tells that managers increasingly receive too many reports containing highly detailed information which they will never use. It is also important to note that managers receive two-thirds of their information from communication with other people, and the remaining comes from documents which are mainly not in electronic form; moreover, managers do not rely on computer-based information (unless the information is proven to be true by some other means).

Also, knowledge is becoming so valuable, as it is the most productive production factor available, that we cannot expect people to give it away. However, this statement is contradicted when too much sharing is taking place, in the sense that people share information which is not relevant to others (Davenport et al., 1992) [3]. For the reasons mentioned above, we do not favour neither too much sharing nor too little: we believe that there is an equilibrium to reach about what to share, to whom, when and how.

The second aspect that we wanted to talk about is authority: we say that decision-making authority should be located where the most appropriate and supportive knowledge resides. Delegating decision-making authority to others in such conditions has several implications:

- managers have to admit that they are not "the only ones who know", using the pretext of authority; IT has increased the knowledge content of work at all levels of the organisation and enhanced decision-making at all levels.
- what is managerial work becoming? will managers still use authority to distribute work and get it done? or will the relationships between managers and workers evolve so that commitment and motivation, rather than obedience, will serve as motives to perform a task?
- will management use technology to reproduce hierarchical relationships, although they give away "physical authority"?

To solve these dilemmas, we propose to involve people at the core of knowledge management approaches: people are users, creators, acquisitors, buyers and sellers of knowledge. We have also to take into account that people have a limit to the knowledge they can handle, even with the support of IT. Therefore, we are promoting a participative approach to knowledge management which would allow people to negotiate the definition, use and needs of knowledge. It is also important to motivate people to be active in the development of knowledge management projects: for too long, it has been assumed that people will start sharing knowledge as soon as they get a right, working and reliable information system; getting people to start sharing knowledge requires more than IT: it requires first to adapt the organisation's information culture.

4 Implications for Information Systems

An other aspect of fragmentation of working time that we have observed is "*introduction of new IT*": the use of IT is increasing at a tremendous rate in organisations. Large investments in IT are made, even if it remains difficult to assess the payback of these investments. Studies indicate that there is no correlative link between IT investments and increased productivity (Roach, 1987) [19]. On an indvidual level, the introduction of new technology translates into periods of adjustments: any new application has to be assimilated before it can be used at its full potential. The use of different applications and their increasing complexity also causes frustration for the workers.

An other aspect worth to be mentioned is the opportunity that IT offers to reach other people and to be reached in a much easier way: for example, it is not rare nowadays that managers receive per day about 50 electronic messages whose content is not crucial for running the business. We are not talking about "information overload" but rather of "noninformation": real and usable information is precious to run any business, but useless information is more dangerous; we tend to categorise each piece of information as "overload" because we lack tools to analyse it.

We stated that managers receive little information in electronic form, compared to other media, but one fact is that the amount of information in electronic format is increasing rapidly and, therefore, IT is likely to have an important role in knowledge management projects.

In conjunction with the knowledge management approach that we proposed in the previous section, we suggest to let individuals (users) have control over the knowledge they use in information systems. In other terms, we believe that there is a lot to be gained if managers can define together the information they need and the tools they want to use (provided that appropriate assistance is available to guide technical choices). However, it is also crucial to keep global some aspects of information, especially for large organisations, so that sharing can be done effectively.

An other aspect of this human-centered approach to information management is that people are trying to get access to a lot of information in order to keep a certain level of control over the work force: in short; they try to replicate the hierarchy in the information system. This attitude has to be corrected, and it has to be done before we bring forward information systems' considerations, that is by changing the information culture.

There are various methods to put this approach into action: among them, the role-based approach can be used (Torvinen, 1997) [21]. The role-based approach aims at constructing an interpretation of the organisation of work by the concept of role: a role is defined as a collection of tasks performed by an individual. In our context (knowledge work and knowledge management), IT is a part of work tasks; they are embedded in work activities. Roles also offers the possibility to study responsibilities of the user vis-à-vis the information that he/she manipulates, and the skills needed to manipulate information. By developing ISs integrated in activities, we build systems which suit better the work of people.

5 Conclusion

In this paper, we have shown how work is changing from different perspectives: work is more integrated, more knowledge-intensive and hierarchical relationships are evoluting since work is more self-coordinated. We have especially paid attention to knowledge work which is emerging as the dominant form of work.

There are implications for improving knowledge work and knowledge management: for the time being, most of knowledge management projects have been merely limited to the implementation of information systems without paying much attention to users. We propose to involve users from the start in knowledge management initiative development so that they could negotiate the definition, use and needs of knowledge in the organisation; involving people in what is more crucial to them, knowledge, will change the way the organisation apprehends information management.

There are also implications for ISs: here again, we suggest to involve owners of knowledge and give them a certain level of freedom to customise their own information environment, so that they could manage and grow in an information-rich environment.

REFERENCES

1. Burchanan D.A., "The Development of Job Design Theories and Techniques", Saxton House, Westmead, 1979
2. Checkland P., "Systems Thinking, Systems Practice", Chichester: John Wiley & Sons, 1981
3. Davenport T. et al, "Information Politics", in Sloan Management Review, Fall 1992, pp 53-65
4. Davenport T. et al., "Improving Knowledge Work Processes", in Sloan Management Review, Summer 1996, pp 53-65
5. Davenport T., "Process Innovation: Reengineering Work through Information Technology", Harvard Business School Press, 1993
6. Davis S. and Botkin J., " The Coming of the Knowledge-based Business", Harvard Business Review, September-October 1994
7. Drucker P., "The Coming of the New Organization", Harvard Business Review, January-February 1988, pp 45-53
8. Earl M., "Knowledge as Strategy: Reflections on Skandia International and Shorko Films", in "Strategic Information Systems: a European Perspective", edited by Ciborra C. and Jelassi T., John Wiley and Sons Ltd, 1994
9. Grant R., "The Knowledge-based View of the Firm: Implications for Management Practice", in Long Range Planning, Vol. 30, June 1997, pp 450-454
10. Grover V. and Kettinger W., "Business Process Change", Idea Group Publishing, 1995
11. Hammer M. and Champy J., "Reengineering the Corporation: A Manifesto for Business Revolution", HarperCollins Publishers, 1990
12. Hammer M., "Reengineering the Corporation", HarperCollins Publishers, 1993, pp 68-71
13. Hedlund G. and Nonaka I., "Models of Knowledge Management in the West and Japan", in Implementing Strategic Processes, Change, Learning and Co-operation", edited by Lorange. et al., Oxford, Basil Blackwell Ltd, 1993
14. Maturana H.R. and Varela F.J., "The Tree of Knowledge", Shambhala Publications, Inc, Boston, 1987
15. Nonaka I. and Takeuchi H., "The Knowledge-creating Company", Oxford University Press, 1995
16. Osterman P., "Impact of IT on Jobs and Skills", in "The Corporation of the 90s", edited by Scott-Morton P., 1990
17. Polanyi M., "The Tacit Dimension", Routledge & Kegan Paul Ltd, 1966
18. Prahalad C. K., "The Work of New Age Managers", in "The Organisation of the Future", edited by Hesselbein F. et al, JB Jossey-Bass Publishers, 1997
19. Roach S., "Technology and the Service Sector: America's Hidden Challenge", Morgan Stanley Special Economic Study, Technical Report, 1987

20. Taylor F, "The Principles of Scientific Management", Bulletin of the Taylor Society, December 1916
21. Torvinen V., "IS Requirements and Work Transformation", TUCS Technical Report No 101, March 1997
22. Zeleny M., "Knowledge as a New Form of Capital, Part 1. Division and Reintegration of Knowledge", Human Systems Management, Vol. 8, pp 45-58
23. Zuboff S., "In the Age of the Smart Machine: the Future of Work and Power", BasicBooks Inc, 1988

A CASE STUDY OF BUSINESS TRANSFORMATION FOR TELECOMMUNICATIONS OF ST.-PETERSBURG

VLADIMIR KVASSOV
Turku Centre for Computer Science TUCS/IAMSR research group at Åbo Akademi University,
Leminkaisenkatu, 14-A, Turku-20540, Finland
E-mail: vkvassov@abo.fi

The paper is dedicated to information technology (IT) - enabled business transformation in the context of transition market economy. We use Russia's telecommunications for the industry context and elaborate on the case of company X providing telecommunications services in the second largest city of Russia - St.-Petersburg. We analyse key factors of current business environment that make a strong impact on the company's activities. We study an impact of IT on the business process transformation, reasoning within the analytical framework presented by Venkatraman (1990). We position the company within the framework and identify the issues that have to be resolved for further development.

Introduction

The paper consists of three sections: in the first section we analyse the telecommunications industry in Russia. We discuss major traits of the telecommunications sector in Russia and identify the driving forces of business environment. We use the case of the real company X to illustrate the current situation on Russian telecommunication market. In this section we stress on the following issues:

- Industry regulation: in this subsection we describe Russian telecommunications industry evolution from the Soviet Era to the latest changes. We stress on such factors as privatisation, licensing and other regulative issues that impact the telecommunications sector;
- Infrastructure analysis: in this subsection we discuss recent developments and present structure of Russia's telecommunication sector. We provide taxonomy of the existing networks and identify recent achievements and bottlenecks in the telecommunications infrastructure.
- Competition: in the subsection we consider the existing competitive environment in Russia's telecommunications sector. We illustrate analysis of competition on the example of St.-Petersburg telecommunications market.
- Company analysis: in this subsection we concentrate on the company specific factors.

In the second section we present an analytical framework for business transformation. We use the framework of IT-enabled business transformation presented by Venkatraman (1990).
The framework include five steps:

- Localised exploitation of IT;
- Internal Integration;
- Business Process Redesign;
- Business Network redesign;
- Business Scope Redefinition.

The framework is based on two dimensions: the range of potential benefits and the degree of organisational transformation. The localised exploitation usually means automation of the routine tasks. Going trough of this stage requires marginal changes in organisation but brings marginal benefits as well. Business scope redefinition requires the most drastic changes in the organisation. In our paper we will go through the all steps and identify what kind of organisational changes are to be done in case of company X on every step of the framework. In this section we also dwell on the complexity issues of business process transformation.

In the third section we will show the application of the business transformation framework to the company X case. We elaborate on analysis of the sales process in company X in terms of Venkatraman's framework. We also show the position of the company on its way to the business transformation. The analysis describes the past experience of the company with IT and current impact of IT on the company's activity. Reasoning in terms Venkatraman's framework we identify the opportunities for further IT-enabled business transformation.

1 Analysis of Russia's Telecommunications

In this section we elaborate on the specific traits of Russia's telecommunications. We also stress on the profile of company X which we are going to consider in our paper.

1.1 Industry regulation

During the Soviet Era telecommunications industry in Russia was completely state-owned. The industry did not get sufficient investments. Before 1992 only 0.15% of GNP was devoted to the development of telecommunication infrastructure (U.S. Chamber of Commerce, 1996). State-owned companies were the only players on the market. Lack of investments and non-competitive business environment has led telecommunications sector to market inefficiency. In this situation new market entrants faced high barriers. The state telephone companies dictated the prices, policy and rules. Neither weak infrastructure nor significant lack of modern services could support a transitional process to market model economy. According to the recent estimate, number of unsatisfied requests for telecommunications services in Russia counts 11 millions (MIS telecommunications review, 1998).

Another factor hindering the development of telecommunication infrastructure was COCOM restriction (the USA's rule prohibiting the provision of modern technologies to the former USSR). It did not permit to apply the advanced telecommunications technologies such as fibre optic, radio relay transmission, and high-speed data transmission in the former Soviet Union. Currently the COCOM restrictions are lifted, so there are some opportunities appeared to develop telecommunications infrastructure with the modern technology.

Privatisation of telecommunications in Russia started in 1992 with investments in the company called Rostelecom that is the only international long distance carrier in the country. The next step of privatisation affected the local public network operators. Currently there are approximately 85 local operators. The recent privatisation allowed to local telecommunications companies to attract investors to the projects in telecommunications industry. Many companies with foreign shareholders emerged on this wave of privatisation and deregulation of the industry. The changes started from the largest cities of Russia such as Moscow and St.-Petersburg. Since we focus on St.-Petersburg in our paper, next we analyse business environment in telecommunications of that region.

The political and legislation issues make the strongest impact on telecommunications market. There are several reasons for that:

a) General political instability;
   The Russian economy is on its way to the market model. This process is accompanied by vague legislation that is full of ambiguities and contradictions. In addition to the country risk that must be evaluated before making any investments, the legislation inconsistency and tax uncertainties must also be taken into account. Such typical procedures as certification and licensing are connected with uncertainty and undefined time frames. The recent financial crisis showed the vulnerability of the emerging markets not only in Russia but worldwide.
b) The infrastructure of public wire-based networks remains under control of formerly state-owned local companies. In St.-Petersburg such a company is a recently privatised AO "Petersburg Telephone Network"(PTN). This company manages the access to the network and its maintenance. At the same time PTN keeps a significant share of the market as an operator. In addition to fixed telephony services, PTN also owns shares in companies operating mobile telecoms sector (standards NMT-450 and GSM-900). These factors make the competitive situation on the telecommunications market hard to predict.

1.2 Infrastructure analysis

The telecommunications infrastructure of Russia is insufficient and not enough advanced technologically to meet the demand. It has insufficient capacity, low penetration and call completion rates, not to mention the lack of modern services.

St. Petersburg has a five million population. It is comparable to the population of Finland, Denmark and Norway. The telecommunication problems of the city are typical for the entire Russia. Next we will study the factors having a significant impact on the telecommunication business in St.-Petersburg:

- Inadequate capacity

The business needs of St.-Petersburg require a high level of telecommunications services. The satisfaction of this demand is connected with the growth of a network across St.-Petersburg. The obsolete wire-based telephone network within the city and outdated switching systems in a number of areas restrain the further growth of telecommunications services in St.-Petersburg.

- Low penetration rate

The penetration rate in St.-Petersburg is 38 telephone lines per 100 people that may seem relatively good in comparison to an average 16 lines per 100 people in Russia (U.S. Chamber of Commerce in Russia, 1996). However we must consider the fact that St.-Petersburg is a large business city placed right on the way from Europe to Russia. The demand for telecommunication services is much higher than the offer. The limitations of the further expansion of the network are mostly technical and financial, such as an obsolete infrastructure of the network, outdated switching systems and a lack of investments. The low penetration rate of fixed telephony is not compensated by mobile telephony like in Nordic countries. The largest operator of mobile telephony (standard GSM-900) counts only 50,000 subscribers. Nevertheless we should notice that the growth of mobile telephony in St.-Petersburg has an explosive character.

- Lack of modern services

This issue logically emanates from the previous one, because it is not possible to provide high quality services when lacking in the modern equipment. This area requires heavy investments in upgrading of the existing switching systems or the in installing the new ones.

In our paper we elaborate on the case of the company providing the wire-based telecommunications services in the second largest city of Russia St.-Petersburg. To discuss the case we need to classify the wire-based networks, which are in use in Russia. Currently we distinct three major kinds of networks:

- Local (Public) Networks

Public networks utilise the infrastructure inherited from the Soviet Era time. In most cases those networks are outdated and have their basis on copper cables and cross bar switching technology. Digital switching technology occupies approximately 5-7% of the total public network switching technology used across Russia (Team Canada Market Research Centre, 1997). In large cities, such as Moscow and St.-

Petersburg this figure is higher but still not sufficient to meet modern business requirements.

The local public telephone network of St.-Petersburg covers almost the whole city with approximately 5 million populations. Most of the communications are constructed on copper cables and built many years ago that makes them outdated. The formerly state-owned largest telecom operator in St.-Petersburg is Petersburg Telephone Network (PTN) that maintains the entire infrastructure.

- Overlay Networks integrated with local networks;

Overlay networks emerged in response to the demand for high quality international communications generated by foreign companies that make business in Russia. These networks are constructed with the latest fibre optic technology and digital switching systems. Operators of overlay networks often utilise gateways to public networks and therefore bring to users the advantage from integration of two different networks. From one hand this step expands a customer base of overlay network operators. From another hand, following this approach makes the operator of overlay network partly depend on the local public network. Part of the network, so-called 'last mile' - the distance between the exchange and the end user still belongs to local public network operator. Therefore success of local and overlay networks integration requires a high degree of collaboration from both operators.

- Dedicated Networks

This is the case of overlay network when it completely bypasses local switched network through the use of micro wave, satellite and fibre optic connection. St. Petersburg has a peculiar example of such network that is constructed through the use of subway tunnels. In difference to the kind of network that we have discussed in previous paragraph this one does not have a gateway to local switched networks. Among the advantages from this approach is complete independence in service provision policy, market policy and technological advantage from entirely new modern network. However this approach has several drawbacks. First, construction of such network requires heavy investments. This issue is quite crucial, especially taking into account vulnerability and high risk of emerging Russian market. Secondly, restriction from public switched network may significantly narrow the range of potential customers

1.3 Competition

Until the privatisation of the telecommunications industry there was no competition on the market. The reason is that the state-owned telecom companies provided services on a monopoly basis. Nowadays the situation has radically changed. New players have entered the especially attractive market segments, such as long distance telephony, wireless telecommunication services and data transmission.

Petersburg Telephone Network is a formerly state-owned PTT, which controls a major share of St.-Petersburg telecom market. The wire-based network is the second largest in Russia (after Moscow). The current switching capacity numbers approximately 1.9 million telephone lines. PTT takes advantage of having been alone on the market for a long time. It carries the largest traffic in St.-Petersburg and holds a major share of the local wire based telephony market.

The competitive situation of the telecommunications market of St.-Petersburg constantly sharpens. Telecommunications business attracts many Russian and foreign companies to enter the market. Relatively recently emerged companies have constructed overlay networks with a gateway to the public network (Peterstar, Baltic Communications Limited), others utilise dedicated networks (Metrocom, Leivo). The latter approach is mostly used for providing data transmission services.

The competition has also increased in the last few years through the expansion of Moscow-based companies. Some of those companies have large and powerful shareholders (Sprint, AT&T), others have a well-developed network in Moscow (Rustel, Combellga), which can offer a good service level in St.-Petersburg through the intercity channels to Moscow.

These factors have a strong impact on the growing competition in the fixed telephony service market segment of St-Petersburg and, primarily, in the sector of long distance telephone services.

We have already discussed in previous sections the problem of the 'last mile' (the distance from the exchange to the customer's premises). There are only a few operators in St.-Petersburg except the local PTT, which have a license to use the existing infrastructure for laying the last mile. It makes the situation in the market hard to predict, because the success in competition often depends on the success in building strong relationships with the dominant provider.

The analysis of the factors we have discussed in this section shows that St. Petersburg has a better-developed telecommunications infrastructure comparing to the rest of Russia but still does not meet Western standards in terms of capacity, penetration rate and the level of modern services. The market is divided between the dominating former PTT and a number of small companies. In addition to this, vague and constantly changing legislation makes it more complicated to enter and operate in the market.

1.4 Company analysis

In this section we will discuss the issues related to the company X which we selected for analysis in our case study. This company emerged on the second wave of privatisation when the formerly state-owned PTTs got a possibility to attract foreign investments. The company was established with foreign investments. The participation of a foreign partner brought not only the money that is necessary for massive investments into infrastructure development. It also assisted in the implementation of "Western Style" in management, with a strong emphasis on a customer service.

The company X attracted heavy investments into the construction of a modern overlay network based on the latest technologies, such as fibre optic and digital switching. The 500km fibre optic network covers the entire city and provides high quality telephony and data transmission services. Modern technology and robustness of the network made the service provided by company X much more reliable than the service provided by its competitors. The company has a dedicated access to the international exchange either through the Russian international operator Rostelecom or through the satellite channels. It also has a gateway to the public local telephone network that makes it possible to expand a customer base. Currently the company provides customers with Western standard wire-based telecommunications services and has approximately 100,000 telephone lines in service. The range of services includes voice telephony, data transmission and private box exchange (PABX) installations. The target customer sector embraces corporate clients as well as medium and small businesses. The company extensively utilises the last mile of local public network to reach its customers around the city. The service is partly provided through the PTT network, that requires a high degree of co-ordination between the two organisations.

1.5 Case issues

In the case we study a business process of service delivery and its evolution from the IT application perspective. Until recently the sales process was supported with a hard copy, paper-based documentation. This approach has many serious drawbacks. First, it took a lot of the sales person's time to fill in customer's request and to deliver it to the proper person in other departments. Besides that, the information had to be duplicated for the different units of the company. As several copies are maintained, some of them inevitably became obsolete.

We aim to study how IT application influenced and motivated changes in the business processes of company X and to identify the impact of IT on the company activities. We will, finally, make an attempt at giving directions for the future development. To evaluate the degree of the business process transformation we used an analytical framework, which we will discuss, in the next section.

2 An analytical framework and complexity issues for business process transformation

2.1 An Analytical framework

During recent years there were a lot of discussion about the impact of information technology (IT) on transformation of business processes. The information systems (IS) evolution has already passed the stage when proprietary information systems could provide a company with a competitive advantage. The case of American

Airlines is maybe the most well known example, which is often used for illustration purposes of proprietary IS advantages (Hopper, 1990).

The last decade brought drastic changes to the IT industry. Extensive research and development in the computer industry combined with mass production made computer systems available for a large mass of people. The fast development of network capabilities makes it possible to reach a tremendous number of information resources.

However we have to be aware of the fact that taking advantage of IT is not a straightforward process. On the one hand, the implementation of IT in business requires a lot of organisational changes and technical co-ordination. On the other hand, a successful IT implementation can facilitate drastic changes in business process and even change the scope of the organisation.

Let us elaborate on the analytical framework of IT-enabled business process transformation. We selected a framework of IT enabled business transformation presented by Venkatraman (1994). This framework was validated by numerous examples from real business environments. The framework is schematically presented in Fig 1.

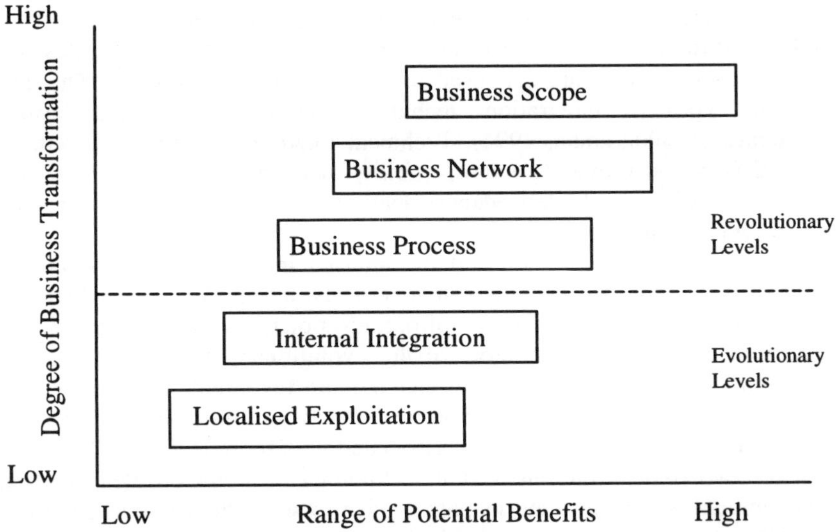

Fig 1. Range of potential benefits and degree of business transformation. Adapted from Venkatraman (1994)

The underlying idea of the framework is that benefits from IT deployment are marginal if they are only imposed on existing processes and strategies of the organisation. The related thesis is that these benefits increase if investments in IT are accompanied by organisational changes (Venkatraman, 1994). We also want to stress the fact that the hierarchy of levels does not represent an evolutionary process. It only shows that any organisation can be identified within the framework of five levels.

Localised exploitation assumes that IT is used for routine business processes. The term "localised" points to the fact that in many cases IS are developed and implemented according to the needs of functional organisational units (e.g. sales, inventory etc). The IS is isolated from the rest of organisation. It pursues only a limited scope of operational targets. In practice, the implementation of IS in this stage requires only minor changes in organisational structure (Venkatraman, 1994). The typical examples of such systems are computerised inventory control systems instead of a paper based card file; electronic mail systems instead of a paper based interoffice memorandum system etc. This level requires only minimal changes to the business process. The weakness of this stage is that the company does not get many of the possible advantages if consider the technical functionality for changing the business process. A technical implementation of this stage usually implies using common IT applications and competitors can easily imitate technical solutions and neutralise the advantages of using the information system. Therefore such an approach cannot provide any competitive advantage for a company.

The level of Internal Integration is an extension of the first level. On this stage all the information systems are supposed to be integrated into one IS. This level involves two types of integration: technical interconnectivity and business interdependence (Venkatraman, 1994). Technical interconnectivity deals with the technological issues of providing interoperability of different applications through one platform. Business interdependence deals with organisational changes in support of an internal integration.

Both the localised exploitation and the internal integration aim to automate existing processes. They fall into the category of evolutionary levels, which does not require fundamental changes in any of organisational conditions. Venkatraman identified main enablers and inhibitors of the evolutionary approach to business transformation (1994). The technological enablers are favourable cost-performance trends and enhanced connectivity capabilities. The organisational enabler is that managers are aware of costs and benefits associated with those levels.

There are two kinds of inhibitors: first, technologies become obsolete. The pace of IT development is very fast and managers have to solve the following problem: either implement the IS now and probably face the obsolescence problem soon or wait for the final standards and then implement the IS. Secondly, managers are concerned with their possible loss of power.

Business Process Redesign (BPR) is a revolutionary level that requires not just adding IT functionality to the processes that currently exists, but using IT for changing the core of business processes. Davenport and Short (1990) define a business process as "a set of logically related tasks performed to achieve a defined business outcome". BPR is defined as "the fundamental rethinking and radical redesign of business processes to achieve dramatic improvements in critical contemporary measures of cost, quality, service and speed" (Hammer and Champy, 1993). This is a method that enables an organisation to make fundamental changes in the way it carries out its business. In addition to those measures, BPR should serve an overall organisational strategy as a response to the pressures of the modern business environment.

Reasoning within the analytical framework, on the stage of BPR an organisation needs to analyse the rationale of existing processes, current changes in their business environment and the threats of not changing the process. The analysis results in an identification of strengths and weaknesses of existing process. BPR provides the IT-enabled capabilities for future competitive advantages of the company rather than the refinement of the outdated business process.

Hammer (1990) considers IT as a key enabler of BPR that implies radical organisational changes and follows the principles:

- Organise around outcomes, not tasks;
- Have those who use the output of the process perform the process;
- Subsume information processing work into the real work that produces the information;
- Treat geographically dispersed resources as though they were centralised;
- Link parallel activities instead of integrating their results;
- Put the decision point where the work is performed, and build control into the process;
- Capture information once and at the source.

IT makes a contribution to BPR by reducing the co-ordination costs across the organisation. We tend to agree with Davenport and Short (1990) who claim that IT and BPR have a recursive relationship, i.e. IT capabilities should support business processes and business processes in their turn should be rebuilt in terms of the capabilities IT can provide.

In contrast to the previous three stages Business Network Redesign (BNR) crosses the organisational boundaries and represents an IT-enabled redesign of the exchange process between multiple participants of a business network. Benefits from BNR are often mistakenly associated with the advantages of a transaction processing between participants of the business network. Venkatraman argues that the potential range of benefits is much broader (1994). It includes:

a) administrative efficiency enhancements from transaction processing;
b) operational efficiency enhancement in inventory movement;
c) the potential for differentiation in the market place through greater coverage of sources of competencies;
d) enhanced learning, which is potentially valuable under highly uncertain circumstances.

Business Scope Redefinition is facilitated by IT functionality. The major strengths of this stage are an opportunity to create more flexible business entities through the use of IT capabilities and the substitution of interfirm business relationships as an effective alternative to vertical integration (Venkatraman, 1994). We can illustrate this stage by the well-known example of American Airline's computerised reservation system SABRE. The role of this system went far beyond the frames of a traditional marketing support tool and became one of the major revenue generating activities of the company.

2.2 *Complexity issues of business process transformation*

In this section we look at the complexity theory from the BPR perspective. The 'complexity' theory, or 'chaos' theory, deals with the non-linear behaviour of the systems. The underlining idea is that a system may have two types of behaviour (Baets, 1998 p.24): chaotic and deterministic (in short term). Genuine chaos has the following properties (Baets ,1998):

- sensitivity to initial conditions;
- existence of random itineraries;
- common occurrence of random itineraries;
- cake-mix periodicity /a-periodicity.

From these properties we can imply that chaotic systems are not replicate the past behaviour. Its behaviour very strongly depends on initial conditions and even small changes in initial conditions, magnified by chaos, may lead to the drastic changes in the system's behaviour.

Recently complexity theory attracted an attention of managers due to its possible implications in a field of management science. We can consider business environment as a complex system, which is characterised by non-linearity and unpredictability of behaviour. This theory also applies to the discipline of strategic planning. If we recognise that the business environment is unpredictable then we can come up to an understanding the reasons of long term plans failures. We can apply this theory considering business organisation as a dynamic complex system that embraces many factors influencing each other. There are several implications of complexity theory for business (CIO Magazine,1998; Baets, 1998):

- In a business context, managers should attend a holistic view of organisation not reducing the entire picture to daily routines;
- We assume that small changes in a system may effect drastic changes;
- It is hard, if not impossible to implement a strategic plan for anything but the short term.

Business process transformation requires the holistic view of and in-depth understanding of the core business processes in an organisation. According to

Malhotra (1996) seventy percent of BPR projects fail because of: (i) lack of sustained management commitment and leadership; (ii) unrealistic scope and expectations; and (iii) resistance to change. Managers often consider a reduced model of a business process out of a context. If we assume that environment is chaotic such a reductionism of the process may limit the success of a BPR project and even lead to a failure.

BPR is characterised by radical and cross-functional process starting from the scratch. In some sense, we can consider it as initial conditions of the complex system. In practice this desire to start from the scratch is very hard to realise. There are many inhibitors: long term nature of BPR projects, need to share the common vision of the process within the entire organisation, daily routines etc. Because of magnifying effect of initial conditions in a dynamic environment, miscalculation of initial conditions may lead to the global failure. Many of BPR projects fail to take into account the non-linearity of the business environment but reduce the business process attributes up to several factors.

However it does not mean that BPR is impossible implement. If organisation can provide a strong leadership in BPR combined with the holistic view of business process then it may lead to the success of such a BPR project.

Those complexity issues can be applied to the telecommunication business environment of St.-Petersburg. In the next section we consider the company X within the analytical framework and figure out how IT enables business process transformation.

3 Positioning the company within the analytical framework

We consider the case of company X in terms of the analytical framework that we have presented in the previous section. We will show how the role of IT has been changing within the company since it started the business. We aim to evaluate an impact of IT on the current company activities and to work out a course of possible actions for the further business transformation. As an example we elaborate on the sales process in the company X.

3.1 Description of the sales process

The model of the sales process in Company X is shown in Fig 2. We describe the process in terms of beginning and end points, interfaces and organisational units involved.

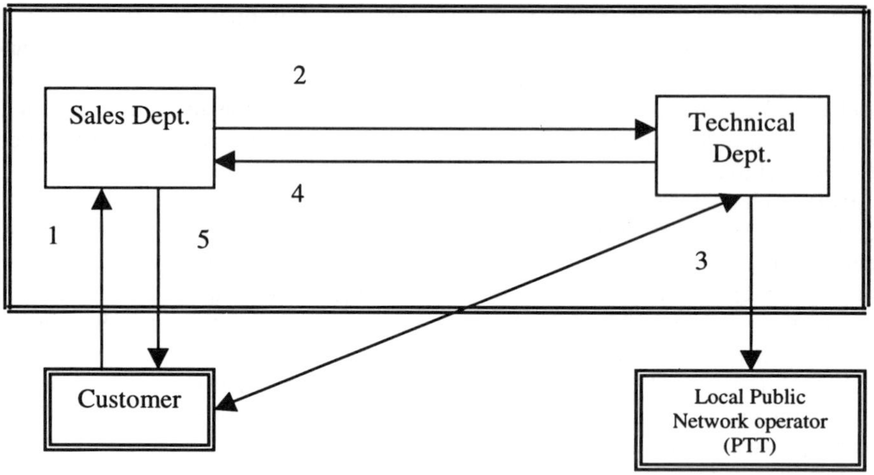

Figure 2: Model of the sales process

Explanation of the symbols:

1. Getting a request from the customer;
2. Sending a request sent for feasibility study to the Technical Department.
3. Feasibility study (checking out customer's premises and local PTT);
4. Delivery of the feasibility study result to the sales person;
5. Signing the order with Customer

As in any other sales process, in the company X it begins with customer's request or by generation of a sales prospect by the sales force. Then the request goes through a technical feasibility study and returns to the sales person with the conclusion about technical possibility of the line installation. The sales person is responsible for negotiating an order. After signing the order the information about the customer details goes to the relevant departments (technical, billing, finance and customer service). Between the start and the end points the process goes through several interim stages and involves different organisational units:
Stage 1 of the process starts with negotiations between the customer and the sales person. Negotiations can be initiated either through a customer's request or be generated by some prospective activity of the sales staff. The sales person identifies the customer's requirements and works out a possible solution. The parties communicate through face-to-face meetings, telephone and facsimile messages. The outcome of this stage is a preliminary agreement for providing the service, or rejection of the offer, which terminates the process.

The Sales Department and the Technical Department are the main participants in stage 2. We identified in the previous sections the problems with checking the feasibility with the local PTT. The company X may solve the technical issues within its overlay network, but in most cases it has to request the local PTT to provide the "last mile". The sales people get the request delivered into the Technical Department. The request form contains information about the customers' requirements and details. That request is the basic document that contains all the necessary technical details. In most cases technicians must check out the customer's premises. In some cases the time for checking out the premises may be reduced if they have been visited recently. The outcome of this stage is providing the Technical Department with information about customer's requirements.

In stage 3 technicians check out the customer's premises for the technical possibilities to provide the service. The main participants in this stage are the customer, the Technical Department and the local public network operator. This is the most crucial stage of the entire process. The network in St.-Petersburg is not so well developed as in most European cities. Therefore the rate of refusals due to technical reasons remains high. In addition to that, we have already mentioned that a major part of the city telecommunications infrastructure network belongs to the local PTT. It means that it is necessary to communicate with the representatives of PTT (in our case with the company "Petersburg Telephone Network") to get approval for an installation. Similarly, for example, in U.K. the major part of the network belongs to British Telecom, but the installation procedure is simpler and less strict - the operator-company does not need to ask for approval of BT to do installation work. This is not the case for St.-Petersburg: the company X has to involve PTN not only in the technical feasibility study, but also in the delivery of the last mile. The outcome of this stage is a conclusion about the technical possibility to deliver the service.

In the 4th stage the result of the feasibility study becomes available to the responsible sales person. The technical feasibility study may last several days. The main purpose at this stage is the delivery of the relevant technical information to the sales person. In case of a positive result of the technical feasibility study the sales person and the customer get together for signing the order in the 5th stage of the process.

In the next section we aim to trace changes of the sales process through the impact of IT. We will elaborate on it to identify its current position and to point out further steps in the business transformation.

3.1.1 Localised exploitation

An implementation of an information system (IS) for managing the sales process was the first step of process automation. The first version of the information system had a limited functionality and basically handled the request entry operations. Its main features were: (i) to input the customer requests, (ii) to trace the request through the stages of the technical feasibility and (iii) information exchange between departments involved in the process. The development of the IT application on this level brought minimal changes to the process itself. However it reduced time spent on handling the customer request and getting it through the process chain.

The IS brought the following advantages at this stage:

- a reduction of the paper based document turnover;
- a reduction or elimination of reproducing lost documentation;
- a reduction or elimination of reprint costs due to factual printing errors;
- it avoided maintaining many copies of the same document;
- it kept the document changes up-to-date;
- it reduced the time for handling the customer request and therefore improved the responsiveness;

There are also obvious drawbacks with the implemented IS:

- a lack of integration with the rest of the company;
- a minimal learning among the mangers about potential benefits and limitations of the IS;
- different departments may use different IS (e.g. sales, customer service, finance, billing) and will therefore use different information, which may lead to confusion and extra work;
- different systems may have different interfaces within the company that decrease the usability of the IS;

To summarise the results of the IS implementation at this stage we can conclude that the implementation of Sales Department IS was a good start. The next logical steps would be to implement an IS with extended functionality that embraces key function of the entire organisation.

3.1.2 Internal integration

Currently the company is in the process of gathering different functional IS into one integrated package. According to Venkatraman there are two types of integration, which must be implemented: technical interconnectivity and business interdependence. Both types of integration should be done, but in practice most attention is paid to the technical IT issues of an integration, that is the technical interconnectivity.

The department pushes a lot of efforts to integrate functional IS within the company. Today there are several information systems to be integrated: the sales department IS, the customer service IS and the billing IS. These systems are built on different platforms and therefore solving the interconnectivity problem remains very important.

A much more challenging task is to make the organisational adjustments. The implementation of a new IS requires strong efforts in changing organisational roles and responsibilities within the company and time for a teaching employees how to use the system. In addition to that, the implementation of new integrated system should not hinder daily business processes.

Currently the company has one database, instead of the previous departmental ones. This helps employees to access the data they need. With the introduction of a customer profile all the relevant data is stored in one place and therefore much more up to date. The current process of internal integration does not only mean, "embedding the outdated process in silicon and software" (Hammer, 1990). Having completed this stage the company is going to solve the following problems:

- to eliminate non value-adding functions;
- to improve the response to customers' requests and therefore to improve competitiveness;
- better customer service, because all customer relevant information is available at every stage of the sales and the after-sales process;
- all involved employees obtain comprehensive accumulated information about the customer and its relations with the company, e.g. financial, technical and traffic generating issues;
- an opportunity to utilise information from the customer database to differentiate the service and to increase the flexibility in service.

An analysis of the current situation shows that the company has moved in its IT application from localised exploitation and is now in the process of internal integration. Next we will identify the necessary actions for a BPR.

3.1.3 Business process redesign

In order to redesign the sales process the company has to assess how its competitors are performing in the same process and to target the best practice in the market. The company X has to identify factors that hinder it from reaching the sales objectives. The transformed process should be redesigned with the help of IT (Davenport, 1993). The sales process as it is represented in Fig. 2 is redundant. It contains steps in which no value is added (e.g. the delivery of information about the technical possibilities from the technical staff to the customer through the sales person). Such operations can be eliminated with a properly designed IS.

An information system that can drastically improve the efficiency has to cross the organisational boundaries. The best option is to avoid dependence on the local PTT because this factor impacts on the company policy. As we cannot change regulations we have to admit that the current sales process will inevitably involve the PTT. The process can be transformed in such a way that non-value-adding stages are eliminated (e.g. checking the customer premises and checking the obtained results with PTT).

The intended IS provides co-ordination capabilities between different organisational units within the company and with the local PTT. It should contain a comprehensive database with information about the technical details in a timely manner. After implementing the IS several stages of the process will be left out (see fig 3). In our opinion, the local PTT would be in favour of such a system, because it will change the paper based communication and therefore increases the effectiveness of its operations.

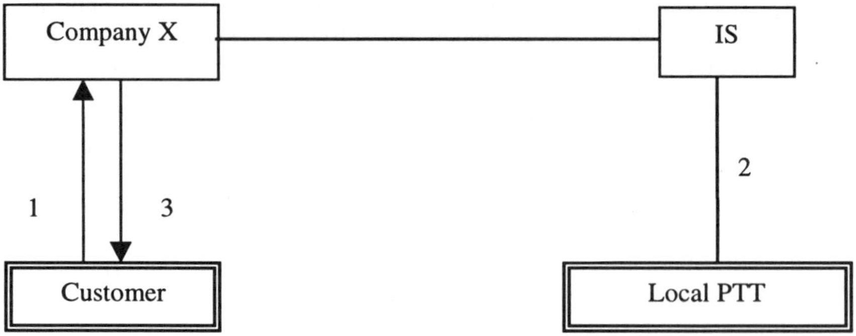

Fig. 3 Intended structure of the process.

1. Getting a request from customer
2. Checking with the IS that includes the information resources of the local PTT
3. Signing the order

As time is a crucial attribute of the process, let us consider time frames of the business process. The technical feasibility study is the process when the technical possibility must be identified for every customer. The duration of this process is relatively long because it involves a third organisation (local PTT). No installation can be completed without its approval. The time spent at this stage can be reduced if the company X implements an interorganisational information system. That process will require a lot of efforts to get organisational and technical issues settled. The advantage of this solution is that the time spent on the clarification of technical problem with PTT in each case will be reduced at the expense of faster and effective information handling through the IS. We can reasonably expect a drastic reduction of the time spent on the feasibility. In this case we exceed the frames of just an automation of the process but go for changing the process itself.

4 Conclusions

In our paper we have studied the business process transformation in the context of the Russian telecommunication industry. We analysed the current situation in the industry and provided taxonomy of existing wire-based networks. Our findings include the factors that make a strong impact on the industry: unstable regulation and inadequate telecommunications infrastructure. We have analysed the infrastructure, the competition and the company that was the subject of our case study.

Our case study showed that the framework is applicable to the case company. We also proposed a course of action for the company to further impose of IT influence on its activities. We have shown the process of changing the role of IT in the vital business process. The case, which we have considered in this paper is quite typical for the rest Russia, so we can assume that the presented analytical framework is valid for other companies.

References:

1. American Chamber of Commerce in Russia, "Key Telecommunications Opportunities and Major Issues in Russia", 1996 (http://www.online.ru/sp/accr/telec/telec1.html)
2. Baets, W.: Organizational Learning and Knowledge Technologies in a Dynamic Environment, Kluer Academic Publishers, 1998
3. CIO Magazine, Enterprise Section 2 – April, 1998, Simple Yet Complex http://www.cio.com/archive/enterprise/041598_qanda-content.html
4. Davenport, T.H. & Short, J.E.: The New Industrial Engineering, Information Processing and Business Process Redesign, Sloan Management Review, Summer 1990, pp.11-27
5. Davenport T.H.: Process Innovation, Harvard Business School Press, Boston, 1993
6. Hammer, M.: Reengineering Work: Don't Automate, Obliterate, Harvard Business Review, July-August 1990, pp. 104-122
7. Hammer, M., Champy, C.: Reengineering the Corporation: A Manifesto for Business Revolution, New York: Harper Business, 1993
8. Hopper M.: Rattling SABRE- New Ways to Compete on Information, Harvard Business Review, May-June 1990, pp.118-125
9. Malhotra, Y.: Business Process Redesign: An Overview. University of Pittsburgh, Katz School of Business, 1996; http://www.brint.com/papers.bpr.htm
10. MIS Telecommunications Review (in Russian), vol. 22 (114), June1998 http://www. mis.ru/1cont.htm
11. Team Canada Market Research Centre, Department of Foreign Affairs and International Trade, "The Telecommunications Market In Russia", April 1997 (http://www.dfait-maeci.gc.ca/english/trade/market/europe/33301-e.htm)
12. Venkatraman, N.: IT-Enabled Business Transformation: From Automation To Business Scope Redefinition, Sloan Management Review, Winter 1994

SYSTEMATIC APPROACH AND COMPLEXITY IN GOVERNMENT POLICY TOWARDS IMPROVEMENT OF FOREIGN DIRECT INVESTMENT ENVIRONMENT - THE CASE OF HUNGARY AND MALAYSIA

VALENTIN GROZDANOV

Inter University Institute of Macao, NAPE, Lote 18, Edf. Tak Ip Plaza, R/C – 3, Rua de Londres – P, Macao

E-mail: valgrozdanov@hotmail.com

This paper discuses the factors influencing Foreign Direct Investment (FDI) attractiveness of two countries Hungary and Malaysia according to their general and specific investment conditions. Both countries applying different government policy towards foreign investment promotion have achieved remarkable capital inflow and economic development. The analysis is based on a model of six group FDI factors and is focused mainly on political and legal conditions, macroeconomic stability and nation specific investment factors, applied in their complexity by the two governments during last ten years. Two different FDI models – the liberal, not restrictive investment environment in Hungary and restricted, narrowly targeted conditions in Malaysia are investigated. Hypothesis is based on believe that only simultaneous and interdependent development and improvement of national investment environment factors could lead to positive results.

1 Introduction

This paper analyses the system of factors influencing Foreign Direct Investment (FDI) environment of two different countries - Hungary and Malaysia according to their economic, political, cultural and social development conditions. Both countries applying sustainable government policy towards promotion of foreign investment have achieved remarkable foreign capital inflow. The investigation is made from the point of view of complexity measures implemented by the two governments, which have different nature and are applied in different environments. Analysis is based on a pattern with six main category factors including - natural conditions, general political stability, government policy towards foreign investment, legal factors, macroeconomic environment and international payment. Within this framework the study researches on comparative base the general and specific FDI conditions conducted in their complexity, which had driven countries to investment success.

One important issue for investigation is how these two countries having different region location, ethnic structure, political system, history and economic development offer conditions that grant them FDI competitive advantage. Thus, will be drawn the overview of critical success factors common for both countries government investment policy as well as the national specificity.

Table 1: System of Foreign Investment Factors
(after Root 1994)

System of Factors	Factor Subsystems
Natural	Geographic location Resources Internal market size
General political stability	Form of government, Ideology,
Government policy towards foreign investment	Political, social, ethnic and other conflicts Past experience of foreign investment, Government attitude, Restrictions and incentives, Investment entry regulations
Macroeconomic environment	The role of government in the economy, Availability of short/long terms plans/programs, Sector distribution of industries, Infrastructure, Main economic indicators, Monetary policy
Legal factors	Enforceability of contracts, Fairness of court decisions, Business and labour laws, Taxation, Patents and trademark protection, Efficiency of public officials
International payment	Conditions regarding international payments, Balance of payment, Foreign exchange position of the country, Repatriation restrictions Exchange rate behaviour

According to recent survey made by the OECD, from 1973 to 1996, the world FDI flows multiplied fourteen times from \$ 25 billion to \$ 350 billion per annum, outstripping growth in international trade. General impact of this growth is the fact that foreign direct investment, together with international trade in goods and services promote economic growth, create employment and rise living standards of the nation states. Achievement of foreign investors' interest is a serious challenge for any government. It is not only question of regulations moreover this goal demands full implementation of comprehensive system of government measures in political and economic area. For the purpose of this analysis a model of minimum required main factors for investment attractiveness is adopted (see Table 1). The model is simplified and includes a range of the most important FDI conditions. Attention will be given to relationships and interdependence between the main and additional factors in this pattern, as well as the differences between the investment strategies implemented in Hungary and Malaysia.

2 System of FDI Factors – The case of Hungary

After the collapse of communist system in Central and Eastern Europe Countries (CEEC) and former Soviet Union in 1989 a new phenomena has appeared on the world's political and economic scene – the countries in transition. Achievement of high foreign capital inflow is considered as a critical factor for the success of the transition process. Thus the government of any CEEC country has to establish a complex system of measures for foreign investment promotion in order to achieve fast, successful and with acceptable social cost market reforms.

Hungarian government has started the implementation of favourite investment conditions ten years ago deeply understanding the strategic importance of FDI for country's economy. Hungary has been the most favourite spot for foreign investment among East European countries during the last few years, accumulating about 17 billion USD until 1998 or 40% of all FDI in Eastern Europe. Its strategic location on the main land routes between Western Europe and the Balkan Peninsula as well as between Ukraine and the Mediterranean basin is one of its major investment advantages. Hence the government has been aware that only this factor will be not enough to speed up transition from totalitarian, planed economy to free market economy and political pluralism. Furthermore, there is no single formula or universal approach that may lead to economic success any of countries in transition.

Countries from this region had different starting conditions in the eve of the transformation process, thus similar economic measures applied in different countries could led to different results. Resource availability in Hungary are poor, some bauxite, coal, natural gas, and fertile soils are available, though they are not sufficient to play an important role for foreign investment attractiveness. According to market size criteria Hungary has 10.5 million population with relatively undeveloped market conditions when the reforms started. The small size of the internal market suggests that main objective for foreign investors will be manufacturing for export on the West and the East, as well as facing the potential demands of the local market. In conclusion the natural factor in Hungary with high importance is – the strategic location, with medium importance – relatively small market size, and with low significance – poor natural resources availability.

2.1 Political and Legal Environment factors

Political stability is discussed in the literature as critical investment factor, for example - Root (1994), Balcerowicz et al. (1997), OECD (1997). All they argue that without stable and predictable political situation within the country high investment inflow is impossible. One remark here is important - political stability not necessary means democracy according to western measures, especially for FDI inflow. The political system in countries like China, Vietnam, Malaysia is far from western model of democracy, but maintain political stability, and hence they enjoy very high investment interest. It is a result of the other important FDI factors we will discuss later in this paper. For transition economies high level of foreign investments is a key success factor. Estrin, Hughes, and Todd (1997) argue that, FDI is playing an important role in transition process for most of CEEC. Foreign investment flow is rather high and appears to be positively linked to progress in transition process and provides an effective vehicle for transfer of knowledge in the widest sense and for a quick restructuring.

Hungary has been a member of the former Soviet Union Block, although its communist system has not been as dogmatic as in other East European countries. During the communist period there were substantial private sector activities mainly in agriculture that together with small businesses amounted to around 25 per cent of the county's GDP in 1990. In comparison other East European countries such as Bulgaria or Rumania had public property of more than 90-95 per cent when the reforms started. The results from the communist regime over all Eastern European economies were weakening agriculture, technological backwardness and inefficient organisation of industry, low development of infrastructure, high armament expenditures, non-competitive economies, and high external indebtedness (Dobosiewicz 1992). In Hungarian case the high external debt is one of the main reasons for the liberal FDI regime.

A key role in the systematic government approach towards FDI environment improvement plays the political and legal reform. First, a group of complex FDI measures was applied in 1988-90 towards political reform and political and legal stability. The Hungarian strategy was clear – establishment of adequate changes in constitution and other laws according to EU legal standards. The transition from communism to democracy was entirely peaceful, accomplished at the negotiating table and violence was not part of the political practice in Hungary. Hungarian political system is based on parliamentary democracy with a freely elected legislative assembly. Achievement majority in the Parliament of one political pro-reform party or establishment of stable coalition is crucial factor for political stability. It had happened in Hungary after 1990 and 1994 general elections that had given the government so-called by Balcerowicz et al. (1997) "political capital"- stable coalition supporting government reforms. In general, the Hungarian government remains committed to a market economy and preservation of political rights no matter which political party has power. Since the beginning of the process of transition towards political pluralism and market economy major political parties have shared identical or similar priorities – EU integration and market reforms. Hungary had achieved historical reconciliation with neighbouring Slovakia and Romania, by signing treaties with each of them in 1995 and 1996 respectively. The main national priorities have been concentrated on fast and radical economic reforms and integration into European Union, NATO, and OECD. The process was supported by involvement of the public in democratic changes with parliamentary debates and consultations based on compromise. Nowadays most of these strategic political and economic targets have been achieved. Hungary has been a member of OECD since 1996, its process of EU integration is intensive, reaching the final phase of direct negotiations for full membership, and its NATO membership, together with Poland and Check Republic, is almost completed. As a result the political, economic, security and investment climate of the country becomes much more favourable.

Foreign direct investments in Hungary are strongly supported by the legal framework that has been developed early, compared to the other countries in East Europe. Process of harmonisation and approximation of the country's legal system towards European Union and OECD requirements is giving additional investors confidence in its policy. First legal steps from all system of measures taken by the government were made in 1988 by adoption of Law of enterprises financed with foreign capital. Main targets of the law were development environments for international economic co-operation, with special regard to the facilitation of the direct appearance of foreign working capital, and the promotion of technological development in the Hungarian economy, providing national treatment for foreign investors free from any discrimination.

Equal treatment is a corner stone in Hungarian FDI policy. Companies with foreign capital have ownership and other rights, trading, wholesale and retail activities, taxation, accounting system, liquidation procedure and legal dispute settlement according to rules applying to other domestic companies. Foreign and domestic private entities may establish and own business enterprises and engage in all forms of remunerative activity, save for those prohibited by law. The companies law guarantees the right of private entities to freely establish, acquire, and dispose of interests in business enterprises. Private enterprises enjoy competitive equality when applied to markets, business licenses, supplies and most importantly credit access. In contrast with Malaysian FDI complexity approach (see Table 3 and 4), Hungary does not impose performance requirements as a condition for establishing, maintaining, or expanding an investment. In the case of foreign investments, there is no requirement for national own shares, that the share of foreign equity be reduced over time or that technology be transferred on given terms. There are no government-imposed conditions on permission to invest, including location in specific geographical areas, specific percentage of local content - be in goods/services or local equity, substitution for imports, export requirements or targets, employment of host country nationals, technology transfer, or source of financing. Investors are not required to disclose propitiatory information to governments as part of the regulatory approval process. The companies are entitled to all tax benefits that are enjoyed by other domestic economic associations.

Restrictions always have poison effect over the investment climate thus as a result of Hungarian government liberal investment policy restrictions are strictly limited. Foreign ownership up to 100 percent is permitted with the exception of some defence-related industries and several other specified companies. Law prohibits acquiring of agriculture land, although the idea for national referendum regarding that issue is under way.

Systems of investment incentives on the other side give the national specificity of Hungarian FDI environment. Priorities are major in modernisation of the processing industry, promotion of investment projects involving new technologies and rippled effects, expansion of development projects aimed at production for export, and stimulation of investments that improve the competitiveness and technical standards of small and medium sized companies.

A composite system of special subsidies and incentives are established and available for mainly industrial investments (see Table 2). During early 1990s a model of tax holidays for new investments has been implemented yet withdrawn in 1993. However, companies that qualified for tax abatements prior to January 1994 retained their benefits, according to court rulings. In place of tax holidays, has been introduced a discretionary new general tax reduction. The new incentive is available in the form of a tax reduction based on a percentage of future taxes to be paid. A common practice in CEEC is to be offered tax holiday for curtain period. It is arguable how efficient is such government FDI measures. If the FDI is in form of "greenfield" investment profit will come after three to five years. If the investment is in privatised company, but with additional substantial investment called by Estrin et al. (1997) "brownfield" investment, then future tax benefits seems also not very efficient incentive. The tax benefits playing role, when FDI leads to profit in short run – one to three years, which is not very often case in CEEC investment practice.

Another key element of the systemic approach toward investment promotion is to secure for local and foreign investor property and trademark protection. Non country could expect modern technology investment without strong intellectual property protection. Thus, government measures in that area have been focused on efficient patent and copyright law. Hungary has adopted modern regulations in these fields and tends to follow strictly the international property standards. The new patent act is in total conformity with the guidelines of the European patent convention. In some areas such as software, audio and videotapes could be found lack of enforcement or court cases of infringement a time consuming. Trademark registration in Hungary is according to EU regulations; however, the speed of the process has to be improved. Serious problem in CEEC as well as in Hungary is the real implementation of the adopted legislation. There are two aspects of this problem. First, provisions of decrees and regulations issued by Council of Ministers sometimes are in contradiction with the main law, which can lead to uncertainty and confusion in foreign investors. Second, a practice of non-implementation of law provisions because of corruption behaviour, and soft institutional and judicial control. These impacts over FDI environment could be reduces with strong government commitment to avoid such practices and establishment of necessary legal infrastructure.

Hungarian government is offered additional business opportunity for FDI by operation in Foreign Trade Zones. Foreign investors may set up operations or acquire interest in companies within a duty-free zone, which is considered foreign territory for customs, foreign exchange, and foreign trade regulation purposes. This is an important instrument for the government to promote FDI inflow oriented towards export using local labour and facilities, though at the same time it is a customs control and corruption challenge.

Hungary's tax, labour, health and safety laws are harmonised with the European Union regulations and do not distort or impede the investment environment. Companies pay 21 percent corporate taxes, which is considerably low comparing with the other countries in Eastern and Western Europe. Value added tax (VAT) in Hungary is one of the highest in Europe - 25 per cent, which reflects directly the level of consumption. Reduction of VAT up to 20 – 22 percent will benefit local and foreign investors increasing level of customer expenditures.

The openness of the country's investment environment is extended also to government procurement. Open tenders for government purchases of goods and services exceeding particular amount require selling to the government. In general, these requirements have led to greater transparency and fairness in government procurement allowing free competition for local and foreign investors. Two other specific factors for Hungarian FDI attractiveness need to be mentioned. First, no cases of expropriation of foreign-owned assets since the early 1950s and after political changes in 1990s have been registered. Second, the local legal environment proclaimed that in case of conflict of interests the dispute could be settled according to international arbitration standards and the rule of law. Corruption is not a serious problem in Hungary as it is in other CEEC, for example Russia or Ukraine. Though the efficiency of public officials is a challenge for any CEEC government targeting high FDI inflows and concrete measures preventing corrupted behaviour need to be improved, especially in the customs and licensing areas.

2.2 Labour

One of the main reasons for foreign investors to allocate capital in CEEC and particularly in Hungary is the competitive labour force (Estrin et al. 1997). Hungarian employee is high educated and skilled, and relatively cheap, a number of times less expensive than in Western Europe, for example average monthly salary for the period 1990 - 97 was between 212 to 310 USD respectively. Moreover foreign investors could find high motivated and productive labour force, in combination with flexibility for adoption to Western style management. Unemployment is a challenge for the Hungary's investment policy. The government is concerned about level of unemployment particularly in the provinces. Unemployment rate is around 10 percent nation-wide, though it varies geographically and among minority groups, for instance in Budapest unemployment is 5-6 percent, but it is about 20 percent in the Northeast and over 75 percent in some Gypsy communities. Measures towards avoid misbalance within the country are in government agenda, though the incentives system targeting FDI inflow in these regions needs to be improved.

Hungary has established the basic underpinnings necessary for a market economy to prosper, which direct determine the investment environment. Stability in fiscal and monetary policy is with crucial significance for the attractive investment climate. After period of high inflation and budget unbalances in 1993–94, Hungary adopted Stabilisation Programme in 1995 providing more coherent and predictable framework for economic development. Since 1989, the private sector in Hungary has grown from approximately 20-25 per cent to over 70 per cent of GDP. The socialist-led government, which won the elections in 1994, has accelerated the privatisation process, with significant progress in 1995 and 1996, notably in energy, banking and telecommunication. Since 1989, the state has liquidated or privatised 75 per cent of its public property and aims to increase this figure to 80 per cent by the end of the century. Government expenditures are expected to be fewer than 50 per cent of GDP in 1998 with plans to increase spending on investment, including on infrastructure (USCC 1998).

Government policy towards privatisation was focused on speeding the process and mostly cash payments with unlimited opportunities for foreign capital inflow. After low privatisation efficiency in first transition years Hungary adjust its strategy with Stabilisation Programme implemented in 1995 (OECD 1997). In contrast with other East European countries Hungary succeeded to sell off large majority of state owned assets in a short time period. With this approach major strategic goals have been achieved – receiving cash necessary for external debt payment, capital for social programmes, technology and management restructuring of the economy on competitive base and free off substantially the public spending burden. For the last decade about one third of all foreign direct investments have gone into privatisation. Government concern towards this national target has made privatisation, particularly in the high value industries such as energy, banking and telecommunications sectors, one of its top priorities. Now Hungary is the only country in the region, which have privatised most of its telecom, banking, and energy sectors mainly with high foreign capital inflow.

In summary, as we have seen from the above discussion the main political and legal factors for FDI attractiveness have been available, established and improved in Hungary in first transition years. It had given the institutional framework for market reforms and foreign investment inflow in particular. Nevertheless, successful investment model could be created only when strategic location, political stability, EU legal standards are implemented together with firm macroeconomic and microeconomic changes.

2.3 Macroeconomic Factors

For successful transition process there are three complex programmes need to be applied simultaneously – Stabilisation, Liberalisation and Institutionalisation in the environment of political stability (see Balcerowicz et al. 1997). Restrictive monetary policy, elimination of the budget deficit, price liberalisation, currency convertibility and restrictive income policy could lead to stabilisation and liberalisation in particular transition economy. In Hungarian case these radical economic targets have been achieved in short time thanks of more favourite starting conditions in Hungary before 1989 and political and legal stability during 1989-97 period. Control over inflation and lowering it is one of the critical success factors for a country's investment attractiveness. Inflation remains problematic in Hungary, even though the government has achieved in 1997 to lower the annual inflation to 18,3%, down from 23.6% in 1996 and 28.3% in 1995 (USCC 1998). The 1998 target is to drive down average annual inflation to about 13.5% and into the single digits from 1999.

In the process of transition it is difficult to predict what would be the effect of all applied legal and economic measures over the real growth of the economy, especially when the economy is open to outside influences. Once a government obtains and maintains positive results it is a clear indicator for successful economic policy. After declining by about 18% from 1990-1993, the Hungarian economy grew by 2.9% in 1994, 1.5% in 1995, and by about 1% in 1996. Absolute GDP for 1996 was still well below 1989 levels. The growth of GDP was 4.4% in 1997, and would be followed by sustainable growth of 4% until year 2000. Another evidence for the efficiency of the government FDI measures is the fact that industrial productivity in Hungary is up 150% over 1988 levels, when in 1995, productivity was just 27.6% over 1988 levels mostly thanks to foreign investment inflow (USCC 1998). The expected growth of GDP is an additional stimulus for investors, which expect increase of home expenditures and growth of middle class customers.

For foreign investor currency convertibility and stabile exchange rate in the host country plays a significant role for strategic planning. Moreover, to be achieved these important FDI conditions, government institutions need to observe closely and delicately regulate the local currency system. Hungary's currency, the forint, is fully convertible for current account purposes. Since 1995, the National Bank has implemented a pre-announced crawling peg foreign exchange regime. The forint is currently devalued at a rate of 1.1 percent per month, which is giving stability and predictability for local and foreign businesses (OECD 1997).

Balance of payment has been one of the chronic problems for almost all East European transition economies. Critical government priority is to cut the state deficit and the importance of high volume foreign investments is directly related to this target. Hungary's balance of payments situation has improved considerably over the last two years. The current account deficit in 1997 was 1.0 billion, down from $1.7 billion in 1996, $2.5 billion in 1995 and $4 billion in 1994. Still the trade balance is negative. For 1997 trade deficit was 2.0 billion, comparing with 3.7 in

1993 and 3.8 in 1994. Exports grew at a faster rate than imports in 1995, 1996 and 1997, but this trend is expected to reverse in 1998. The complete elimination of the import surcharge on July 1996, which the government initially imposed at 8% in March 1995 (and was gradually reduced in 1996 and 1997), may cause imports to grow at a faster rate. The government still anticipates that FDI will cover the current account deficit.

Regarding price policy in Hungary - the liberal approach has driven to elimination of state subsidies and price controls for many products, allowing prices to be determined by market forces. Although the state still supports some basic services such as utilities, mass transit and gasoline, though price increases are bringing rates in line with costs. An exception from the overall total economic liberalisation is the fact that the government still continues to subsidise pharmaceuticals, though this practice has a short run application. The volume of "grey economy" is estimated to be about 30% of GDP, which is less than in other East European countries, but stills a high figure (USCC 1998). Thus, to reduce the importance of grey economy resulting in unfair competition and tax evasion will be another challenge for the future Hungarian FDI policy.

When the reforms started Hungarian foreign debt was 21.3 billion USD in 1990. It reached its highest level of 31.7 billion in 1995 and has been reduced to 22.1 billion USD by the end of 1997. In literature high foreign debt is assessed as a crucial factor determining a country's economic policy and foreign investment policy, particularly for small countries. Hungary is a clear example of this vision. Foreign indebtedness of Hungary is one of the most important reasons for the country's open FDI policy. On one side, without the pressure of high external debt in the eve of reforms, perhaps the government would not implement such a liberal and open FDI environment. Hence, on the other side, without liberal FDI model the country would not achieve technology improvement and received the high amount of FDI. This would delay the process of transition and EU integration, and would made the economy less competitive with serious difficulties in debt payments.

2.4 Other FDI Factors

Foreign investors are keen to find no barriers to capital transfer of their profits, thus repatriation restrictions would damage the investment image of the country. In Hungary capital export and outward direct investment have been allowed since 1974, with prior permission of the Hungarian government. Currently the Ministry of Finance and the Ministry of Industry and Trade review license applications for capital export. Hungary has emerged successfully from a period of destabilising exchange rate uncertainty.

According to Hungarian foreign investment agency the foreign capital investments have been fulfilled with favourable sector structure, and in addition new sectors (e.g. car and automotive parts manufacturing) have been established as well. The latest data shoes that sector distribution of all the foreign capital invested up to the end of 1997 is the following - 48.5 per cent of FDI was made in industry, 14 per cent in telecommunication, 12 per cent in energy, 10 per cent in commerce, 6.5 per cent in finance and 9 per cent in other sectors. Considering the regional structure of all the FDI invested up to the end of 1997, Budapest continues to be the main attraction in the eyes of investors, as the city attracted 46 per cent. The underdeveloped eastern provinces are strongly promoted by the Hungarian government by offering more favourable incentives there. Nevertheless foreign investors preferred the western and middle part of Hungary and Budapest.

Infrastructure conditions in the host country are directly related to investor's interest. Having strategic geographic location and only land borders Hungary has to improve and maintain its infrastructure according to Western standards. The Hungarian transportation system is relatively well developed compared to other countries in the region, however, highways system needs to be improved. The government has expansion plans in this field as well as new projects for European Corridor from Venice, Italy to Kiev, Russia, passing trough Hungary. Infrastructure improvement is targeted in country's railway and airport system towards needs of the growing economy. Government investment measures towards fast privatisation are reflected on country utilities. Hungary has privatised nearly all its gas and electric generation and distribution facilities. The foreign energy companies, that bought the generation plants plan to modernise and expand the plants' capacities in order to meet the expected further surge of energy use and the retiring of some plants (USCC 1998). Investment in new infrastructure and renovation of existed will be one of the major government's short turn priorities.

Table 2: Summary of Hungarian Export Promotion Measures
Source: OECD, 1997

Hungary implements the following measures to promote exports:

- Officially supported export credits (according to the OECD agreement).
- Other interest subsidies for SMEs that are not specifically aimed at promoting exports but that may results in export substitution: SMEs can obtain a loan of the maximum amount of HUF 100 million, with a maturity of less than one year with a subsidy equal to 50 per cent of the interest payment.
- Import duty may be reclaimed by re-exporters, although this possibility will be reduced starting from 1 July following agreements with EU and CEFTA on the determination of rules of origin.
- Preferential tax treatment on dynamic exporters: if a company invests more than HUF 1 billion and its exports increase by at least 25 per cent then the corporate tax is halved.
- Agricultural price subsidies: like many other countries, Hungary developed a system for subsidising the export of several agricultural products and food items. In general, the higher degree of transformation of the product, the lower the unit subsidy (with the important exception of meat products).

2.5 System of Incentives

Export and volume benefits are offered for foreign investors operating in Hungary> It is a typical government policy in east European countries, though the final positive effect is not always clear. Further EU integration process will lead to harmonisation Hungarian promotion measures with European practice (see Table 2).

2.6 Impacts of Hungarian FDI model

The results from the systematic approach and complexity application by the Hungarian government towards improvement of FDI environment could be summarise as follows:

Positive consequences:

- Political and economic integration in EU and OECD
- New technology
- Competitive goods and services, lower prices and higher quality
- New and efficient management skills and practices were implemented
- Achievement of growth in export, import and greater import substitution
- Growth of competition
- Higher labour profitability
- Better country-wide investment structure
- R and D improvement
- Adopted new (for the country) marketing concepts and tools
- Overall growth of the middle class and widening the social base of reforms

Negative consequences:

- Dual type companies – foreign invested firms (high technology, easy capital access, new management) and local firms (old equipped and managed, capital shortage, not competitive)
- Disintegration between foreign and local firms with difficulties for local to became subcontractors of the others
- Short/long term gain and perspectives are different and not equal for all citizens
- Dependence on the influence of factors and processes external (global or regional) for the national FDI environment system

In conclusion we have seen how Hungary has applied successful systematic approach towards foreign investment attraction. The base of this success is simultaneous development and implementation of main investment factors –

political, legal and economic stability, according to Western standards combined with location advantage and adjusted government policy when necessary to be improved some conditions. The initial starting conditions and other national specific FDI factors have played additional positive role in that process. Thus, Hungarian foreign investment model could be used as a pattern from other CEEC, though implemented according to national specific environment of any particular country.

3 System of FDI Factors - The case of Malaysia

Malaysia is a different type example of FDI promotion model. In contrast with the liberal, little restricted Hungarian strategy, the Malaysian government tends to regulate the FDI inflow in the country favouring only foreign manufacturing investments, especially in high-technology industries. The high foreign investment inflow is one of the main reasons for continuos high GDP growth. Since 1985-95 Malaysia has been one of fastest growing economy in the world - with 7.8 percent average annual increase of GDP. Currently Malaysia is actively wooing foreign investments in the multimedia and information technology industries to facilitate the establishment of the so-called Multimedia Super Corridor project. However, it does not actively seek foreign investment in service industries or foreign participation in agriculture or construction. Investment is also restricted in the oil and gas industry (USCC, 1997). Malaysia offer continuos political stability and strong government support towards its own foreign investment priorities. The ethnic factor plays an important role making the country an original example for successful FDI policy with strong ethnic business requirements and limitations. The Malaysian government focuses on narrow strategic targets and builds its FDI model towards these achievement goals.

The Malaysian FDI system is a balance between typical investment factors critical for any country and strong with specific focus FDI regulations. As a result the foreign investment model is successful, though highlighted by national specificity. With 21.2 million population the country is attractive internal market for investors, though Muslim cultural and religious specificity plays a constrain role comparing with the other markets with same size and purchasing power. An important investment factor is the fact that Malaysia is rich in natural resources and has long been a leading producer and exporter of commodities such as natural rubber, palm oil, cocoa, timber, pepper and tin, and is a net exporter of petroleum and natural gas (USCC 1997). Nowadays, Malaysia is recognised as one of the world's leading exporters of electronic semiconductors, room air conditioners, audio-visual equipment, and products based on the country's natural resources such as rubber products (gloves, threads, catheters), palm oil products (soaps, margarine, oleochemicals) and timber products (plywood, mouldings, furniture). According to first group of factors in our pattern Malaysia has comparative advantage with strategic location, medium size internal market and reach natural resources availability.

3.1 Political and Legal Environment factors

Malaysia is strategically located in the heart of South East Asia - one of the world fastest growing regions. The country is well serviced by all major air and shipping lines. The easy access to the rest of the world has today made Malaysia an attractive centre for trade and foreign investment. From the point of view of ethnic structure the population of the country is a very interesting example with its 21.2 million citizens – unique mix of Malay, Chinese, Indian and other ethnic groups. In Malaysian ethnic spectre Malay and other indigenous people are 58%, Chinese - 26%, Indian - 7% and others - 9%. This ethnic and religion specificity directly influenced foreign direct investment environment in the country.

Malaysia gained independence from United Kingdom in 1957 through peaceful negotiations. Malaysia is a constitutional monarchy, its head of state is one of the Malay Rulers elected for a term of five years by his brother Rulers, who has to act in accordance with government advice. The Cabinet headed by the Prime Minister consists only of members of the legislature and is collectively responsible to Parliament. Such form of political establishment is different then the Western style democracy, but it has secured political stability, which the important issue for foreign investors. As we have discussed above Malaysia is an example of country where political stability is more important for investors' then real democracy.

Crucial economic changes in the country's government objectives had happened in 1969. After the ethnic violence in 1969 against ethnic Chinese - controlling a large share of the national commerce, the government of Malaysia has launched the so-called New Economic Policy. This programme had targeted restructuring of society and full involvement of bumiputras (ethnic Malay) in the country's economic life. The new policy has two key focuses: to fight poverty and to prevent monopoly on a particular profession by any one ethnic group. Until 1970 the Malay share of national wealth was 2.3%, the Chinese had 25% and foreigners had 70%. The new economic policy resulted in radical change of wealth distribution in Malaysia in 1990s. In 1995 bumiputra reached 20.6%, ethnic Chinese 41%, and foreigners 28% national wealth share. Comprehensive system of privileges and quotas has been established as a background of this policy (FEER 1998).

In contrast with Hungary the Malaysian government has established a complicated, though economically efficient system of FDI regulations combined with ethnic restrictions and close government supervision. Proposals for a manufacturing license, either foreign or local, are screened by the Malaysian Industrial Development Authority to determine whether they are consistent with the government strategic policies. The legislation is a combination of British and European laws with strong influence (copping) from laws of Singapore. The legal framework is determined by the government strategic objectives and is used as a tool towards strict regulation of political and economic activities.

Foreign investors established in Malaysia are generally accorded national treatment in all but equity limits. The government policy limits foreign equity to a minority 30 percent share, but 100 percent foreign ownership in manufacturing is permitted in certain instances for export-oriented. Foreign companies, which are granted Multimedia Super Corridor status, are also permitted 100% ownership (see Table 4). Malaysia also promotes the holding of economic assets by bumiputra (ethnic Malays) and usually requires foreign and domestic firms to take on bumiputra partners (usually 30 percent) and to have a workforce, which reflects Malaysia's ethnic composition. Malaysia offers a number of fiscal incentives to foreign manufacturing investors, which may be linked to performance requirements.

Investment regulations are specified in the Promotion of Investments Act 1986 and the Industrial Co-ordination Act 1975. Different laws and government institutions are involved to control investment policy. On acquisitions, mergers and take-overs, the Securities Commission and the Foreign Investment Committee implement the regulations specified in the Malaysian Code on Take-overs and Mergers. The Foreign Investment Committee also formulates policy guidelines for foreign participation in the non-manufacturing sector.

In addition to the standard 30 percent cap on foreign equity, certain industries have additional barriers to entry. For example, the government severely restricts establishment of entities in the financial service industry. Restrictions are implemented in banking, and insurance sectors, and ownership of agricultural land is restricted to Malaysian citizens. The national monopoly is established in crude oil and gas deposits. Over time a number of foreign firms selling their products and services into the domestic market have received licenses with limited-term exemptions to the standard limit of 30 percent foreign equity. As these exemptions expire or the licenses come up for renewal, government agencies require foreign firms to demonstrate substantial progress towards meeting the foreign equity limits. Private entities, both foreign and domestic, have a right to acquire, merge and take over business, however, the acquisition or disposal of 5 percent or more of interests in any local financial institution requires the prior approval of the Minister of Finance. Although public enterprises rarely compete with private enterprises for market or credit access, bumiputra firms may be given preference in securing government contracts when there is equal competitiveness (USCC, 1997).

Malaysia has an effective legal system and adequate legislation to protect private property. Foreign investors are keen to find strong intellectual property protection in any host investment country. Thus in order to be world producer of electronics Malaysia has adopted a comprehensive regime for protecting intellectual property rights. The range of national laws and regulations covers Western standards in intellectual property rights. In 1997, four new pieces of legislation have been passed which extend intellectual property rights protection equally to new cyberspace applications. The laws cover such problems as computer crimes, digital signature, tele-medicine. The Copyright Act includes enforcement provisions allowing Government officials to enter and search premises suspected of infringement, and to seize infringing copies and reproduction equipment.

3.2 Performance Requirements and Incentives

Fiscal incentives, granted to both foreign and domestic investors, are subject to performance requirements, usually in the form of export targets, local content requirements, and technology transfer. Performance requirements are often written into the manufacturing license of both local and foreign investors. If a firm (foreign or domestic) fails to meet the terms of its license, it risks losing any tax benefits it may have been awarded. In extreme cases, the firm could lose its manufacturing license. The government has stated that, over the long term, it intends to eliminate gradually most of the fiscal incentives now offered to foreign and domestic manufacturing investors. For tax purposes, local and foreign enterprises are treated on essentially the same footing. The corporate tax rate is 30 percent, except in petroleum production, which is taxed at 40 percent. Regulations in domestic banking system contain restrictive conditions for FDI, though after Asian economic crises in 1997-98 deregulation and foreign investment in that area is due to established.

3.3 Labour

Factor condition particularly cheap labour force is a main attraction in this region. Though there are contradiction, especially in Malaysian labour policy. On one side the government favourite high technology manufacturing and use of skilled work force. On the other side, there is a shortage of skilled employees in some industries and restrictions for use foreign work force. The official unemployment rate in Malaysia is low - for 1996 was 2.6 percent (down from 3.9 percent in 1992). The low unemployment rate is caused mainly of two reasons – sustainable high economic growth lasts ten years and significant annual FDI inflow. Problems for companies are difficulties in obtaining and retaining workers at all skill levels, caused by the gap between wage rates climbing faster than productivity rates and efficiency.

As a result of its FDI policy Malaysia no longer seeks labour-intensive industries and reserves its fiscal incentives for high value-added projects. It encourages labour intensive industries to move offshore. Threats from unions are low – they account less than 10 percent of the workforce and negotiation or arbitration system of settlement strikes is established and control by the government. More then that the Malaysian government prohibits the formation of a national union in the electronics industry, a major employer, but it does allow for in-house unions in the electronics industry. For foreign investors strong foreign employee recruitment restrictions a sensitive issue. In Malaysia restrictions are implemented regarding the number of expatriate personnel employed by foreign and domestic firms (Ball et al. 1996). In addition, the government monitors hiring practices to ensure that all employers strive to meet guidelines designed to ensure a racial balance in employment. Due to shortage of professionals, scientists and academicians, Malaysia has responded by simplifying the approval process of work permits for these categories of foreign workers (USCC 1997). From 1997 the new procedure have been implemented towards relaxing recruitment of foreigners.

3.4 Other FDI Factors

Special measures have been implemented towards corruption prevention. Malaysia reacts quickly to allegations of corruption and has set up the Anti-Corruption Agency to combat corruption by officials. In 1997, the Government directed senior state-level officials to declare their assets to the agency upon taking office. Foreign businessmen are asked to report any individuals who ask for payment in return for government services (USCC 1997). Surveys by media groups have placed Malaysia in a category of "less corrupt" in Asia. Singapore is generally rated as the least corrupt in the region. The agency probes are reported in the newspapers. According to recent survey of Transparency International Malaysia and Hungary are occupied similar place – 22 and 25 respectively.

Table 3: OECD Multilateral Agreement on Investment (MAI) - Principles
Source: OECD

MAI Principle Requirements	Hungary	Malaysia
National treatment	Implemented	Equity, Area and Labour restrictions
Transparency	Implemented	Implemented
Transfer of funds	Permitted	Permitted
Entry and stay of key personal	Permitted	Permitted with some restrictions
Performance requirements	No	Yes
Expropriation	Yes - MAI standards	Yes - MAI standards
Dispute resolution	Yes - MAI standards	Yes - MAI standards

Malaysia has a relatively open foreign exchange regime. Payments, including repatriation of capital and remittance of profits or dividends, are freely permitted and are transacted on a timely basis. All payments to other countries may be made in almost any foreign currency, through authorised foreign exchange dealers. Non-residents are also allowed to open foreign currency and ringgit accounts freely. Resident exporters and approved Operational Headquarters are allowed to retain a maximum of US$10 million in export proceeds in foreign currency accounts. A traveller may carry any amount of ringgit or any foreign currency when he leaves or arrives in Malaysia. There is no indication that there are any discrepancies between Malaysia's stated exchange policy and its implementation (USCC, 1997).

The government policy towards expropriation is based on the international standards - expropriation in Malaysia may only be undertaken for a public purpose and subject to prompt, adequate and effective compensation. In case of legal problems foreign investors could settle them also according to international regulations in this field and with the help of domestic legal system, which is open and accessible for foreign and local businesses.

Table 4: Special Policy on Malaysian Equity Distribution
Source: USCC, 1997

Foreign equity distribution	Local equity distribution
From 70% to 100% foreign equity	30% or less reserved for Malay
Less then 70%	30% for Malay, the rest for other Malaysians
JV with not Malay	30% for not Malay, any other for Malay

A sophisticated system of equity regulations and FDI incentives are the specificity of Malaysian investment environment. The main conditions for government equity distribution policy is highlighted by ethnic protectionism and high technology and export promotion orientation. No equity conditions are imposed on projects, which export at least 80 percent of their output. For projects exporting from 51 to 79 percent of output, majority foreign ownership (51 percent) is permitted. This can be raised to 79 percent ownership under certain circumstances. For projects that export between 20 and 50 percent of output, 30-51 percent foreign ownership is allowed. For projects exporting less than 20 percent of output, maximum foreign ownership is 30 percent.

A comprehensive system of investment incentives is adopted and applied in Malaysia (see Table 5). The principal incentives for the manufacturing sector are contained in the Promotion of Investments Act of 1986 and the Income Tax Act of 1967. These incentives apply to companies subject to Malaysia's 30 percent corporate income tax. Some of these incentives are also available for investments in agriculture, tourism, R&D and technical training.

Table 5: Incentives for FDI in Malaysia
Source: USCC, 1997

Status	Incentives
Multimedia Super Corridor	Pioneer status for 10 years or 100% investment tax allowance; No duty on multimedia equipment; Special foreign currency transactions; No restriction on recruitment of expatriates; Funding for R&D for SMEs
Manufacturers Pioneer Status	Full or partial tax exemption for 5 to 10 years incentives, dividends tax free; type dependence
Not Manufacturers Pioneer Status	60% reinvestment allowance
Investment in Infrastructure	Allowance of 100%
Research and Development	Five years tax free and permission to carry tax-relief period losses forward
For raw materials or not local components; export oriented only	Customs exemption
Export	Deduction – 5% of the FOB value of export sales
Investment more than 2 mln. USD	5 foreign employee automatically ; more - on request

3.5 Macroeconomic Factors

The strong economic performance with stable and high growth for last ten years has been critical for Malaysian investment success. It has been combined with high growth in all major economies in the Asia Pacific region. This is an additional specific advantage for Malaysia FDI policy - to be in the heart of fast developing region. The Malaysian economy, a mixture of private and public enterprises has achieved a remarkable record of 7.6% average annual growth in 1988-96 (The Economist, 1998). This growth has resulted in a substantial reduction in poverty and rise in real wages. The average inflation has been kept low – about 3.8 percent during the 1989 – 1996 period. In 1996 and 1997 manufactured goods exports expanded less rapidly than in previous years because of the global slump in electronics; nonetheless, foreign investors continue to commit large sums to the economy (USCC, 1997).

As a result of economic crises spread in Asia in 1997 government and investors have faced new economic challenges such as sharp devaluation of national currency - more than 50 percent, rise of inflation, serious problems in banking system, increase internal and external indebtedness (IMF 1998). The restrictive measures offered by IMF to local governments, particularly to Malaysia are an additional important investment factor. The crises will have two-direction effect. On one side, low labour and production cost will make export more competitive. On the other side, the local market had shrunk, and import goods became more expensive, especially these from developed countries. Foreign debt currently is 34.4 billion USD and represents 42.6 % of GDP. This difficult financial situation will reflect on continuing government investment promotion policy. One of the measures could be relaxed of strict equity restrictions for non-Malays and foreigners, which may lead to increase the level of foreign capital inflow.

With combination of favourite political, economic and country specific factors the government FDI policy had helped Malaysia to reach change of economic structure from agriculture dominated in the past to high percent in GDP of industry and services today. Country's GDP composition by sector is: agriculture - 14%, industry - 45% and services - 41% (1997). Unemployment rate was maintained low 2.6% with tendencies to increase. One of the important FDI impacts of Asian crises over Malaysian economy is the exchange rate devaluation of the national currency - ringgit. Currently the exchange rate is 3.85 ringgit per 1USD, down from 2.5-2.7 ringgits per 1USD for more than 10 years. Serious challenge for the future economic policy will be the negative growth figures for 1998 and predictions for 1999. The external and the internal debt servings are government concern as well. Foreign and domestic businesses need to operate in changing environment with rising interest rates and a credit shortage. The possessions of Asian crises will force the government to place particular emphasis on the tightening of monetary policy, on improvements in corporate governance and fiscal transparency, and on the early implementation of commitments to deepen structural reforms. There is expectation that this policy will not effect the interest of foreign investors in Malaysia, and if is applied deregulation in foreign equity restrictions the country will become even more foreign capital attractive.

Nevertheless, as a result of implemented measures and substantial maintenance promotion of FDI policy for decade the government has succeeded to attract many global multinationals to allocate capital and technology, management and know-how in Malaysian economy. On a manufacturing project approval basis, total new foreign investment flows in Malaysia jumped 86 per cent from USD3.7 billion in 1995 to US$6.8 billion in 1996, and US$3.5 billion in 1997. The first three major investors in the country are Singapore, Japan and USA. The major foreign investors in Malaysia are main export destinations at the same time – the export share for USA is 20.7 percent, for Singapore is 20.3 percent, and for Japan about 13 percent of total. The three countries made up nearly three-quarters of total foreign investments in Malaysia. This is indicator for the confidence of the world business in Malaysian economy and in the system of FDI environment particularly. It is an unwritten rule that the volume of foreign investment is a vote of confidence in country's policies. At the same time the government has to react with adequate changes in the national FDI factors, subsystems and components towards more friendly and attractive investment environment.

4 Conclusions

As we have seen from these brief analyses the systematic approach of Hungarian and Malaysian governments towards improvement of country's FDI environment are different, but successful. Both countries enjoy strategic geographical location, which is a crucial precondition for general FDI success. Stabile political and economic conditions and government commitment towards attraction of foreign capital are also common characteristic for two countries.

Hungary has a different history and political system, which influence the country investment environment. The implementation of political and economic standards for fast EU, NATO and OECD integration has played a critical role in its transition development. Hungarian FDI strategy based on open, liberal model has been more competitive than Czech Republic or Poland investment policy on the way to achieve capital inflow and technology renovation.

Fast cash privatisation has been one of the country's FDI competitive advantages in comparison with other CEEC and Malaysia in particular. Restrictions of the model are strictly limited and the system is adaptable to new European and world political and economic realities. The relationships between components of the Hungarian FDI system are strong and balanced, although some factors such as the rate of inflation and the level of unemployment will demand special government attention. The successful government management on complexity investment environment gives Hungary competitive advantage against east European, Asian and Western countries in the global battle for more FDI.

Malaysian FDI systematic approach is coloured by strong national specificity. Factor with critical importance is the implementation of the New Economic Policy in 1969, which has determined the future economic development of the country. The system of foreign equity restrictions has changed the wealth structure of the country towards poverty avoidance and enhancement the economic standing of the Malay ethnic majority. Malaysia has tolerated high technology foreign investments in manufacturing and has achieving remarkable results. Except the equity limitations and export requirements its system of main FDI environment factors is stable and work efficiently. All investment conditions are focused on a narrow target, although restrictions are not a serious barrier for high foreign interest and capital inflow. We do not discuss in deep fairness of such investment model, hence future deregulation of FDI factors in Malaysia may increase the total volume of investment

Competitive advantage is created and sustained through a highly localised process. Differences in national values, culture, economic structures, institutions, and histories all contribute to competitive success (Porter 1990). All analysed success factors in the adopted investment pattern act in cumulative interdependence and in complex national environment. For example it is impossible to reach favourite FDI conditions in particular country without a stable political system and sound legal framework; without firm macroeconomic characteristics (growth of GDP, low inflation, stable exchange rate, balanced budget); and stable social environment. These measures have to be implemented as composite government actions at the same time.

Host country investment factors strongly influenced each other. Negative or positive changes in some of them could improve or destroy the investment environment attractiveness, depending on the impact and response of all major agents within the investment system. The process of investment improvement is a hard and delicate work for government and business organisations involved, though the results could benefit all citizens.

Hungarian and Malaysian models of foreign investment environment have pointed out that with different national FDI policies, starting conditions and approaches combined with specific advantages could be achieved similar outcomes – high level of foreign direct investment and sustainable prosperity of the nation state.

References:

1. Balcerowitcz L., Blaszczyk and Dabrowski M. 1997. The Polish Way to the Market Economy. In W. Woo, S. Parker and J. Sachs, eds., *Economies in Transition: Comparing Asia and Europe*. Massachusetts Institute of Technology.
2. Ball D., McCulloch W. Jr. 1996. "International Business – The Challenge of Global Competition", Irwin.
3. Daniels J.D., Radebaugh L.H. 1998. "International Business, Environment and Operations", 8th edition, Addison-Wesley.
4. Dobosiewicz Z. 1992. Foreign investment in Eastern Europe, Routledge London and New York.
5. Estrin S., Hughes K., Todd S., 1997. "Foreign Direct investment in Central and Eastern Europe: multinationals in transition", Pinter, London and Washington, Royal Institute of International Affairs.
6. Far Eastern Economic Review (FEER), May 28,1998, p. 23.
7. International Monetary Fund, Press Information Notice (PIN), No.98/31.
8. Porter M. 1990. The Competitive Advantage of Nations, HBR, pp. 73-93.
9. Organisation for Economic Co-operation and Development 1997. Economic Survey - Hungary, Paris, OECD.
10. Root Fr. R. 1994. "Entry Strategies for International Markets", Lexington Books.
11. United States Chamber of Commerce 1998. Country Commercial Guide – Hungary, USCC.
12. United States Chamber of Commerce 1997. Country Commercial Guide – Malaysia, USCC.
13. The Economist 1998. Pocket World in Figures, Profile Books Ltd.
14. The Financial Times, May 12, 1998, p. 4.

STRATEGIC FRAMEWORK FOR COPING WITH THE TUMULTUOUS ARENA OF THE CIS: A CASE OF TWO JOINT VENTURES

JONATHAN MENUHIN

Warwick Business School, University of Warwick, Coventry CV4 7AL, United Kingdom E-mail: phd97jm@wbs.warwick.ac.uk

A Strategic framework is developed in order to have a better understanding of the practical process of strategy formulation in the tumultuous arena of the CIS. This frameworks is based on the sustainability of the firm as the key factor, while the conventional strategic framework focus on market positioning or competences development process. A strategic framework for coping with the environment risks, is developed with six operational guidelines. The framework is applied to two case studies; Joint Ventures in Kyrgizstan and Moldova.

1 Introduction

The risks a company faces can be divided into two types: business and environmental risks. The first type of risk results from the nature of its business; the second type of risk is associated with changes in the business environment and related to issues such as politics, economics and cultural differences. This distinction is required in order to understand the nature of investment in Eastern Europe and the CIS. While usually the dominant risk to investors in the West is business risk, investors in the CIS are exposed to both types of risk. Mueller and Mueller (1997) have explored the dimensions of foreign investment risk in Eastern Europe as perceived by foreign business managers there. They have identified four clusters of risk type:

1. Political - regarding establishment of democratic government, political stability, and the development of a banking system and a stock market
2. Macro-economic stability - regarding inflation, the tax regime, membership in trade agreements (e.g. KU, EFTA) and currency convertibility.
3. Business infrastructure - regarding the education and skills of the local workforce, legal framework and other infrastructure.
4. Ethnic conflict risk - regarding ethnic conflict and border disputes.

None of these clusters are part of the conventional business risks but constitute environmental risks. Mueller and Mueller intend to focus on the additional risks to the regular business ones; of course, the conventional business risks are also exist in Eastern Europe.

However, regarding investment in this market, there are some advantages for risk-takers. Luo and Peng (1998) have showed that there are significant advantages for first movers in this arena and detailed the risk-return tradeoff between first mover and late mover strategies. Estrin *et al.* (1996) found that investment can be beneficial, with certain limitations, if the market entry is part of a well-conceived and financed global strategy. Margolis (1995) in her research on the real estate industry in Poland showed that foreign investors can gain high returns in this high-risk environment.

According to World Bank data (1997), the main economic activity in Kyrgizstan (40% of GNP) and Moldova (60% of GNP) is agriculture. The level of production there is low (2.5 tons per hectare of winter wheat in Kyrgizstan compared with 7-10 tons per ha in France and England). Lee and Orpen (1992) have identified that foreign investors usually come to the countries of Eastern Europe to utilize the low-cost workforce and business opportunities. Considering these characteristics, the industry as general in these countries, is low technology.

It is possible to characterize the Eastern European market as: (i) a high risk - High complexity environment, (ii) high return to risk takers and (iii) low-tech industry.

These characteristics require an appropriate strategic framework. This framework will be developed and discussed in comparison to the traditional frameworks. This framework result in six guidelines for successful implementation. This paper presents two joint ventures: the Kyrgizstan Agribusiness Company (KAC) and the Moldavian Wine Company (MWC) to illustrate the use of these guidelines.

2 Theoretical background

2.1 Appropriate strategic framework for CIS

Recent discussion in the literature on strategic approaches differentiates the strategic framework of a company according to the characteristics of the market. Chakravarthy (1997) claims that the Porter (1980, 1985), Hamel and Prahalad (1989, 1993) and D'Aveni (1994) strategic frameworks are distinguished by their attitude to two factors: the complexity of the firm and environment dynamics. The complexity of the firm is measured by the number of competitive configurations, that the firm has to consider in shaping its own strategy to cope with the environment complexity; the dynamics of the environment is the rate at which these configurations change over time.

Chakravarthy criticizes the Porter framework for the hidden assumptions of low complexity of the firm and low dynamic of the environment. Moreover, in the Eastern European arena, since foreign companies have tend to be first-movers and thereby have relative monopoly power, it is difficult to apply the Porter framework to these circumstances. Chakravarthy agreed that the Hamel and Prahalad framework focuses on the notion of change (moderate dynamics) but criticizes their work for regarding the complexity of the firm as low. Trying to apply their framework to agri-business projects in Eastern Europe fails because the environment of these projects has a high complexity, while the dynamics of these projects are low. The problem with D'Aveni's Hypercompetition world is, according to Chakravarthy, lies in its ignoring environment complexity.

In his paper, Chakravarthy characterizes the strategy approach in the environment that he called turbulence. Chakravarthy describes a turbulent environment as industry in which structures change dramatically including: falling entry and mobility barriers, increasing economics of scale and innovation dynamics. All these business risks arise from R&D development, while in Eastern Europe the main business risks are those that arise from the socio-economical-political circumstances as well as the environmental risks.

This paper follows Chakravarthy's arguments to a degree by describing the nature of strategy in the market that can be described as possessing a lot of complexity, as with the market in Kyrgizstan and Moldova with the agro-industry that has a low level of dynamics. Usually the term used for these markets is "turbulent", but in order to avoid confusion with Chakravarthy's work, this paper describes these markets as "tumultuous".

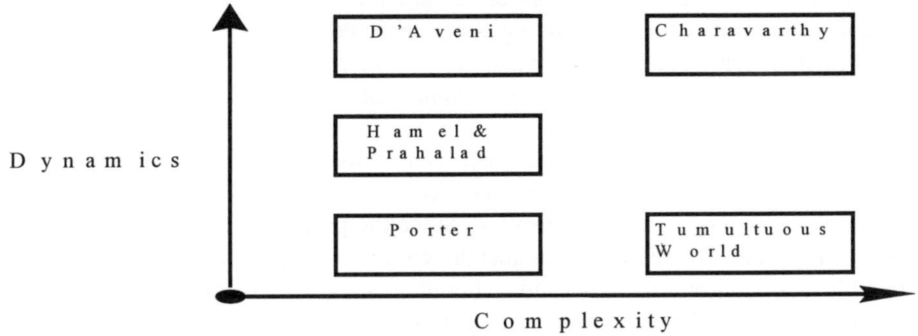

The characteristics of the East European market discussed earlier show the issue of sustainability that a foreign firm faces. The main problem is not to gain a rent, as a foreign investor could gain profit above normal in this market, but to form the venture in a way that could survive in the tumultuous environment, namely to reduce the risks. In order to succeed two strategies need to be present: to minimize the business risk and to minimize the environmental risks.

2.2 *Minimizing risks*

2.2.1 Minimizing the business risk

Das and Teng (1996) define two types of risk in an inter-firm alliance structure: relationship and performance. Joint venture is one type of inter-firm alliance and the preferred mode of market entry of foreign firm into the Eastern Europe and New Independent States as Shama (1995) identify. Performance risk involves the prospect of achieving the strategic goals of the alliance. The relational risk is concerned with whether the cooperation among the partners will go smoothly. The inter-firm alliance is unique since it involves both these kinds of risk and not only the performance risks.

In the tumultuous arena, reducing performance risks is connected with two issues; economising and creating flexibility. Economising focuses on ensuring cost efficiency in production, distribution and organisation (Williamson ,1991) in the firm's operation. The idea underlining economising is to achieve the ends with the minimal resources. As Teece *et al.* (1997) have suggested "economy is the best strategy". Flexible organisation is concerns with the ability to continually adapt and master the continual change. Flexible organization can easily develop strategies that adapt to new market realities, and shift all aspect of the organization so that they are congruent with the new strategy (Overholt, 1997). In the two joint ventures presented in this paper, these principles are translated into three guidelines: (i) rehabilitation and not re-creation e.g. since these counties are not in a fallow field, rehabilitation has economical advantage over re-creation. The venture should be based on the rehabilitation of existing business processes and not on the re-creation of new processes (ii) The second guideline is to focus on the weak link in the chain of the process. For it was this link which caused the collapse of the system. This is the reason for its importance in the value chain and the need to rebuild it. (iii) The last one is a general caution suggestion. Since the project architects can not foresee everything, and the reality is subject to continuous change, the plan should be flexible and the investment decision, should be made with cautious.

Overholt (1997) claims that the company should integrate the activities of diverse functional departments within the firm, and develop and implement an internally consistent set of goals and functional policies. These collectively define the company's position in the market. This issue is even more important in the case of a joint venture, where the shareholders have different interests. In order to reduce relational risks, an explicit reinforcing set of goals and functional policies is needed to counter the centrifugal forces that can lead functional departments and shareholders interests in separate directions. The fit between all organizational components should be consistent with the chosen organizational design and goals. Moreover, Lorenzoni and Baden-Fuller (1995) raised the issues concerning building consensus on the goals, for structuring the relationship, and building trust between the partners. In the tumultuous arena, the guidelines that lie beneath the idea of minimizing the relational risks are: (i) strive for control in order to counter the centrifugal forces, and (ii) the importance of consensus building. These two guidelines are complementary; that is if the company does not have full control it should build a strong consensus mechanism.

2.2.2 Minimizing the environmental risk

Environmental risks is connected with the complexity of the environment. The environmental risk is reduced by network building with all stakeholders, such as: government, banks, partners or clients. This view of the environment and its influence on the firms comes from the resource dependence theory (Oliver, 1991) that emphasizes the relationship between the firm and its external stakeholders Hickson *et al.* (1986) suggest that strategic acts are susceptible to influence from stakeholders, who have different concerns and attempt to achieve their own ends. Differences are resolved through bargaining, negotiation and compromise.

The idea of a web includes the notion of bi-directional links between the components. In order that a link is sustained, a mutual interest should be built in. In the tumultuous arena, company strong web links, may reduce the risk in the unstable environment.

To summarize, six principles have been presented to aid survival in the tumultuous arena:

Performance risk:	*	rehabilitation rather than re-creation.
	*	focus on the weak link.
	*	the general caution suggestion.
Relational risk:	*	strive for control.
	*	building consensus.
Environmental risk:	*	create a web with your external stakeholders.

3 The research objective and methods

The analysis of the strategy formulations of two international Joint Ventures in the agro-business sector in Kyrgizstan and Moldova is used in order to develop the strategic framework to fit the tumultuous environment. Because Moldova and Kyrgizstan are among the least developed economies in the CIS, many foreign investors have been hesitant to enter these markets. Some entrants, such as the two joint ventures under study here, have responded to what they perceive as a fallow field by developing comprehensive business strategies. Analytically, it is the very comprehensiveness of these strategies that enables us to examine a whole range of responses to the tumultuous environment, thereby yielding new insights into strategy formulation.

The field work for the KAC was carried out within the period February - November 1994; for the MWC it was March - August 1996. It included in-depth interviews with investors, consultants, government officials, local companies and representatives of international organizations such as the World Bank and the EBRD. During March - May 1998 a second phase of interviews was held with the project consultants of the two companies.

4 Description and analysis of the case studies

This section includes presentation and analysis of the two case studies. The analysis is based on the six guidelines stated above.

The ventures have been formulated with the assistance of international organizations (EBRD in Kyrgizstan and the World Bank in Moldova) in order to facilitate the entrance of foreign companies to these countries. The ventures are based in the agro-industry sector; KAC was established in 1995 and MWC was designed during 1996 - 1997 and is taking its first steps of implementation currently.

4.1 The Kyrgizstan Agribusiness Company (KAC)

KAC provides a full range of services for producing and marketing wheat for selected farms. These services include: inputs distribution (seeds, fertilisers, herbicides, pesticides, fuel and machinery maintenance), acquisition and maintenance of farm machinery and irrigation equipment, professional advice and assistance in critical periods (sowing, harvesting and diseases) up to marketing the outputs.

The company shareholders are a group of four foreign investors (55%) and a group of four local companies (45%). The foreign investor group consists of two agro-chemical companies, an agricultural project management company, and a non-sector related company that has business activities in the country. The local companies are a grain elevator and milling company, an agricultural machinery supplier, and two agro-chemicals distributors.

The company investment capital of $20 million is made up of $6 million from the foreign investors in cash, $5 million of the local investors in-kind investment, and the rest in loans mostly from the EBRD ($7 million). The equity capital is planned to be disbursed on a phased basis in accordance with the needs of the project.

KAC has established long term relations with the farms, which were carefully selected according to criteria of operational efficiency, management leadership and economies of scale. With the selected farms, KAC signed a contract for three years. The inputs were calculated in order to achieve a production norm of 6.0 tons per ha, out of which the company will take 3.6 tons, with the rest (2.4 tons) to be retained by the farms, for their contribution of land and labor. In this scheme of payment, KAC is sharing the operational risks with the farm and finances the farms working capital.

4.2 Analysis of the Joint Venture

4.2.1 Rehabilitation and not recreation

- The main costs in agriculture are usually the cost of purchasing the land, the infrastructure requirement and the variable costs of labour. KAC chose to refrain from investing in these costs and to specialise in the complementary services of agriculture. Since Kyrgyzstan is an agricultural country, KAC uses the existing facilities (mostly the farms) in order to achieve its goals.
- This scheme enables the company to develop marketing channels for the foreign agro-chemical companies, since some of the local partners are distribution companies. KAC sells chemicals to other farms, that are not participants in the venture, via these channels.
- The principle of rehabilitation of existing economic process enables KAC to receive financing from the EBRD that it would not be possible to get for an agro-chemical distribution company. KAC will distribute the agro-chemicals (approximately $10 million per annum) for the land project.

4.2.2 Focus on the weak link

The severe problems of agriculture in Kyrgyzstan are based on economic problems that cause a lack of funds to invest in working capital and maintain the facilities. The continuation of the economic difficulties cause use of the existing working capital for consumption and not for investment. KAC invests 40% of its funds in working capital and 55% in purchasing and fixing agriculture machinery.

The banking system in the country could not provide the farms with the funding due to lack of capital and the lack of legal reinforcing instruments (land can not be taken as collateral). KAC is providing the farm with funding for their operations in cash and goods that they could not have received from elsewhere.

4.2.3 The general caution suggestion

The company project area is 70,000 ha in two parts of the country. The implementation is spread over four years; 10,000 ha in year one as a pilot and the remainder over the next three years. In year 3 the company will consider whether to diversify to higher value crops (oilseeds). The pilot enabled the company to revise its calculations.

4.2.4 Strive for control

The KAC shareholders are a group of foreign investors (55%) that are organised as a single off-shore entity. This entity has the power to appoint the KAC chair of the board and the management team.

The owners of KAC are seven companies that have different objectives. In order to prevent dispute between the partners, they have signed an agreement that specifies the company mission, the management and board of directors' role, and a mechanism to settle disputes.

4.2.5 Consensus building

The KAC has one clear, single objective. All the company's activities are based on serving this purpose. The company architects reject business opportunities that pose the danger of contradicting to this mission. A clear and agreed mission is essential to the consensus building.

In order to prevent mistrust between the partners due to issues of transfer cost (that the mother company will gain its profit by selling its goods above the market price), it was agreed that the KAC management have autonomy in purchasing its goods and that the farms needs and the market price will be the only criteria.

4.2.6 Create a web with your external stakeholders

The company's dominant external stakeholders are: the EBRD, the government and the farms. There follow a summary of the links of the KAC with its stakeholders:

The EBRD - The EBRD's objectives for the loan are: involve the private sector, to encourage real investment in the country, have no negative environmental effects, and create sound and profitable projects. KAC was design according these guidelines. The connection with the EBRD was mutual interest. This project was the first for the bank in the country (the EBRD has political as well as economical objectives) and the first of the private sectors in the region. The foreign investors wanted the loan from the bank not only for financial reasons but also as a political insurance. They thought that the government would not act against the interests of the EBRD and so would not legislate or impose regulation that would harm the company.

The Government - The investors were concerned with issues such as appropriation, imposing new regulations on hard currency and new taxes. The Kyrgiz government wanted this project for various reasons, including to ensure the wheat supply to the country (Kyrgyzstan imports wheat), to encourage foreign investment (this project was the second capital investment in the country) and to rehabilitate the agricultural sector in the country. The Government signed a guarantee that defines its obligation to the project.

The farms - The KAC decided to look at the farms as partners rather than clients. This attitude is implemented through several procedures:

- the farms' selection process; one of the criteria was the leadership type of the farm.
- establishment of long-term contracts, in order to build commitment from the two sides.
- investment in the farm premises and maintenance of the machinery.

The farms were enthusiastic to participate in the project, since it would increase their revenues from 2 tons per ha, before the project, while they purchased all the inputs, to 2.4 tons per ha within the project and the KAC is paying for the expenses, bar labor cost.

Table 1 Present a summary of the guidelines and the implementation for the KAC.

Table 1. Summary of the guidelines and the implementation for the KAC

The guideline	Implementation
Rehabilitation rather than recreation.	• Investment in working capital and rehabilitate eroded facilities. • Local existing distribution channels to market agricultural inputs.
Focus on the weak link.	• Funding the farms.
The general caution suggestion.	• Pilot project.
Strive for control.	• Appointing the company CEO and Director. • Company agreement document. • Dispute solving mechanism.
Building consensus.	• One agreed company objective
Create a web with external stakeholders.	• Establishment of mutual interest relation with the stakeholders

4.3 The Moldavian Wine Company (MWC)

The MWC is a joint venture set in one of the biggest wineries of Moldova. The objective of the Joint Venture (JV) is to grow grapes and produce quality wine for Western markets. The JV will lease existing vineyards, lease land for grape plantations introduce new varieties of grapes and sell agricultural inputs for other vineyards. The Winery will be renovated, modernized and upgraded to western standards. The JV vineyards will be cultivated by land owners who that are part of the winery's local shareholders.

The joint venture was established between one foreign company (minor shareholder), local bank (minor shareholder) and a locally owned winery (major shareholder). The foreign investor is an agro-chemical company. The local bank serves as the financing body and provides local business managers for the JV.

The local winery has suffered in recent years from a lack of funds to maintain its assets, and part of the facilities need to be upgraded, but as a whole the general condition was acceptable. The investors estimate that an investment of $500,000 will be sufficient to upgrade the winery to western standards.

The MWC was designed after studying the initial results of the KAC operation. Therefore some similarities can be found, such as the investment of a foreign investor in a company to serve as a distribution channel for its products. The new feature of the MWC is the mechanism for ensuring foreign currency for the payment of the imported agricultural inputs. The foreign agro-chemical company will receive payment for its products directly from the revenue of the wine exported to the West.

4.4 Analysis of the Joint Venture

Since the MWC was formulated after studying the KAC experience, some similarities exist. In order not to repeat unnecessary details, the similarities will be presented only shortly.

4.4.1 Rehabilitation rather than re-creation

- To upgrade the winery is much less expensive than to establish a new one.
- The usage of existing facilities for building distribution channels of agro-chemicals for a whole range of consumers not only the participants in the projects.

4.4.2 Focus on the weak link

The project architects found two weak links in the process: (i) the wine's marketing to the West, and (ii) the grapes supply to the winery. The first caused by lack of experience; while the second caused by a lack of funding for purchasing fertilisers and herbicides to the vineyards.

In order to built capability to export wine to the West, the MWC investors decided to create an in-house wine marketing consultant. A western-owned wine producing company that manages several wineries in Romania and markets their wines in the West will serve as the in-house consultant. It's role will extend from improving production processes, introducing new varieties for the new vineyards, and to be in charge of the wine marketing in the West.

The farmers' working capital problem is dealt with a similar solution in the case of the KAC. By providing the farmers with their needs and a barter payment scheme, they can pay at the end of the harvest.

4.4.3 The general caution suggestion

The foreign investors have two caution steps in the formulation stage: the first one regards the level of the investment. At the initial stage, when the foreign investor considers gaining control of the company (more than 50% of the ownership), a large scale investment program was considered. This program was revised and reduced due to the governing issue.

The foreign investors held a market research test in the West; only after receiving the results was the business plan approved.

4.4.4 Strive for control

The shareholders discussed the issues of the company mission, each member role's and the mechanism for dispute solving. At the end of the process an agreement should be sign.

4.4.5 Consensus building

The company has three shareholders each responsible for different parts of the value chain. The MWC business is based on a win-win concept; all the participants would gain from the venture if successful , and if it collapses all the participants will suffer.

4.4.6 Create a web with your external stakeholders

The company views its external stakeholders as the World Bank and the Government. It is important to see that the EBRD credit line is excluded from this list. There follows a summary of the links of the MWC with its stakeholders:

The World Bank - The investors decided to participate in the World Bank political insurance scheme, protecting the foreign investor governmental action. The World Bank's political insurance program is managed locally by agency with an affiliation to a Dutch bank. The participation in the scheme in on commercial basis; that is the foreign investor pays an insurance fee to local agency.

The EBRD - Although the MWC is entitled to get credit at special terms from the EBRD credit line, the investors are still considering whether to apply. The main reasons are the long-term approval procedure and their financial solidity.

The Government - The World Bank's political insurance reduces some of the investors concern about possible governmental action; nevertheless, the investors decided to establish close relations with the government. The foreign investor consulted with the authorities for both their official and non-official approval for the projects. Since Moldova was one of the two wine manufacturers in the former USSR (the second was Georgia), the Government sees the recovery of the industry as highly important in the development and reconstruction of the country's economy.

Table 2 Present a summary of the guidelines and the implementation for the MWC.

Table 2- Summary of the guidelines and the implementation for the MWC

The guidelines	Implementation
Rehabilitation rather than recreation.	• Investment in working capital and in rehabilitating eroded facilities. • Local existing distribution channels to market agricultural inputs.
Focus on the weak link.	• Building the capability to market the wine to the West (in-house consultant). • Funding the farmers required working capital.
The general caution suggestion.	• Revising the investment depending on the level of control. • Market research.
Strive for control.	• Discussion about company founding agreement. • Dispute solving mechanism.
Building consensus.	• One agreed company objective. • Win-win business concept.
Create a web with external stakeholders.	• Establishment of mutual interest relations with the stakeholders

5 Final comments

This paper has addressed on two main issues:

1. The unique strategic framework to cope with tumultuous Eastern Europe and the CIS arena for foreign investors within low dynamic industry; as the agro-business contexts.
2. The guidelines and the implications of this framework in reality.

This paper aims to fill the gap in the existing literature on strategic frameworks that seems to be inadequate in the tumultuous arena. This arena is characterized by a high level of business and environmental risk. The risk factor is the focal point in doing business in Eastern Europe and CIS countries, and has not received sufficient coverage in the conventional strategic frameworks.

The essence of the strategic framework developed here is that the framework should be adjusted to the characteristics of the environment and the configuration of the firm (as in Chakravarthy, 1996); in the tumultuous world it is necessary to identify the possible risks and try to minimize them according to the suggested guidelines.

The second issue was how to apply the strategic framework in the cases of the two joint ventures. It should be noted that careful formulation does not guarantee full successful implementation. The two companies today face some difficulties. In the

KAC case the Government has imposed a new set of regulations, and the company has some disagreements with the farms. The MWC implementation process is slowing down due to a lack of total commitment by some of the partners (possibly because of the management of the relational risk). This paper did not aim to give a recipe for success but, rather, to explore the guidelines of the business formulation in the CIS arena.

Reference List

1. Mueller James D., and Mueller Rene Dentiste. 1997. Perceptions of Foreign Investment Risk in Central and Eastern Europe and the Former Soviet Union. *International Journal of Management* 14, no. 4: 598-606.

2. Luo, Yadong, and Mike W Peng. 1998. First mover advantages in investing in transitional economies. *Thunderbird International Business Review* 40, no. 2: 141-63.

3. Estrin, Saul, Kirsty Hughes, and Sarah Todd. 1996. Investing in Eastern Europe: Two case histories. *Business Strategy Review* 7 , no. 1: 24-37.

4. Margolis, Adrienne. 1995. Rewarding risk-takers. *Central European* 5 , no. 9: 25-27.

5. Lee, Choong Y, and C Orpen. 1992. Marketing strategies for British firms entering Eastern Europe. *Journal of International Marketing and Marketing Research* 17, no. 2 : 81-88.

6. Chakravarthy, Bala. 1997. A new strategy framework for coping with turbulence. *Sloan Management Review* 38, no. 2: 69-82.

7. Porter M.E. 1985. *Competite Advantage*. New York: Free Press.

8. Porter M.E. 1980. *Competite Strategy: Techniques for Analyzing Industries and Competitors*. Yew York: Free Press.

9. Hamel, Gary, and C. K. Prahalad. 1989. Strategic Intent. *Harvard Business Review* 67, no. 3: 63-77.

10. Hamel, Gary, and C K Prahalad. 1993. Strategy as stretch and leverage. *Harvard Business Review* 71, no. 2: 75-85.

11. D'Aveni R.A. 1994. *Hypercompetition: Managing the Dynamics of Strategic Maneuvering*. New York: Free Press.

12. Das, T K, and Bing-Sheng Teng. 1996. Risk types and inter-firm alliance structures. *The Journal of Management Studies* 33, no. 6: 827-43.

13. Shama, Avraham. 1995. Entry strategies of U.S. firms to the Newly Independent States, Baltic states, and eastern European countries. *California Management Review* 37, no. 3: 90-109.

14. Williamson, Oliver E. 1991. Strategizing, Economizing, and Economic Organization. *Strategic Management Journal* 12: 75-95.

15. David J Teece, Gary Pisano, and Amy Shuen . 1997. Dynamic capabilities and strategic management. *Strategic Management Journal* 18 , no. 7 : 509-33.

16. Overholt Miles H . 1997. Flexible organizations: Using organizational design as a competitive advantage. *Human Resource Planning* 20 , no. 1: 22-32 .

17. Lorenzoni, Gianni, and Charles Baden-Fuller. 1995. Creating a strategic center to manage a web of partners. *California Management Review* 37, no. 3: 146-64.

18. Oliver C. 1991. Strategic Responses to Institutional Processes. *Academy of Management. The Academy of Management Review* 16, no. 1: 145-79.

19. Hickson D.J, Bultler R.J., Cray D., Mallory G., and Wilson D.C. 1986. *Top Decisions - Strategic Decision-Making in Organization*. Oxford: Basil Blacwell.

THE COMPLEXITY OF MANAGEMENT IN THE CASINO INDUSTRY

Y. LINDENBOIM
Warwick Business School, University of Warwick, Coventry, CV4 7AL, England
E-mail: lindenya@aol.com

The gaming operation differs from other service operations in its objectives. The traditional services (such as banks, hospitals or public transportation) seek to process customers as quickly as possible, both for 'throughput' reasons and customer convenience reasons, whilst the casino's objective is to 'process' the customers as long as possible. This unique aspect of the service leads to a set of complex management decisions, in regards to the planning of the service design.
For several months the casino industry was researched. In particular, two British casinos were analysed in detail. The study analysed the gaming operation in those casinos, and explored the different aspects of operation, including the demand conditions, transformation process, layout design, performance measurement and human resource management. The financial implication of the current service design were explored and proved to be fascinating.
This paper suggests a fresh systemic approach for analysing a traditional gaming operation by identifying underlining factors in customer behaviour and business infrastructure. A new managerial model for handling the complexity of the casino business is presented, emphasising the need for developing a learning process.

1 Introduction

The aim of this paper is to explore the nature of the unique casino operation, assess its current performance and suggest ways regarding how this performance could be improved in the future.

Levitt (1980) argues that the core product or service sooner or later becomes a commodity within an industry, as competition increases and the industry matures. As a result, in order to have the edge over competitors, performance on the supplementary services is usually emphasised. An important implication of servicescapes (the impact of physical surroundings on customers and employees) is the need for managers to understand their power in communicating environmental cues (e.g. layout, music, lighting, smell and colour) to consumers. With this knowledge and understanding, environmental cues can be strategically planned in order to communicate the correct image of the business to the consumer.

This complex task requires management to develop a set of new expertise. Applying contemporary approaches to traditional gaming business should enhance the casino performance and establish their competitive advantage. Careful scheming of the customer-server interaction needs the understanding of the underlined factors of the punters' expectations and the business resources. The majority of research into servicescapes has been conducted in restaurants and department stores (Bitner 1982,90,92 Milliman 1986,88). However, research into servicescapes in the casino industry has been neglected.

This paper explores the literature of service design and servicescapes, analyses the gaming operation in two specific casinos, and offers conclusion and recommendations for potential enhancements in this operation.

2 Industry Background

The term "Gambling" covers two distinct forms of activity: betting and gaming. Betting covers gambling in which the participant can use some degree of skill to determine the outcome of the event on which he/she is betting, e.g. a horserace. Gaming, on the other hand, covers gambling in which the outcome is determined by statistical odds and the participant can have no control over the outcome, e.g. the National Lottery.

The gambling industry in Britain has to operate within a tight regulatory framework. The legislation is designed to allow the individual to gamble in a fair and free manner, while at the same time not encouraging the practice.

The industry has to be protected from criminal elements, while at the same time be allowed to flourish as far as possible. The government has an incentive to maintain the health of the industry, given that it is a major source of revenue in the shape of gaming duties and tax. The Home Office is ultimately responsible for the regulation of the gambling industry, although on a day-to-day basis, The Gaming Board of Great Britain, which was established in November 1968, oversees operational matters for casinos. Section 10(3) of the Gaming Act provides that "it shall be the duty of the Board to keep under review the extent and character of gaming in Great Britain".

Gambling has become one of the fastest growth areas in consumer spending and households have set aside 42% more in the past year for gambling, mainly attributable to The National Lottery. However, growth has been hampered by very restrictive legislation, which has failed to keep pace with this demand. During the last 3 years, however, some beneficial deregulation has taken place, with more currently in the consultation process.

The betting and gambling industry is unique among consumer markets, in that consumers, in the long run, tend to receive some of their expenditure back in the form of winnings. Spending on betting and gaming, therefore, is measured in two ways: gross money staked and net expenditure; that is, stakes less winnings, or net losses.

2.1 Strengths

- Casinos are relatively unaffected by the health of the domestic economy.
- The introduction of two new games which have appealed to continental gamblers.
- The reduction of the 48-hour cooling-off period down to 24 hours.

- The increase in the number of gaming machines allowed in casinos.

2.2 Weaknesses

- Casinos can be affected by external events, such as exchange rates, due to a high percentage of overseas players.

2.3 Opportunities

- The allowance of some print advertising, which is currently in the consultation process, will allow casinos to advertise in holiday guides, hotel brochures and telephone directories.
- The increased licensing hours for consumption of alcohol in casinos from 12.00am to 2.00am in the Provinces and 3.00am in London.
- Players will be allowed to play with debit cards.
- For the industry to campaign for the allowance of credit in casinos.
- The potential acceptance of postal applications for casino membership makes it more available to overseas visitors, as they can apply before arriving in the UK.

2.4 Threats

- The major companies building too many casinos in the 21 new locations where casinos are to be allowed.

3 Literature Review

Following the initial review phase of the casino industry, a theoretical framework for analysing this operation was chosen. Fascinated by this unique service operation, a decision was made to explore the literature covering service design and learn more about the importance of servicescapes.

3.1 Service Design

When defining a service concept, perceptions and expectations of the service must be established in the minds of customers, employees, shareholders and lenders. Then to target the right market with the service concept, the service organisation must position itself competitively in the environment for that market. This can be done, Michael Porter argues, in three ways; by overall cost leadership, differentiation and market focus.

A service package will generally consist of many services, not just the one. In the casino industry, one customer will visit a casino with the intention to gamble, have a drink and a bite to eat: the gaming floor is the core product and all the others are supplementary. Another customer may visit the casino just to dine in the restaurant, in which case the restaurant is the core service. Eigler and Langeard (1984) argue that it is sometimes difficult to define the core service in a service organisation, hence it is imperative that no service within the operation is ignored. Sasser, Olsen and Wyckoff (1978) define the components of the service package as,

- Physical items or facilitating goods.
- Sensual benefits or explicit services.
- Psychological benefits or implicit services.[1]

As argued by Levitt (1980)[2], the core product or service sooner or later becomes a commodity within an industry, as competition increases and the industry matures. As a result, in order to have the edge over competitors, performance on the supplementary services is usually emphasised.

When a service operation is being designed, there are many areas that must be considered. Those are mainly the result of customer participation in that service: the different amounts of interaction between customers and employees affect the design and co-ordination of services.[3] During the design process, all other key areas within operations must also be considered in order to meet customers' needs, such as marketing, organisational behaviour, human resource management and operations management.

Thus, the end result is dependent on several factors:

- What is the 'service package' or 'service packages'?
- Who are the customers for the 'service package'?
- What are the standards for the 'service package'?
- How can the 'service package' be given to the customers? [4]

Schmenner (1986)[5] suggest that two elements can be used to describe and classify service businesses - the levels of labour intensity and customer-server interaction. He classifies services into either "high" or "low" with respect to these two factors and from this is able to basically categorise services into four major groups. Not all businesses fit clearly into these extremes, however, as casinos, for example, lie somewhere between 'Mass Services' and 'Service shops'.

Subsequently, a manager's challenge changes, depending on the degree of labour intensity and customer interaction in a service organisation. If the labour intensity is high, there are many human resource activities involved, such as recruitment, training, development, and employee welfare. The operational issues are also major tasks for managers, including scheduling of the workforce and managing growth.

Lovelock (1992) analyses a case whereby components of a service organisation are also rated on their levels of customer-server contact and subsequently the amount of communication this interaction brings about. This is then compared against the power held by front line staff. This analyses concluded that customer satisfaction improved when the staff were given more customer contact skills and customers could choose when they wanted the service to be carried out, rather than being dictated to by the organisation. Staff being trained further in the communication skills necessary for diagnosing problems, gathering and disseminating information, managing waiting periods and resolving conflicts further enhanced this satisfaction.

Customers always have expectations when they approach a service organisation, either based on past experience, word-of-mouth or advertising. The common areas of thought, on which judgements are made for the expectations, are those on which managers should focus, so that customers are delighted when their expectations are fulfilled positively.

Customers' expectations of the whole service package are inherent in the design stages, so that they are at least met, to make them satisfied, if not surpassed to delight them. Service design should aim neither to narrow or close, but to reverse the gaps among customer expectations and perceptions of service quality.[6]

However, the standard for the service package should be clearly defined in the design stages, so those customers do not expect more than the service can deliver. Service breakdown occurs when one's experience of service falls short of expectations. The truly successful service companies are those that focus on the details of service design and deployment and then work equally hard at recovery when things go wrong, as they do the first time out.[7]

Much of the research into service design has been conducted in a methodical way, as it has been for manufacturing design, but service design has the added perspective of increased customer contact and interaction. Several methods are given for classifying the service organisation, but not many take into account the physical attributes of the customer (i.e. sex and age) and relate this to the classification context. Also, the ability of individuals to control and regulate their own behaviour in response to service delivery has not been researched in great depth, so the extent of this influence is not really known clearly.

Consumers' internal feelings brought into an environment are not really taken into account, but the extent of their satisfaction on experiencing the service is well documented, with methods of adjustment depending on their reaction. So the real influence of the service delivery system and design on their feelings cannot be accurately assessed.

3.2 Servicescapes

Bitner defined servicescapes as 'the impact of physical surroundings on customers and employees.' The more people are satisfied with servicescapes, the more they are likely to be satisfied with the complete service encounter. Servicescapes involve those controllable internal and external environmental elements that people experience with at least four of the five senses: Kotler identified these as *visual* (colour, brightness, size, shape), *aural* (volume, pitch), *olfactory* (scent, freshness), and *tactile* (softness, smoothness, temperature).[8] To improve customers' perceptions of quality and satisfaction, management must focus on these controllable physical cues as well as the tangible and intangible components of a business.[9]

In the design and management of servicescapes, the aim should be to build long-term customer relationships. There is evidence to show that when customers are satisfied, their perception of service quality is enhanced, which in turn increases customer loyalty.[10]

3.2.1 Elements of The Servicescape

Layout

When an interior space works, it is no accident. Research shows that the physical surroundings of the environment can influence behaviour, so servicescapes have to be carefully designed. Developing and delivering a well-designed casino, for example, takes unparalleled talent and experience. It isn't enough anymore just to offer the public a few tables and machines, a place to sit, and free drinks. A certain lifestyle is communicated and required.[11] In the design process, the first step is to identify desirable customer/employee behaviour and what the company hopes to achieve from its physical facility.

Investigations into retail crowding in particular layouts have shown that under dense conditions, shoppers experienced less satisfaction with the environment than when it was quiet. There is also a significant correlation between time pressures, stress and satisfaction with the environment.[12]

The responses to these can all in turn influence whether or not people stay in and enjoy an environment. It is well documented that the comfort of seats in restaurants for example, subconsciously or consciously affects the amount of time spent by customers in these establishments. In casinos, customers are more likely to stay longer, wager more and be more likely to return, if their seats are comfortable, the gaming equipment convenient, the atmosphere pleasant and the service first-rate. When they walk away miserable, customers often think that something interfered with the game itself. That's why it is important for companies to continually dedicate themselves to the experience of the player.[13]

Music

A link has been discovered between music and productivity in the workplace. Where background music is played in a retail environment, shopping time and the number of shoppers who make purchases increases. Shoppers have a better image of stores, which play music and tend to equate it with better customer service.[14]

Experiments have been carried out to test the influence of background music on consumer behaviour. Although research has been more focused on the effects within department stores and restaurants, it can be related directly to the casino environment. Following interviews with consumers in-store, it showed that music did not directly affect their moods or unplanned purchases, no matter what time of the day, but perception of shopping time varied depending on age and type of music:[15] music preference is most strongly influenced by age rather than regional or racial factors.[16]

Gorn (1982) concluded that product preferences could be conditioned when paired with specific types of background music. Allen and Madden (1985) however, on attempting to replicate these results, were unsuccessful. They speculated that although product preferences could not be conditioned reliably by exposure to a particular piece of music, this did not mean that affective conditioning never occurred.[17] On some occasions, particular products may dominate peoples' minds to such an extent that attempts to influence their choice by means of music are a waste of time.[18]

Background music has such an influence on customer behaviour that it is an important consideration within businesses. Studies have shown that music, which compliments the objectives of a restaurant, for example, will contribute to its profitability. In one study, a restaurant added slow, middle-of-the-road instrumental music to change the atmosphere. Although this resulted in a reduction of table turnover, more beverages and desserts were consumed, thus resulting in a higher profitability. Another restaurant increased the table turnover at lunchtime by having faster tempo music and bare tables. Most people do not know that they are being influenced by atmospherics. An explanation of this effect is that the careful and deliberate choice of particular colours and music will provide harmony or resonance.[19]

Another study looked at the affect on customer behaviour of different tempos of background music, on six variables: 1. serving time, 2. time between serving and leaving, 3. number of customers leaving before being seated, 4. $ of food purchases per bill, 5. $ of bar purchases per bill, and 6. the estimated gross margin for each type of music. The results showed that customers stayed at the table longer with slow music and also purchased more from the bar, resulting in a higher gross margin for slow music. There was no significant difference between slow and fast music for the other variables.[20] In summary, background music can influence the amount of time and money, both real and/or perceived, a customer spends in an establishment. While research has identified relationships between specific behaviours and specific musical characteristics (e.g. tempo, volume, mode), companies should be cautious when attempting to manipulate specific aspects of their background music. A safer and more effective strategy would be to choose background music that reflected the musical preferences of targeted market segments.[21]

The effect of "music loudness" showed that significantly less time was spent in environments where music was loud and the rate of spending was greater.[22]

The effect of music-induced pleasure and arousal on consumer's desire to participate in buyer-seller interactions have been investigated, by varying background music in terms of pleasure and arousal (low, moderate and high). The results suggested that consumers had a higher desire to interact in environments associated with more pleasure and more arousal; pleasure had a stronger positive impact under low and high arousal than under a moderate level, and arousal had a stronger effect under low and high pleasure compared to moderate level.[23]

A time interval can appear longer or shorter depending on the stimulus information encountered with it and the internal state of the perceiver. The perception of time is important in a service encounter, so background music is being used to fill these time intervals in order to manipulate the states of listeners. It has been found that soothing music has a positive effect on consumers during a time interval, contrary to arousing music.[24]

Lighting

It has been found that lighting is an extremely important determinant of the environment because brightly-lit rooms are more arousing than dim ones. Within the casino environment, dynamic and themed lighting, combined with interior design and acoustical effects, stimulates the guests' senses; glittering lights and flashing graphics keep the patron from feeling tired.[25]

Upgrading both the exterior and interior lighting can provide a marked improvement on the profitability of a casino[20]: exterior lighting to attract players could be enhanced in entry driveways, for example, or give long distance visibility by way of floodlights and laser graphics. Inside, combinations of colour and lighting can be used to create moods and emotions to calm or stimulate the central nervous system.[26] Proper lighting, both interior and exterior, can be used to raise the perception of safety, which is critical in any environment. It can also be used architecturally and theatrically to enhance the environment.[27]

Smell

Consumer behaviour is determined by the relationship between several linked factors. One of these is the effect of smells; they can reduce stress, induce a feeling of well-being, or evoke an emotion that will influence behaviour.[28]

The presence of inoffensive ambient scents (e.g. certain floral scents) and offensive scents (e.g. overripe vegetables) has been investigated, to evaluate consumer behaviour in these environments. Findings suggested that consumers in ambient conditions perceived that they spent less time in that environment than those in no-scent conditions: queuing times appeared shorter, thus enhancing the positive image of the environment.[29]

Retailers are looking into the use of fragrances to increase sales. The theory is that because smell is believed to be most closely linked of all the senses to memory and emotions, stores can use it to put people in the mood to shop. Relating this to the casino environment, research has shown a 45% increase in slot machine usage in scented casinos. This use of subliminal stimulation has been criticised by consumers, but businesses argue that it is all part of the ambience, like background music or lighting. The scent has to be right for the target audience, however, as people might go out of their way to avoid shopping in a store that was scented in a way that didn't appeal to them.[24]

Colour

Tests have been carried out on people to see which colours they associate with such words as expensive, happy, love and dependable. The results have shown that while some colours have cross-cultural consistency, others have different meanings in different countries.[30] For marketers today, colour - in branding, advertising, packaging and stores - will be a major competitive resource.[31] Because colour is likely to enhance the attractiveness of objects and environments, consumers are more likely to have favourable attitudes towards stimulating alternatives.[32]

The effects of red and blue have been studied in a purchasing context - chosen because of their opposite colour properties. It was found that red is perceived as negative and tense, as well as physically arousing. Blue, on the other hand, was identified as calm, cool and positive: people tend to stay longer in a blue environment. Although at first the results might indicate a reaction of arousal to the perception of colour, this may be due to the general positive and negative perceptions of those two colours.[33]

Warmth, technology and nostalgia drive paint colour choices, and with such a diverse choice, any taste can be accommodated. Yellows and reds create a warm, comforting atmosphere, while green is popular for those with environmental concerns. Browns and 'earthy' hues are being used to reconstruct nostalgic eras, while metallic paints are being chosen by those seeking to invent futuristic environments.[34]

Consumers have also become more sophisticated about the use of colour and have higher expectations than in the past. Colours actually send out signals about the physical characteristics of a product. e.g. red or black signify strength and durability.[35]

Subliminal perception

Subliminal perception is a person's ability to perceive and respond to stimuli that are below the level of consciousness. Some people believe that subliminal cues appear deliberately, while other maintain that they are there by chance. Dudley (1987) believes it is unlikely that subliminal stimulation is an effective method of controlling consumer behaviour. With claims being supported by little published evidence, advertising and communication practitioners still continue to use subliminal messages. While these messages may encourage people to act in a way that they had already decided to, the possibility that they make people act in a completely different manner is questionable.[36]

The paper written by Hawkins in 1970 on subliminal stimulation has been referred to for many years as real evidence that subliminal advertising can affect consumer-related behaviour. Even though there have been many theories and papers since then counteracting that argument, there is enough evidence and interest in this subject that it cannot be dismissed as irrelevant.[37]

Mood

Mood states has been the subject of much recent research, with respect to psychology and marketing, particularly in the effect of persuasion and consumer behaviour.[38] In tests, researchers have attempted to induce 'positive' and 'negative' moods in consumers by the use of different environmental stimuli, to examine the influence of those moods on behaviour. In general, it was found that positive moods increased confidence, risk taking and decision making.[39]

In pleasant environments, with respect to condition, information rate and layout, it was found that customers' moods improved as time went on, whereas they deteriorated in unpleasant environments. Subsequently, customers tended to spend more money in pleasant environments, spent longer in them and had more intention of returning.[40]

Image

Images displayed by businesses have long been recognised as determinants of their success and have also been used as a positioning and differential tool. Much work has been carried out in image research, with the view of predicting consumers' overall impressions of businesses by their physical attributes. The value of this is to help managers identify which image dimensions are most important in the formation and modification of consumers' impressions, and pre-planning of costs involved and strategies required for these modifications.[41]

The perception people have of products is often more important than reality. Companies have began to realise that the sensory details of products - the way they sound, smell or feel - influence consumers in a big way, giving them 'feelings' of quality. The little details are becoming more important.[42]

In one particular investigation, combinations of specific elements in a store environment were assessed to examine how they influenced customer perceptions of goods and service quality, and how these perceptions influenced the overall perception of the store image. Of the two stores used, one had a prestigious image, the other a discount image. The results indicated that ambient conditions and salespeople are more likely to affect customer perceptions of goods and service quality than design factors such as colour, display, trim, store and merchandise layout. Classical music, soft lighting and many salespeople created a prestigious store atmosphere, resulting in customers perceiving the goods and service to be of a higher quality than in a discount store environment.[43]

Consumers rely on external and internal environmental cues to categorise and hence make perceptions about products and services. Tangible cues are used to pre-judge expectations of a service. Thus managers are recognising this and spending more on exterior and interior remodelling of their businesses.[44]

Physical surroundings also influence customers' reactions to service failures. In the event of a service failure in an organised, professional environment, the customer may blame the company less and see the incident as unintentional and temporary. But in a disorganised environment, customers take the physical cues to mean incompetence and inefficiency, and expect the same type of problem to occur in the future.[45]

Gender

Pubs and casinos have, in the past, always been associated with male domination, with particular areas focused on a purpose, i.e. drinking or gambling. Women are increasingly becoming a major element in these environments and companies are having to accommodate this factor through the use of servicescapes. Women place more emphasis on the importance of physical surroundings than men, so signals from the physical surroundings should indicate that both men and women are welcome and should feel comfortable.[46]

The impact of consumer behaviour to music (tempo and loudness) has been studied with respect to males and females. It was found, for example, that females react more positively to music at lower volumes. The implications for this study are significant for application through differential targeting of men and women through music, in advertisements and retail environments.[47]

Price Acceptability

The impact of servicescapes is also an issue in consumers' price acceptability. A pleasant experience created by clever use of colour, layout, architecture, scents and temperature, also sends information cues to the consumer as to the image of a business. An important issue is the identification of specific environmental cues which consumers use to judge the acceptability of price. In general it was found that customers more readily accepted goods at a particular price in businesses with a high level of customer service and perceived 'higher-class' environment.[48]

Pleasure induced by certain environments also appears to be a strong reason for consumers spending more money in them than they intended. Pleasant environments intensify feelings and arouse consumers, contributing to their increased emotional behaviour.[49]

3.2.2 Summary

It is evident from research carried out that physical environments have the ability to influence consumer emotions and subsequent behaviour, from the evaluation of these environments and formation of perceptions. However, the strength of these effects varies with the ability of individual consumers to control their own behaviour: managers cannot expect servicescapes to have the same affect on all consumers.[50]

3.2.3 Managerial Implications

An important implication of servicescapes is the need for managers to understand their power in communicating environmental cues to consumers. With this knowledge and understanding, environmental cues can be strategically planned in order to communicate the correct image of the business to the consumer. In forming strategies for designing servicescapes, managers must realise the needs of the ultimate user and also the requirements of different functional groups.[51]

Just as managers can attempt to create distinctive environments that differ from their competitors by using different servicescapes such as store layout, display and colour, they can also use other methods such as scent as a way of differentiation. Customer satisfaction and purchasing behaviour can also be influenced via mood manipulation and overall perceptions of service quality can be enhanced.

3.2.4 Literature Pitfalls

The majority of research into servicescapes has been conducted in restaurants and department stores. Although the findings can be related to other environments, direct research must be carried out to confirm these assumptions. Also, the ability of individuals to control and regulate their own behaviour in response to environmental stimuli has not been researched in great depth, so the extent of this influence is not really known.

Research into moods doesn't appear to take into account consumers' internal feelings brought *into* an environment, but concentrates more on the apparent moods that environment creates. So the real influence of physical factors on moods cannot be accurately assessed. The evaluation of price acceptance, although correctly basing it to some degree on consumer's perception of environmental factors, doesn't talk about the influence of individual's knowledge on competitors' prices and general price awareness.

4 The Gaming Operation

The gaming operation is the core business of a casino. The analysis of these operations will be at the core of the following chapters. The objective of this chapter is to review the performance of the gaming operations and to suggest measures that may enhance it in the future.

Firstly, the overall financial results of the gaming operations, in the observed case, will be covered, followed by a more detailed analysis of the revenues and profit for each game. Secondly, the customer demand will be explored. Thirdly, the transformation process will be discussed. Fourthly, the layout design will be examined in detail. The distinct gaming cells will be inspected along with the positioning of these cells on the gaming floor. Finally, all the major findings will be summarised and a set of recommendations will be offered.

It is expected that this order of analysis will allow the reader to become familiar with the nature of the casino gaming operation, before the making of recommendations regarding how this operation may be improved. This order represents a top-down approach, which will characterise all the sections in this chapter.

5 Methodology

Information was collected from a wide range of sources to provide the basis of analysis in this report. Company accounts were received from the management accounting department, gaming figures were collected from the cash desk, and attendance levels were drawn from the reception database. For a period of four weeks, the gaming operation was investigated through observation and interviews with managers, staff and customers. For competitors benchmarking, newspapers, journals and investor reports were used, three provincial casinos were visited, and five telephone interviews were made with the duty managers of other casinos.

The use of both quantitative and qualitative analysis enabled a deep and comprehensive understanding of the operation. Operations management books were consulted during the analysis, in addition to the literature review presented in the last chapter. Recommendations as to how this operation may be improved were made on the basis of this analysis.

6 Financial Results

6.1 Average customer spend

The average-spend per customer sways seasonally. The drop per head is at the maximum during the winter (December-January) and at the minimum during the summer (June-July-August). The explanation for this seasonal effect is probably the fact that gambling expenses are by nature part of the customers' disposable income. This income is highest in December (the most common time for annual bonuses) and lowest in the summer, when most people take (and pay for) their annual vacation.

The average spend is also influenced by the number of customers in the club. A negative correlation was found between the attendance levels and the average-spend. As the number of customers in the club increases, the 'drop per head' decreases, and vice versa.

Looking at the same data on a daily basis [Figure 6-1], a distinction can be made between the weekdays and weekends. During the weekends, many customers attend the club, but the average gaming spend is low, while during the weekdays, attendance is lower and the drop per head is higher.

Figure 6-1: DPH vs. Customer Attendance (daily)

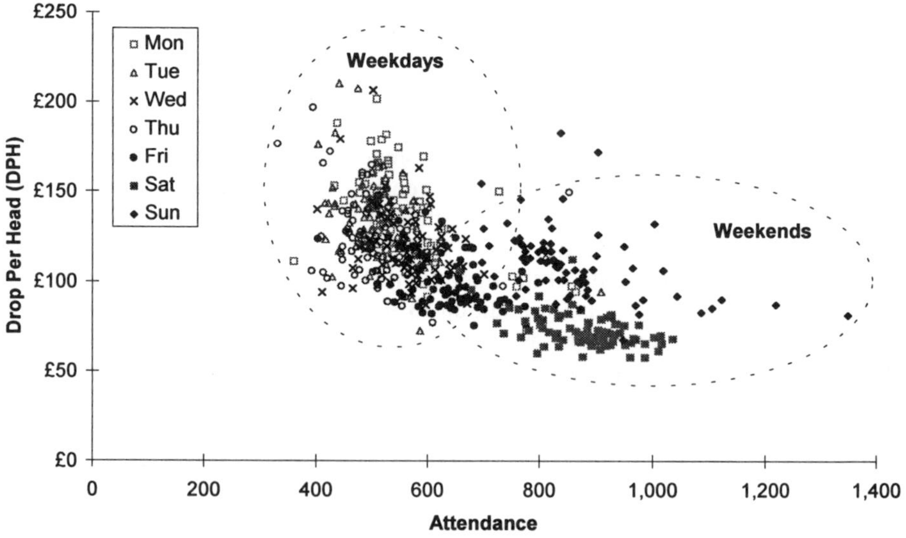

6.2 Game Profitability

The profitability of the games varies according to the chance of the game, the punters' skills, and the cost of operating staff. The Roulette, Blackjack and Casino Stud Poker games employ one dealer per table and one inspector per two tables. The Punto Banco employs one dealer and one inspector for the one table available in the club, and the most expensive game to operate is Dice, with three dealers and one inspector. Table 6-1 shows the gross and net margins for the different games, as was calculated in the case study.

Table 6-1: Gross and Net Margins per Game

	Gross Margin	Net Margin
American Roulette	16%	13%
Blackjack	24%	18%
Casino Stud Poker	25%	17%
Punto Banco	15%	13%
Crap (Dice)	19%	11%

Both Punto Banco and Dice generate low gross margins and are more expensive to operate, resulting in a marginal net contribution to the overall club revenue. Roulette, Blackjack and Casino Stud Poker have similar operating costs, but differ in their share of income and margins. Roulette, therefore, is the 'cash cow' of the gaming operation, Blackjack and Casino Stud Poker are the 'nice but small' contributors, and Punto Banco and Dice are the 'problem children'.

It is concluded, from this distinction, that the American Roulette, Blackjack and Casino Stud Poker operations can be seen as the core products of the gaming operation, with Punto Banco and Dice acting as supplementary services. However, it is thought that the contribution of the last two games for the attractiveness and thereby success of the club is crucial. Punto Banco is very popular among the oriental customers; these account for about one fifth of the attendants, more than a quarter of the gaming income. For this reason, the availability of Punto Banco in the club is important. Dice's major contribution to the club lies in the atmosphere it creates; a noisy and exciting game; it attracts attention and keeps the area around it alive.

7 Demand

7.1 Introduction

The demand analysis covers the general customer attendance levels in terms of frequency of visits per customer and general daily load in the club. Following this, the customer flow into the club is explored, with an emphasis on the operational implications to the reception department. The differences between weekday and weekend demand are discussed, with a specific emphasis on the effects of card room activities on the results. The section will be closed with the marketing and operational implications.

7.2 Data

Customer attendance in the club is recorded on the reception database every time a member enters the club. This process is required by law, and helps the club not only as a security measure but also as a tool for monitoring customer attendance.

7.3 Annual Demand

Over the last year, 18,000 members visited the club (all visitors must be member or guests of members). From them, around 10,000 visited only once or twice and an additional 6,000 visited the club up to 12 times (i.e. once a month on average), leaving around 2,000 'loyal' gamblers. These figures may be explained simply by the fact that the casino and gambling experience does not necessarily appeal to all people. In this respect, the people who visit the club only occasionally should not be regarded simply as 'lost' or 'dissatisfied' customers. Increasing customer loyalty may be achieved by using loyalty cards, which are used in many service operations, including the casino industry. Special loyalty cards, that offers benefits in return for appearance in the club, may help not only in bringing the customers to the club but, also, in bringing them at specific times which are characterised by a low level of demand (i.e. midweek).

7.4 Daily Demand

The daily attendance levels in the club deviate from 350 to 1350 per night over the year. There is a clear differentiation, as mentioned in the previous section [Figure 6-1], between the levels of attendance during the weekdays and weekends.

During the week the club accommodates 'professional gamblers', while over the weekend most of the attendees may be regarded as 'social gamblers' who spend less money on average (as well as a small minority who enter the club only to use the bar).

7.5 Hourly Demand

British Casino clubs are open to customers from 1:00 PM to 4:00 AM. During these hours, customers flow into the club at different rates. In the observed casino, during the afternoon, about 20 customers pass through reception, most of whom are registered members, and only need to present their membership card for a new computer controlled system without the need for signing in. Most of the customers arrive at the club after 7:00 PM. This is the time when the main restaurant starts to serve and, therefore, customers arrive for their booked dinners. Typically, the parties include a mixture of members and guests, requiring a 'signing in' procedure to take place, in which each guest enters his/her name, address and signature into a special book. Although this is not a very long activity, it does causes queues, especially during weekends, where up to 80 guests may enter the club within an hour.

Different customer groups have different patterns of attendance. The Chinese, who generate an extensive part of the casino's income, tend to arrive at the club after midnight (once the restaurant businesses, which many of them run, have closed). They tend to use the club during the week, especially on Mondays and Wednesdays for the Mah-jong competitions.

8 The transformation process

All operations produce goods or services or a mixture of the two, and they do this by a *process of transformation*. They use their resources to change the state or conditions of something to produce *outputs*. Put simply, operations take a set of input resources, use them either to transform something, or to be transformed themselves, into an output of goods and services. The inputs to the operations can be conveniently classified as either *transformed resources* (the resources that are treated, transformed or converted) or *transforming resources* (the resources that act upon the transformed resources).[52]

In the case of the gaming operation, the transformed resources are the customers and their money, and the transforming resources are the gaming staff and equipment.

The transformation process is closely connected with the nature of its transforming resources (i.e. materials, information or customer processing). Casino clubs, as customer-processing operations, are concerned with transforming the *psychological* state of their customers, and the *possession* of the money they bring with them. The gaming transformation process, as with most service operations, involves a high level of contact between the customers and the operation. Customers flow into the gaming cell, are processed, and flow out of the cell.

The gaming operation differs from other service operations in its objectives. The common services (such as banks, hospitals or public transportation) seek to process customers as quickly as possible, both for 'throughput' reasons and customer convenience reasons, whilst the casino's objective is to 'process' the customers as long as possible. Put simply, a dealer's goal should be to keep the customers on the table as long as possible, so they will spend more money. This, of course, should be done by as many cycles (game rounds) as possible. Dealers also need to balance their game with the customers' optimal pace; that is, quick enough to be profitable, but slow enough to keep customers entertained and not irritated (some of the slow players tend to prefer the inexperienced new dealers for this reason).

It is the objective of the service provider, the casino, that customers should leave the place with a smile on their face, although this is a mission that is virtually impossible when the client has lost more money than planned. The casino wants the customer to win sometimes and take money home because the overall long-term revenue from any specific customer will be positive; the statistics of the game are to the casino's benefit, and winning customers strongly tend to re-visit the club. The ideal scenario for the casino, is when customers win during the game, but stay and play till they lost everything.

9 Layout Design

9.1 Introduction

The layout of an operation is concerned with the physical location of its transforming resources: its facilities, equipment and staff. Before considering the various methods used in the detailed design of layouts, it is useful to consider the objectives of the activity. To a certain extent, the objectives will depend on the circumstances but there are some general objectives which are relevant to all operations, such as inherent safety, length of flow, clarity of flow, staff comfort, management co-ordination, accessibility, use of space and long term flexibility.[53] The layout is one of the factors used to control the profitability of a business, either by reducing the costs of the operation or by increasing the revenues generated by it, or both. In the case of the casino gaming floor, it is the layout of the cells that could reduce the costs, and their positions in the room that may increase revenues.

9.2 The Gaming Cells

The extended nature of the cells can best be described by examining the amount of the direct and indirect resources located within each cell. From the customers' point of view, the service cell is a gaming table. However, from operations' management point of view, it is the unit of operation with the optimal resource utilisation that should be considered as a cell.

Each table requires one dealer and a fraction of the supervision for its operation. The Great Britain Gaming Board recommends that the supervision of the games should be done by inspectors, who should inspect not more than two games at the same time. The Gaming Board also advises that inspectors should supervise only one type of game simultaneously. Thus, two tables, two dealers and one inspector constitute the minimum cell of operation for Roulette.

9.2.1 American Roulette Tables Cells

Three common cell types are available for the American Roulette tables: Side-by-Side [Figure 9-1], Back-to-Back [Figure 9-2] and the L-Shape type [Figure 9-3]. The different cell types vary in their use of space, staff and customer comfort and accessibility.

Figure 9-1: AR Table cell - Side by Side

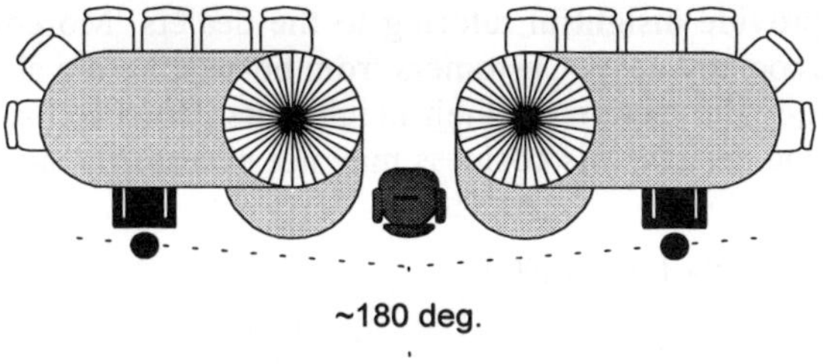

The Side-by-Side cell is the least comfortable for both staff and customers. The inspector needs to turn their head very often from table to table in an angle that is close to 180°, an act that is irritating and time consuming. In this setting it is impossible for the inspector to see both tables simultaneously, so while watching one game, he/she listens to the other[a]. The second major disadvantage of the Side-by-Side setting is the intimidating effect it has on customers. As the customers are sitting opposite to the inspectors, they may experience a strong sense of confrontation (i.e. 'me against the casino').

---

[a] From their experience, inspectors can identify mistakes in payout without looking at the bet, as there are some combinations, which are either rare or practically impossible.

Nevertheless, since many casino clubs operate in buildings that were not originally built for that purpose, that cell type is used in many of them, as a result of floor constraints (e.g. a narrow rectangular shape). In some cases, the angle between the tables is a little less than 180°.

Figure 9-2: AR Table cell - Back to Back

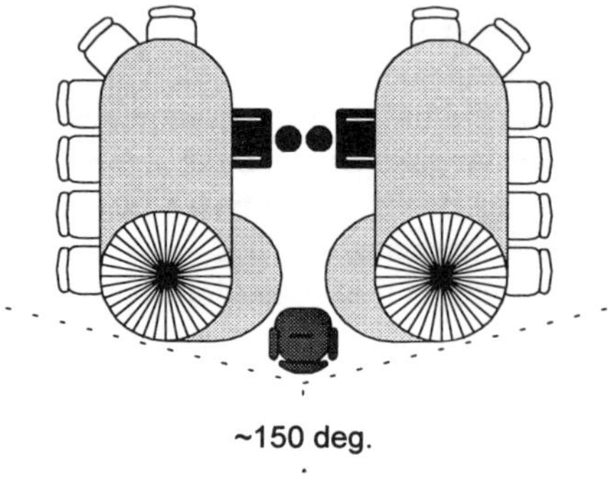

The Back-to-Back cell is more comfortable than the Side-to-Side cell for both staff and customers. This setting allows the turning angle for the inspector to be approximately 150°, having the dealers all the time in his sight. Watching the dealers constantly not only improves the level of inspection, but also the ability of the inspector to provide insightful tutoring to the dealers. Moreover, the Back-to-Back cell is more socially cosy. Customers from both tables are able to see and hear each other and enjoy the 'action' of their neighbours. However, as customers differ in their attitudes and desires, this cosiness may not necessarily suit all the punters in the club.

Figure 9-3: AR Table cell - The L-Shape (90°)

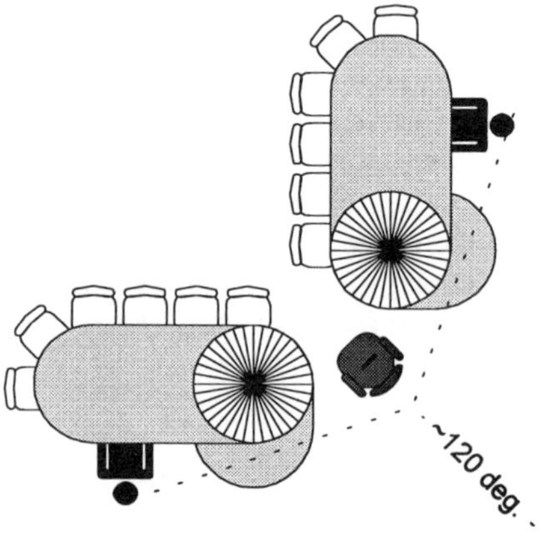

The third common American Roulette setting, is the L-Shape cell. This is the most comfortable for customers and staff. With the observing angle reduced to approximately 120°, inspectors are much more comfortable in their seats, are able to concentrate on their job, with the ability to spot exceptional events in the corner of their eye (or by ear). For customers, this setting provides a protected feeling and a sense of choice. As they approach the cell (most likely in a 45° angle), they can see both dealers and tables and choose to join the table which is more or less crowded (according to their personal preferences) or the dealer they perceive as more advantageous for them[b]. The choice factor is highest with this setting, as the tables are closer to each other. This setting is often used in room corners and in groups of 2 or 4 L-Shape cells.

9.2.2 Blackjack and Casino Stud Poker Tables Cells

The two common cell types available for BJ/CSP are Side-by-Side [Figure 9-1], and Back-to-Back [Figure 9-2]. These cells are mostly used for the same game (i.e. two BJ tables or two CSP tables).

Figure 9-1: BJ/CSP Table cell - Side by Side

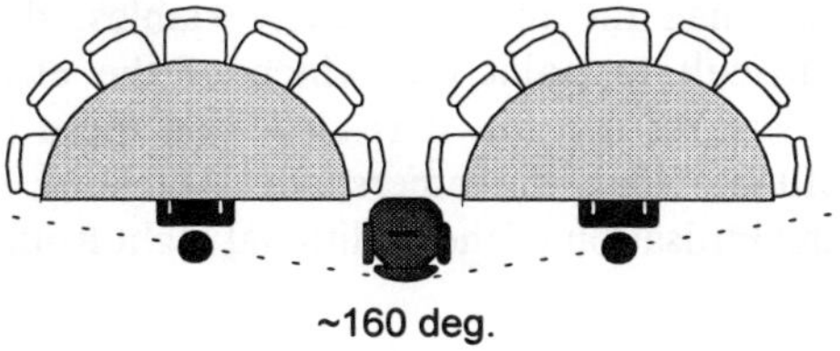

Figure 9-2: BJ/CSP Table cell - Back to Back

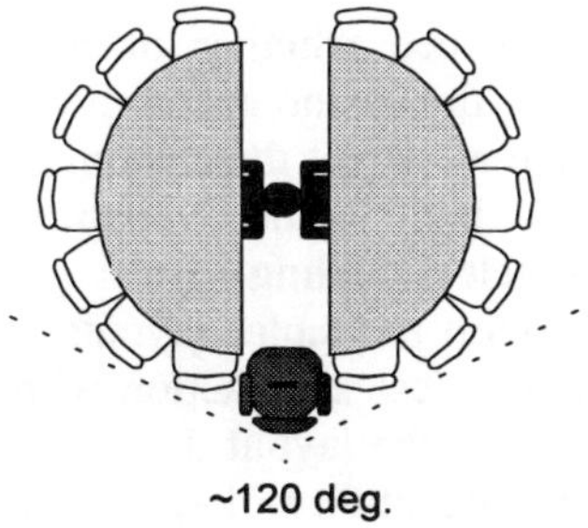

---

[b] Some gamblers consider particular dealers to be more 'lucky' for them than others.

As with the Roulette cells, the BJ/CSP Back-to-Back cell design is more practical and efficient than the Side-by-Side design. The only difference between the AR cells and the BJ/CSP cells, is the option of operating two different games in the same cell.

9.3 Table Positioning

9.3.1 The Stars

From a detailed analysis of the particular results for each table in the club observed, a very interesting picture emerges. Out of the 12 American Roulette tables, the core revenue generator, 4 tables account for 62% of the total drop (STDEV=7%). These four 'star' tables have several distinguishing characteristics. First, they are located in the centre of the gaming floor, in front of the stairs facing the cash desk. Second, they have an L-shape layout and are located opposite to the Dice and Punto Banco tables that are 'noisy' and attract interest from the club's visitors. Last, and probably most importantly, these are 25p chip tables, which attract more customers than the more expensive tables; this allows a generally longer game. It is difficult to point out which of the elements have the most influential effect on these results, but it is clear that these 'star' tables are much more financially profitable than the others are. Adjusting the layout or pricing policy may improve the utilisation of the additional eight Roulette tables.

9.3.2 Servicescapes

As referred to in the literature review above, many elements influence customer preferences and behaviour in the interior of the service place. The responses to the environment can in turn influence whether or not people stay and enjoy an environment and, in the case of the casino spend more money. The music, lighting, smell, colour, mood and image all affect the customers' attitude toward the service place. In a casino, these also include the availability of couches for relaxation during the game, and the timeless feeling resulting from the absence of clocks. Investigations into crowding with particular layouts have shown that under dense conditions, customers experience less satisfaction with the environment than when it is quite. Therefore, in respect to the layout design of the gaming floor, one of the key issues to reflect on is the availability of space around the tables for people to feel a less crowded atmosphere.

10 Conclusion and Recommendations

10.1 General

In order to achieve long-term profitability, it is suggested by Doyle (1994) to consider two main routes of action: (i) increase volume, or (ii) improve productivity.[54] It is useful to consider these as alternatives because they can conflict. A focus on volume requires cash and investment; a focus on productivity should generate cash. In the following section, suggestions are made, with the use of this theoretical model, with regards to improving casino performance.

10.2 Increase Volume

10.2.1 The principle of unstimulated demand

As mentioned earlier, some individuals can become addicted to gambling and it can cause others financial and other problems even if they are not addicted. A small minority of people gamble in a way, which compromises, disputes or damages family, personal or recreational pursuits and a significant proportion of these commit illegal acts to pursue their gambling. To a large degree, the extent and potentially adverse consequences of problem gambling have been controlled in the past in Britain by the restrictive regime applied to gambling generally, and to gaming in particular. Of special relevance to this has been the principle that facilities should be no more than adequate to meet the <u>unstimulated</u> demand for them and the controls on advertising and promotion, on access to gambling facilities and on gambling on credit.[55] Because of these restrictions, some very obvious marketing tools that can be used in many industries are not relevant to casino clubs. Expanding the market by <u>stimulating</u> demand is legally as well as ethically restricted. However, customers who have visited a club in the past, either as members or as guests, may be regarded as 'unstimulated demanders' of gaming. Therefore, by approaching them to 'remind' them of the availability of this service, the club may gain an increase in volume without breaking the law.

10.2.2 Word-of-mouth

Besides the direct mail tool, the most common and effective measure of communicating with customers (as suggested by an industry expert[56]), it is the word-of-mouth of the current customers that can expand the customer base. According to the marketing literature, only a small portion of dissatisfied customers complain to the management when they are not happy, but a large portion of dissatisfied customers tend to create negative publicity. Hence, encouraging current members and guests to promote the club, not only direct customers to tell their friends about the club, but also encourage them to feedback any discomfort to prompt management. This should reinforce the already existing listening attitude among managers.

10.2.3 Customer Training

Some of the customers do not understand all the games and stick to the easy American Roulette. Since the profitability of Blackjack and Casino Stud Poker is much higher than Roulette, it should be in the casino's interests to promote these games. This could be done by teaching the customers privately or in groups. Special training sessions could be conducted, by training officers, allowing the customer to play 'dry' until they understand the game and are ready to play in the club. The benefits here are twofold: the club should increase its profitability and customers should become more satisfied with the overall caring attitude of the club toward them.

10.3 Improve Productivity

10.3.1 Cost reduction

Most expenses in the casino are fixed costs. Staff salaries account for more than two thirds of the total expenses, and the use of part-time employees is very limited. In order to reduce costs it is suggested that a trade-off be made between fixed and variable costs. This will allow lower costs in months of lower demand (during the summer). Practically, it may be done by holding a group of 'on call' dealers that get paid per shift, and may be reached within hours. From conversations with employees in several clubs we learned that many 'young mums' consider this arrangement as beneficial, because they can spend time with their children during the day and get paid during the night).

10.3.2 Capacity Management

In terms of general demand management, a summer promotion may be considered as a way to reduce the annual seasonal decline. The changing level of demand needs to be met with the assistance of part-time employees and the adjustment of shift switching according to the sway in demand during the hours of the day, the days of the week and the months of the year.

10.3.3 Increase Prices

The option of increasing prices is relatively limited, as the customers in the casino do not pay a direct fixed price for the service. Moreover, they get a free service of soft drinks, coffee, tea and sandwiches on the gaming floor in order to keep them going. The possibility of charging them for this service is irrelevant, as this is a standard of the gaming industry. Nevertheless, the minimum stakes at the gaming table may be regarded as 'price'. This price has a high elasticity (i.e. percentage of volume <u>reduced</u> as a result of price <u>increased</u>), resulting in different levels of table profitability. High prices tend to lead to low volume, but it is the objective of a business that their multiples will be high.

10.3.4 Rationalise the product mix

The casino product portfolio was investigated in section 6, resulting in the definition of American Roulette as the 'cash cow' of the gaming operation, Blackjack and Casino Stud Poker as the 'nice but small' contributors, and Punto Banco and Dice are the 'problem children'. It was concluded, from this distinction, that the American Roulette, Blackjack and Casino Stud Poker operations may be seen as the core products of the gaming operation, with the Punto Banco and Dice acting as supplementary services. However, it is thought that the contribution of the last two games for the attractiveness and thereby success of the club is very important. Nevertheless, the product portfolio should be re-evaluated on a periodic basis and measures of cost savings (e.g. small dice tables, which are operated by one dealer instead of three) should be considered together with the contra-effect they would have on demand.

10.4 Long term Profitability - Crossing the chasm

As mentioned above, staff attitudes and commitment should contribute to a business's ability to innovate and provide a better service to its customers and to achieve higher long-term profitability. Dealers and inspectors should be involved in gaming activity decisions, reception personnel should be consulted with regards to changes in their systems, the chefs should have an influence on the menu, and the casino managers should have a say with respect to the overall club management. It is also recommended that club managers undergo professional management courses from time to time, so that they can be fully updated with recent industry developments as well as general management methods.

11 Summary

"We are what we repeatedly do. Excellence, then, is not an act, but a habit."

-- Aristotle

This study analysed the gaming operation in the casinos industry, and explored the different aspects of operation, including the demand conditions, transformation process, and layout design. Several suggestions have been made with regards to enhancing performance in the future.

It is argued that casino management should strive to provide a high quality service for their customers, and that the excellence should be their primary goal. Managing a gaming operation, it is concluded, is not a game after all.

12 Acknowledgements

I thank Marie Davies for her contribution to the literature review and Mike Shulver for his critical review and helpful insights on the subject.

Reference

1. W. Earl Sasser, R. Paul Olsen and D. Daryl Wyckoff (1978), *Management of Service Operations*, 8-10.
2. Levitt, Theodore (1980), "Marketing Intangible Products and Product Intangibles", *Harvard Business Review*, May-June, 94-102.
3. Rikard Larsson and David E. Bowen (1989), "Organisation and Customer: Managing Design and Coordination of Services", *Academy of Management Review*, Vol.**14** Iss.2, 213-233.
4. Colin G. Armistead (1985), "Design of Service Operations", *Operations Management in Service Industries and the Public Sector*, 51-64.
5. Schmenner R.W. (1986), "How Can Service Businesses Survive and Prosper?" *Sloan Management Review*, Vol.**27** Iss.3 Spring, 21-32.
6. Nicholas C. Georgantzas and Christian N. Madu (1994), "Service Designs for Quality: Integrated-Process Necessity", *The Mid-Atlantic Journal of Business*, Vol.**30** Iss.1 March, 55-80.
7. Bell, Chip R. and Zemke, Ron E. (1987), "Service Breakdown: The Road to Recovery", *Management Review*, Vol.**76** Iss.10 Oct, 32-35.
8. Charles S. Areni and David Kim (1994), "The influence of in-store lighting on consumers' examination of merchandise in a wine store", International Journal of Research in Marketing, Vol.**11**, 117-125.

9. Hutton, James D. and Richardson, Lynne D. (1995), "Role of the facility and physical environment on consumer attitudes, satisfaction, quality assessments, and behaviours", Health Care Management Review, Vol.**29** Iss.2 Spring, 48-61.
10. Bitner, M.J. (1990), "Evaluating service encounters: the effect of physical surroundings and employee responses", *Journal of Marketing*, Vol.**54** April, 69-82.
11. Carry, James (1996), "Winning Spaces", *World Gaming and Leisure Industry Review 1996/7*, Sterling Publications Ltd, 78-82.
12. Eroglu, Sevgin A. and Machleit, Karen, A. (1990), "An Empirical Study of Retail Crowding: Antecedents and Consequences", *Journal of Retailing*, Vol.**66** Iss.2 Summer, 201-221.
13. Weinert, Joe (1996), "Setting the Table", *International Gaming and Wagering Business*, Vol.**18**, No.8, 1997 (August).
14. Matthews, Virginia (1994), "If Muzak be the food of love", *Marketing Week*, Vol.**17** Iss.21 Aug 5, 24.
15. Yalch, Richard and Spangenberg, Eric (1990), "Effects of Store Music on Shopping Behavior", *Journal of Services Marketing*, Vol.**4** Iss.1 Winter, 31039.
16. Stipp, Horst (1990), "Musical Demographics", *American Demographics*, Vol.**12** Iss.8 Aug, 48-49.
17. Kellaris, James J. and Cox, Anthony, D. (1989), "The Effects of Background Music in Advertising: A Reassessment", *Journal of Consumer Research*, Vol.**16** Iss.1 Jun, 113-118.
18. Pitt, Leyland F. and Abratt, Russell (1988), "Music in Advertisements for Unmentionable Products - A Classical Conditioning Experiment", *International Journal of Advertising*, Vol.**7** Iss.2, 130-137.
19. Milliman, Ronald (1988), "The Sweet Sound of Profits", *Restaurant Management*, Vol. **2**, Iss.7 July, 53-54.
20. Ronald E. Milliman (1986), "The Influence of Background Music on the Behaviour of Restaurant Patrons", *Journal of Consumer Research*, Vol.**13** September, 286-289.
21. Herrington, J Duncan and Capella, Louis M. (1994), "Practical applications of music in service settings", *Journal of Services Marketing*, Vol.**8** Iss.3, 50-65.
22. Ronald E. Milliman (1986), ibid.
23. Dube, Laurette, Chebat, Jean-Charles and Morin, Sylvie (1995), "The effects of background music on consumers' desire to affiliate in buyer-seller interactions", *Psychology and Marketing*, Vol.**12** Iss.4 July, 305-319.
24. Kellaris, James J. and Mantel, Susan Powell (1996), "Shaping time perceptions with background music: The effect of congruity and arousal on estimates of ad duration", *Psychology and Marketing*, Vol.**13** Iss.5 Aug, 501-515.
25. Israel, Chip (1996), "Don't Gamble When it Comes to Proper Lighting", *World Gaming and Leisure Industry Review 1996/7*, Sterling Publications Ltd, 78-82.

26. LaFollette, Shirley (1996), "Interior Design: a global language", *World Gaming and Leisure Industry Review 1996/7*, Sterling Publications Ltd, 78-82.
27. Gallegos, Patrick (1996), "Casino Magic at Biloxi: the evolution of lighting", *World Gaming and Leisure Industry Review 1996/7*, Sterling Publications Ltd, 78-82.
28. Bidlake, Suzanne (1992), "Scents of Real Purpose", *Marketing*, Oct 15, 21.
29. Eric R. Spangenberg, Ayn E. Crowley, & Pamela W. Henderson (1996), "Improving the Store Environment: Do Olfactory Cues Affect Evaluations and Behaviours?", *Journal of Marketing*, Vol.**60** April, 67-80.
30. Miller, Cyndee (1993), "Scent as a marketing tool: Retailers - And even a casino - Seek sweet smell of success", *Marketing News*, Vol.**27** Iss.2 Jan 18, 1-2.
31. Woudhuysen, James (1994), "Add local colour and understand your consumers", *Marketing*, June 23, 7.
32. Joan Meyers-Levy and Laura A. Peracchio (1995), "Understanding the Effects of Color: How the Correspondence between Available and Required Resources Affects Attitudes", *Journal of Consumer Research*, Vol.22 September, 121-125.
33. Bellizzi, Joseph A and Hote, Robert E. (1992), "Environmental Colour, Consumer Feelings, and Purchase Likelihood," *Psychology and Marketing*, Vol.**9** Iss.5 Sep/Oct, 347-363.
34. Anonymous (1997), "Colours for the new millennium", *Modern Paint & Coatings*, Vol.**87** Iss.1 Jan, 90.
35. Triplett, Tim (1996), "How consumers pick colours", *Marketing News*, Vol.**30** Iss.20 Sep 23, 20-21.
36. Dudley, Sid C. (1987), "Subliminal Advertising: What is the Controversy About*?", Akron Business & Economic Review*, Vol.**18** Iss.2 Summer, 6-18.
37. Beatty, Sharon E. and Hawkins, Del I. (1989), "Subliminal Stimulation: Some New Data and Interpretation", *Journal of Advertising*, Vol.**18** Iss.3, 4-8.
38. Schultz, Roberta J. (1996), "The effects of mood states on service contract strategies", *Journal of Professional Services Marketing*, Vol.**14**, 117-135.
39. Ronald Paul Hill and James C. Ward (1989), "Mood Manipulation in Marketing Research: An Examination of Potential Confounding Effects", *Journal of Marketing Research*, Vol.**26** Feb, 97-104.
40. Kordelia Spies, Friedrich Hesse and Kerstin Loesch (1997), "Store atmosphere, mood and purchasing behaviour", *International Journal of Research in Marketing*, Vol.**14**, 1-17.
41. Amirani, Shahrzad and Gates, Roger (1993), "An attribute-anchored conjoint approach to measuring store image", *International Journal of Retail & Distribution Management*, Vol.**21** Iss.5, 30-39.
42. Dogar, Rana (1997), "Marketing to the senses", *Working Woman*, Vol.**22** Iss.4 April, 32-35.

43. James C. Ward, Mary Jo Bitner and John Barnes (1992), "Measuring the Prototypicality and Meaning of Retail Environments", *Journal of Retailing*, Vol.**68** Iss.2 Summer, 194-220.
44. Barry J Babin and William R. Darden (1995), "Consumer Self-Regulation in a Retail Environment", *Journal of Retailing*, Vol.**71** Iss.1, 47-70.
45. Mary Jo Bitner (1990), "Evaluating Service Encounters The Effects of Physical Surroundings and Employee Responses", *Journal of Marketing*, Vol.**54** April, 69-92.
46. Ruth A. Schmidt and Roger Sapsford (1995), "Issues of gender and servicescape: marketing UK public houses to women", *International Journal of Retail & Distribution Management*, Vol.**23** Iss.3, 34-40.
47. Kellaris, James J. and Rice, Ronald C. (1993), "The influence of tempo, loudness, and gender of listener on response to music", *Psychology & Marketing*, Vol.**10** Iss.1 Jan/Feb, 15-29.
48. Dhruv Grewal and Julie Baker (1994), "Do retail store environmental factors affect consumers' price acceptability? An empirical examination", *International Journal of Research in Marketing*, Vol.**11**, 107-115.
49. Robert J. Donovan, John R. Rossiter, Gilian Marcoolyn and Andrew Nesdale (1994), "Store Atmosphere and Purchasing Behavior", *Journal of Retailing*, Vol.**70** Iss.33, 283-294.
50. Barry J. Babin and William R. Darden (1995), "Consumer Self-Regulation in a Retail Environment", *Journal of Retailing*, Vol.**71** Iss.1, 47-70.
51. Bitner, Mary Jo (1992), "Servicescapes: The Impact of Physical Surroundings on Customers and Employees", *Journal of Marketing*, Vol. 56 (April 1992), 57-71.
52. Slack et al, (1995) *Operation Management*, Pitman Publishing, London. 11-19.
53. Slack et al, (1995) ibid. 237-259
54. Doyle P. (1994), Marketing management and strategy, Prentice Hall, 117.
55. Report of the Gaming Board, p.**8**.
56. Romero John, "How casinos use direct marketing", Direct marketing, Vol.**59**, No. 10, 2/97, 54-55.

HOW AN OIL GIANT LEARNS: A CASE STUDY OF BRITISH PETROLEUM

MR. JIMMY C. HUANG
Warwick Business School, University of Warwick, Coventry, England, CV4 7AL
E-mail: phd96ch@wbs.warwick.ac.uk

DR. ELENA ANTONACOPOULOU
Warwick Business School, University of Warwick, Coventry, England, CV4 7AL
E-mail: irobea@wbs.warwick.ac.uk

A growing interest in learning and knowledge management is reflected in the current management literature. However, in contrast to individual and organisational learning, learning at the team level is an underdeveloped area of investigation. On the other hand, the separation of both organisational learning and knowledge management research has led to an incomplete understanding of the relationship between learning and knowledge. Companies are increasingly adopting virtual teamwork in their work design, but there is little research to illustrate the pitfalls such arrangements can cause. This paper attempts to investigate both learning and knowledge management in the context of the virtual setting, and seeks to reveal the interrelationships between these three areas.

1 Introduction

A growing interest in learning and knowledge management is reflected in the current management literature. However, in contrast to individual and organisational learning, learning at the team level is an underdeveloped area of investigation. There can be no doubt, however, that team learning is important. Senge (1990), for example, argues that team learning is one of the five key elements involved in building a learning organisation. In addition, McBain and Kusy (1997) indicate that team development is a powerful way of implementing learning, as well as an essential method for improving organisational effectiveness. Therefore, this paper seeks to show that it is essential to take the concept of team learning into account when examining learning issues.

Companies are increasingly adopting virtual team arrangements designed to achieve various benefits: cost cutting (Hamblen 1998), flexible working (Suttling and Wood 1997), increased productivity (Hamblen 1998; Solomon 1998) and the effective collaboration of global knowledge workers on a variety of tasks (Warkentin, Sayeed and Hightower 1997). This trend has created a unique arena for investigation. What is particularly obvious at present is that the advantages of virtual arrangements are strongly promoted by the advocators, but their pitfalls and impacts are not yet very well understood.

If we seek to incorporate the virtual dimension into discussions on learning and knowledge management, there are currently obvious gaps in the literature. In particular, two key questions need to be addressed: firstly, precisely how does an individual or a team learn within a virtual environment? Secondly, how do various levels of learning within a virtual organisation construct the general process of knowledge management? The purpose of this paper is to begin to fill these gaps in the literature by extending the discussion of knowledge management and learning to the virtual dimension.

Huber (1991) distinguishes among four organisational learning processes: knowledge acquisition, information distribution, information interpretation and organisational memory. This paper adopts his theoretical model as a basis for investigating how learning processes take place within a virtual organisation, especially in the area of team learning. Various levels of learning, including individual, team and organisational learning, are examined, and a theoretical model of team and organisational learning processes is generated. We then explore the dynamic interrelationships between these learning processes. Finally, we illustrate how various levels of learning can facilitate knowledge management within a virtual organisation.

The remainder of the paper is divided into four sections. Section One provides a review of the current literature and some gaps within this body of work are identified. These gaps form the research questions that this paper seeks to investigate. Section Two describes how a case study in British Petroleum Knowledge Management Centre was conducted to investigate issues in relation to learning processes and knowledge management within a virtual environment. Section Three presents an analysis of the empirical findings to help the respond to the gaps illustrated above. Conclusions are offered in Section Four.

2 Knowledge and Learning within Organisations: Current Perspectives and Debates

It is first of all essential to clarify the concepts of knowledge and learning. The two concepts are clearly related, but their precise meanings differ and have been interpreted in various ways. The following literature review will concentrate on knowledge and then turn to the subject of learning.

2.1 The vitality of knowledge

Bell (1973) points out that knowledge is a central feature of post-industrial societies. Team organisation has become of crucial importance, and employees generally need to be managed as knowledge workers (Zuboff 1988). Concepts such as knowledge society (Drucker 1968, 1993; Toffler 1990), intelligent enterprise knowledge (Quinn 1993), intangible resources (Hall 1992), knowledge assets, knowledge work (Fisher and Fisher 1998); knowledge-intensive firm (Starbuck 1992) and intellectual capital (Allday 1998; Brandley 1996; Brooking 1996; Edvinsson and Sullivan 1996; Probert 1996; Roos, et al. 1997; Steward 1997) all highlight the fact that knowledge is not a fancy term or merely a managerial fad, but is one of today's most vital assets (Davenport and Prusak 1998; Sveiby 1997) and resources (Kamoche 1996; Prahalad and Hamel 1990).

2.2 Knowledge: the definition

The current management literature investigates not only the value of knowledge to organisations, but also attempts to create more comprehensive theories to explain knowledge. Debates about what knowledge is can be traced back to the ancient Greeks. Few thousand years later, no consensus has been achieved, but more thoughts on the matter have emerged. We adopt the idea of Sanchez and Heene (1997): 4, who define knowledge as "the set of beliefs held by an individual about causal relationships among phenomena". However, we would like to expand this definition to incorporate the knowledge of a team or even an organisation, and to focus on the shared understanding and beliefs held by individuals, teams and organisations. For the purposes of our analysis, knowledge is defined as 'the set of shared beliefs held by individuals, teams and the organisation as a whole about casual relationships among phenomena for making effective decisions'.

2.3 Different types of knowledge

Based on the previous debate about the nature of knowledge, as well as the theoretical approaches taken, many different typologies of knowledge have been proposed in the literature. These include: tacit and explicit knowledge (Polanyi 1958; Nonaka 1994; and Nonaka and Takeuchi 1995); embrained, embodied, embedded, encoded and encultured knowledge (Blackler 1995); know-how, know-what and know-why (Sanchez 1997); dictionary, directory, recipe and axiomatic knowledge (Sackmann 1991); symbol-type and encultured knowledge (Collins 1993) and procedural and declarative knowledge (Cohen and Sproull 1996; Kogut and Zander 1992, 1993 and 1996). Although the typologies vary, there is a high degree of overlap among these concepts. This paper draws on the distinction between tacit and explicit dimensions of knowledge, not simply because this has been widely accepted, but also because, in their discussions, Nonaka and Takeuchi (1995) have paid particular attention to dynamic interactions between these two dimensions. By contrast, other theories have been less successful in explaining the interrelationships between different types of knowledge. Reasons to understand interactions and interrelationships between tacit and explicit knowledge are the dynamic nature of knowledge can be best described through their interactions, and the process of interaction determines the formats of knowledge as well as its locations. This point is illustrated in the following paragraph.

2.4 The locations and formats of knowledge

The locations and formats of knowledge vary enormously. It could be located in experience (March, Sproull and Tamuz 1991; Levitt and March 1988), routines (Cyert and March 1996; March and Simon 1958; Nelson and Winter 1982), organisational memory (Stein and Zwass 1995; Tuomi 1996;Walsh 1995; Walsh and Ungson 1991), culture (Cook and Yanow 1996; Sackmann 1991; Schein 1992, 1993 and 1996), individuals (Kolb and Fry 1975) and organisations (Sparrow 1998). All the formats of knowledge listed above can be explained and are created through the interactions between tacit and explicit knowledge, and various formats of knowledge may contain both types of knowledge. For example, experiences that an organisation's members possess may be recorded in documents as explicit knowledge, and they may also exist within individual members' brains as tacit knowledge. This paper seeks to investigate both types of knowledge from three perspectives. First of all, we examine tacit and explicit knowledge located in experiences and routines. Secondly, we revisit the concept of knowledge creation through the interaction of both tacit and explicit knowledge in relation to learning processes. Finally, we attempt to explain how both types of knowledge may be managed.

2.5 Knowledge management

As illustrated above in the discussion of the vitality, types and locations of knowledge, there is no shortage of opinions and theories about knowledge. Currently there is still no agreed framework of analysis, however, the varying practices of knowledge management in companies have demonstrated the need for more systematic research to enhance our understanding of these complex issues. The field of knowledge management brings together a range of issues, including: knowledge creation (Giunipero 1997; Nonaka and Takeuchi 1995; Vicari, von

Krogh and Roos 1996), knowledge transfer (Howells 1996; Richter and Vettel 1995, Wathne, Roos and von Krogh 1996), knowledge sharing (Martinez 1998; Trussler 1998) and knowledge and competence building (Boisot, Griffiths and Moles 1997; Klavans and Deeds 1997; Sanchez and Heene 1997; von Krogh and Roos 1996). Unfortunately, there are few theoretical developments which illustrate how knowledge can be managed. Some of the works listed above share a common assumption in terms of the way that knowledge can be captured. They assume that knowledge is something solid which can be 'easily' located in different sections of the organisation. We run the risk of not acknowledging that knowledge is dispersed (Tsoukas 1994). Our attempt to overcome the pitfalls of previous research will involve exploring individual accounts of how knowledge may be conceptualised, created, shared and selected by members of teams working in the same organisation. Another assumption the knowledge management literature generally has made is that knowledge is something already existing within the organisation. There is little explanation of how knowledge is articulated in the first place, and by what approach knowledge is accumulated.

After our discussion of the concept of knowledge, we now have a brief understanding of the vitality of knowledge; different types of knowledge; where it is located and by what formats. We also know something of the application of knowledge to other business functions. It is critically essential to understand how knowledge is acquired and developed in the first place. Undoubtedly, the main approach, if it is not the only approach, is through learning. In the next section, we discuss some of the current literature on the concept of learning.

2.6 The concept of learning

Fiol and Lyles (1985): 811 define learning as "the development of insights, knowledge, and associations between past actions, the effectiveness of those actions, and future action". Instead of concentrating on the idea of a permanent change of behaviour, as Wilson and Rosenfeld (1990) suggest, this paper adopts the viewpoint of Fiol and Lyles, which sees learning as a process of development and growth (Antonacopoulou 1995). As similar definition can be found in Kolb (1984): 38 who suggests that "learning is the process whereby knowledge is created through the transformation of experience". These two definitions underpin the essential role of knowledge within the learning process. Learning can be located at three different levels - for example, individual learning, team learning and organisational learning (Cangelosi and Dill 1965; Kasl, Marsick and Dechant 1997; Senge 1990). In terms of the content of learning, a few distinctions may be illustrated: single-loop and double-loop learning (Argyris 1982; Argyris and Schon 1974); adaptive and generative learning (Senge 1990); reactive and proactive learning (Miles and Randolph 1980); foundation, formation, continuation, transformation and transfiguration learning (Jones and Hendry 1992); interactive and informational learning (Gnyawali, Stewart and Grant 1997) and exploration and exploitation learning (Boisot, Griffiths and Moles 1997; March 1991). What these typologies have very different theoretical foci, as well as different emphases in terms of what organisations ought to learn, there are some degrees of similarity in their assumptions.

The first three typologies are based on the assumption that there is only one type of learning which produces positive outcomes while the other type inhibits effective learning. In this case, only double-loop learning, generative learning and proactive learning are valuable to organisations. Although Jones and Hendry (1992) do not indicate that only transfiguration is good for the organisation and others inhibit effective learning, they argue that only through transfiguration can a learning organisation be built. These four typologies all support a more radical approach to learning.

The last two typologies share the same assumption that different situations need different types of learning, and demonstrate that both radical and incremental learning are valuable to organisations. This research agrees with the later view and considers both approaches to be beneficial to our understanding about the learning concept. Further, understanding both approaches may provide a more holistic view when one investigates learning. Similar opinions about the holistic approach can be found. For instance, Dodgson (1993) argues the essence of the integrative approach to understand learning concept. This perspective also reflects Miner and Mezias' (1996) argument about the importance of investigating learning in a convergent manner.

At individual level, learning has been primarily viewed from the psychological perspective primarily, and various schools of thought can be found in the literature which reflect a number of assumptions about how people learn (see Antonacopoulou 1997 for a review of the relevant literature). It is beyond the scope of this analysis to provide an overview of learning theories. However, individual learning within organisations has so far been focused on training activities and the way in which individuals learn on the job (e.g. Burgoyne and Hodgson 1983)

In terms of organisational learning, significant research issues include the following: the creation of learning organisations (Garvin 1993; Jones and Hendry 1992; Stein and Pinchot 1995), the building and continuous improvement of learning enabling capabilities (Barrie and Pace 1997; Beer and Eisenstat 1996; McLaughlin and Kaluzny 1997; Parkinson 1995; Upton and Kim 1998), learning and the formulation of strategies (Barnett and Hansen 1996; Chatterji 1996; Epstein and Roy 1997; Kim 1996; Sinkula, Baker and Noordewider 1997) and the relationship between learning and culture (Antal, Dierkes and Hahner 1997; Bartlett and Ghoshal 1998; Hodgson 1998; Keys, Wekks and Denton 1998; Schein 1996). Learning approaches vary accordingly, since organisations can learn in many different ways. For example, they learn from experiences (Barkema, Bell and

Pennings 1996; Burgoyne, Brown, Hindle and Mumford 1997; Luxhoj, Riis and Stensballe 1996; Verganti 1997), mistakes (Makridakis 1997), benchmarking (Liedtka, Haskins, Rosenblum and Weber 1997) and system thinking (McKenna 1992; Senge 1990). Furthermore, the work of Marsick and Watkins (1994) on informal and incidental learning has provided new avenues about the ways in which learning can be understood within organisations. Miner and Mezias (1996) termed inter-organisational learning as learning within populations of organisations. This may be accomplished through strategic alliances (Doz 1996; Khanna, Gulati and Nohria 1998; Lei, Slocum Jr. And Pitts 1997; Medcof 1997; Simonin 1997) or networks (Powell, Koput and Smith-Doerr 1996). More recently, researchers have begun to tackle the difficulties of measuring organisational learning, and a number of tools have been developed. Examples are the Organisational Learning Survey (Goh and Richards 1997), the Learning Organisation Assessment Framework (Gephart, Marsick, Van Buren and Spiro 1996) and Organisational Learning Capacity Measurement (Hult and Ferrell 1997).

Compared with the amount of attention given to individual and organisational learning, the concept of team learning remains underdeveloped. Senge (1990) argues that team learning is one of the five major disciplines involved in the formation of a learning organisation. Sashkin and Franklin (1993) illustrate three approaches to team learning: crisis problem solving, learning for the improvement of current activities and anticipatory learning. The various factors which enable and inhibit team learning are depicted by Purser, Pasmore and Tenkasi (1992) and Meyers and Wilemon (1989). Brooks (1994) explains the relationship between team learning, power among team members and the difficulties which teams are facing. McBain and Kusy (1997) indicate that team development is a powerful way of implementing learning, as well as an essential method for improving organisational effectiveness. McKenna (1995) argues that team learning requires a change in attitudes as well as a change in assumptions and paradigms. All these works either assume that team learning is merely a smaller scope of organisational learning, or consider team learning to be no more than the sum of a group of individual learning. Some fundamental issues, such as how a team learns differently to a group of individuals, nature of the processes in relation to team learning and the interrelationships between different levels of learning, are not fully discussed in the current literature. Furthermore, a large amount of research into teamwork has explored the importance of teamwork to the organisation and its wide acceptance by both practitioners and academia. Few works, however, provide insights into team learning to explain how a team interacts with the knowledge possessed by team members as well as by the team as a whole.

The underdevelopment of the team learning concept is an indication of the incompleteness which characterises the current organisational learning literature. This is why this paper seeks to make the case for adding a team learning perspective to the research agenda. In addition, this work also attempts to explore the impact of virtual settings on various levels of learning, as well as to investigate how virtual settings affect the general processes of knowledge management. At present, the complex relationships between knowledge management and learning are not clear. This paper attempts to respond to this gap by providing evidence of the relationship between both fields.

3 Research Method

The issues raised in the previous section form the focus of a research project which is currently in progress. The aim of the study is to explore the concept of team learning in general. More specifically, this study firstly seeks to examine the concept of learning at the individual, team and organisational levels, as well as the interrelationships between each level of learning. Secondly, it attempts to generate a learning process model which will be able to depict the processes underlying team and organisational learning, and the dynamic interrelationships between each process. Thirdly, this study endeavors to illustrate the role of knowledge in relation to the concept of learning. Finally, the research strives to examine the effect of virtual arrangements on both learning and knowledge management.
The methods employed in the research design are based on two major considerations:

- the nature of the research gaps which concentrate on the exploration of the concepts of team learning, knowledge management and virtual arrangements, as well as their interrelationships, and
- data needed to generate answers to existing gaps are the insights from learners and their perspectives on the three areas for investigation.

This type of inquiry, according to Van Maanen (1983), and Morgan and Smircich (1980), is qualitative research. Yin (1993): 9 argues that " a "how" and "why" question is being asked about a contemporary set of events over which the investigator has little or no control", and the most appropriate strategy is case study. On the other hand, other considerations, including the validity and reliability of data, are taken into account. After deciding to apply the case study method, another critical decision involved the sampling procedure of the case company. Criteria included not only the implementation of virtual teamwork across geographic boundaries, but also the awareness and efforts devoted to both organisational learning and knowledge management programmes. Of course, the major consideration was the case company's willingness to collaborate with the research. British Petroleum (BP) fitted all the criteria and also demonstrated its willingness to invite researchers to investigate these issues.

The paper is thus based on a five-month case study of BP. The main data collection methods used were interviews and documentation. Interviewees are members of the BP Knowledge Management Team based at various sites in England, Scotland and the USA. The interviews were mainly face-to-face, except in the case of one member in the USA who was interviewed through a video-conferencing system. Findings from the data analysis stage are presented after an introduction to the case company.

3.1 Introduction to the British Petroleum Company plc.

The British Petroleum Company plc. is one of Britain's biggest companies and one of the world's largest oil and petrochemicals groups. BP's main businesses include BP Exploration, BP Oil refining, BP Oil marketing, BP Chemicals, BP Solar, BP Marine and Air BP. BP has operations in over 70 countries. Its key strengths are in oil and gas exploration and production; the refining, marketing and supply of petroleum products; and the manufacturing and marketing of chemicals.
A corporate-wide change programme, "Project 1990", launched in March 1990 by the Group's chair, Mr. Robert H. Horton, aimed to change BP's organisational structure and culture as well as people's visions and values concerning BP. A company-wide survey was carried out in October 1989 to understand employees' perceptions about BP, and the following is a summary of the main results:

BP is secure and healthy but overcomplex and bureaucratic. Staff are loyal and proud to work for the company but feel that it cares too little for them. (Project 1990 Report: 2 from BP Archival Centre)

According to the report, two key targets of the project, based on the outcome of the company-wide survey, were to change BP from a complex bureaucracy with a high degree of uncertainty into a simple, teamwork and trust organisation; and to integrate vision and value into day-to-day jobs.
This change project became a landmark in BP's history and provided a blueprint for major changes in the company's organisation and way of working to improve efficiency and flexibility. This general project also led to a collaborative culture which enabled the introduction of projects such as: the Virtual Teamwork Project, the Organisational Learning Project and the Knowledge Management Project.
The focus of the Virtual Teamwork Project was to increase the level of communication internally and externally through the use of video-conferencing technology. This project aimed to minimise the time and geographical constraints arising from the company's global operation, and enabled the connection of people inside and outside BP through more convenient communication channels. It provided a valuable platform for later projects such as Knowledge Management, by

establishing a sophisticated communication infrastructure, and promoting virtual collaboration. Greater detail about each project will be given below. Their influence on learning and knowledge management will be examined in the following discussion.

3.2 The structure of learning and knowledge management in BP

The main goal of knowledge management in BP is, according to one of the project consultants, "to re-use learnt knowledge so that business process may be optimised and that the correct decisions are made. The knowledge management vision is 'BP knows what is knows, learns what is needs to learn, and uses that knowledge for overwhelming competitive advantage'." According to most of the interviewees, "both tacit and explicit knowledge offer key leverage for supporting decision-making and improving business operations. Therefore, learning from past experiences and transferring the learning outcome into knowledge assets for future reuse are the keys to knowledge management in BP." This is why knowledge management tools were introduced to BP in the first place. Various knowledge management tools, which are essential mechanisms for learning, were developed mainly for teams and various groups to articulate what knowledge they have learnt. The concepts of these tools, as well as the techniques by which to use them have spread. More essentially, these tools are coached by Knowledge Management Centre, described by three interviewees: "they can ensure that people not only aware these tools, but also have the experience to operated them." These tools operate and are practised at many different events. This research investigated the three main knowledge management tools which are commonly used in BP. Due to the requirements of confidentiality, we label these tools A, B and C.

According to the company's brochure, tool A should be used immediately after each identifiable event. The rationale for this view is that, at this time, one's memory is fresh and unvarnished, participants are still available and learning can be applied straight away. An event must have a beginning and an end, and can be a complete small action or, alternatively, one part of a large action.

The time to use tool B is when a business unit is facing a challenge and where the experience and insight of others might be valuable. It is also appropriate to use tool B when the diversity of views external to the business unit can broaden the range of options under consideration.

Tool C is used upon completion of a piece of work. Meetings are organised in order to capture the lessons learnt during the project. The benefits of tool C are that the valuable lessons are identified, there is enhanced openness and co-operation in the team and closure at the end of the project is achieved.

One of the essential concepts of knowledge management, according to most interviewees, is "to share what they know from experiences with others." Another important factor to knowledge management, indicated by most interviewees, is an "open organisational culture." Explained in more detail by these interviewees, one of the most essential elements of an open organisational culture is that "different opinions are tolerated." Other critical elements, illustrated by most interviewees, include "trust to share and trust to accept other opinions" and "the willingness to share and to help other people's problems".

4 Findings

In this section, we discuss how learning and knowledge management takes place in BP. In addition, the relationship between different levels of learning and knowledge management is elaborated. A team and organisational learning process model is formed according to the findings from this case study. Problems and issues in relation to various levels of learning and knowledge management are illustrated. Before presenting the research findings on learning and knowledge management, it is essential to distinguish data, information and knowledge.

4.1 Data, information and knowledge

Distinguishing between data, information and knowledge is always difficult. A commonly shared story in BP provides an interesting interpretation of how they achieve this differentiation. A few interviewees say that "using the example of travelling, a clock or a watch provides the time, and time is treated as data. A bus timetable is considered as information, and knowledge is that the bus on Monday morning is always ten minutes late." Both data and information are essential for planning and decision-making; knowledge can provide insights which may not be achieved only by data and information, and allows correct decision-making (e.g., when to leave the house and walk to the bus stop).

4.2 Different levels of learning in BP

The BP learning model, according to some interviewees, contains three stages: learning before doing, learning during doing and learning after doing. The model applies to learning at the individual, team and organisational levels. It is organised according to the consequences of time for learning, and it represents the core structure of the learning concept in BP. Learning tools are designed to serve different learning purposes as well as to target various users on specific occasions. Some project consultants indicate that "most business activities require massive capital investment, and that trial and error is extremely costly". They therefore need to plan carefully before taking actions and learn from each action in order to feed information and knowledge back to the next project. An example provided by one of the project consultants is as follows: "In the Gulf of Mexico, they put a certain valve into the sea bed. A valve goes down to the bottom of the sea, and it does not cost just £100 to put it right. In Scotland, they use the same valve. If the lesson learnt from Mexico can pass to Scotland, they can avoid the same problems". A more detailed explanation and application of this three-stage learning model will be discussed below as part of the analysis of the BP team and the organisational level learning processes.

4.3 Individual learning

Responses from interviewees show that individuals learn from "education, training courses, operation manuals, past working experience and all sort of sources". The purpose or motivation for individuals to learn is mainly to ease problem-solving and to support decision-making. The learning sources described above provide useful background knowledge for individuals to handle their basic day-to-day work. However, most project consultants indicate that "the knowledge learnt from training courses, operation manuals and so on may not be sufficient to handle some unique problems or to make highly complex decisions".

Learning from others, according to some interviewees, "either through informal personal networks or by seeking advice via the intranet," is perceived to be the most efficient learning source. Two interviewees state that "individuals may hesitate to ask for help from others, because it will be considered as a weakness". Some people may therefore prefer to tackle their problems alone, or try to solve problems before anyone is aware of them. This not only implies that time has been wasted by different people trying to solve the same problems, but that delays in the progress of the project result from the failure to share knowledge. In order to overcome this barrier, people are encouraged to speak to others and to learn from others through various communication channels. Moreover, people are advised to investigate who has faced a similar problem before. Instead of devoting time to solving problems alone, people are changing their habits by starting to ask for help and advice.

In BP, both tacit and explicit knowledge are vital resources to the company's business activities and people have access to both types. One interviewee explains the relationship between these two types of knowledge:

It is like learning how to cook. You may learn it from a cookery book and make lots of mistakes. Alternatively, you can go to an evening class, by which you are taught how to do it properly. So you need both tacit and explicit knowledge. That is why we provide one-to-one coaching, as well as reading backups. So when they go away, they have got something to read and can remember what is what.

Most interviewees assert that "tacit knowledge is powerful", but on the other hand, difficulties in terms of articulating and transferring tacit knowledge are inevitable. The wide availability of video-conferencing in BP allows the transfer of tacit knowledge at a distance; often from one country to another. Furthermore, BP Connect, the nature of which will be explained later, helps people to find the person who has faced the same problem before. People are equipped with various facilities to learn from others or people can learn from face-to-face interaction. Information technology (IT) facilities in BP include e-mail, groupware, phone, video-conferencing and so on.

Inevitably, an individual's ability to learn from others relies on the personal willingness of the other person at the end of the communication channel. In addition, cultural boundaries between business units might restrict the individual's ability to learn from other people in different units. Feedback from individuals shows that they are relatively satisfied with the help they receive from others. Most interviewees agree that "there are more than enough people who would like to help, and they have very good experiences in terms of learning from others".

Delegating authority to individuals has motivated individuals to learn continuously due to their work responsibility. Individuals are not only able to learn from others before doing a project, but also while they are doing a project. After accomplishing the project, individuals discuss what they have learned, make the learnt lessons conscious and explicit, and record it for the benefit of others. Learning from others has enabled individual learners to solve their problems more efficiently by integrating various sources of knowledge.

4.4 Team learning

Huang, Newell and Antonacopoulou (1998):1 define team learning as "the dynamic and collective process of acquiring or creating knowledge through formal and informal interaction to build up a shared understanding and to achieve the common purpose and goal for the team". The main goal of team learning is to create a higher level of shared knowledge than that which can be achieved by individuals. Senge (1990) argues that the value of team learning is indeed greater than the sum of individuals' learning. Empirical data gathered from BP confirms the validity of the concept of team learning and underlines the value of team learning as advocated by Senge.

As indicated in "Project 1990", teamwork has become a prevalent working arrangement in BP's various business units. Learning from each project in the team context to benefit other similar projects has become an essential concern in BP. As knowledge management tools A, B and C are introduced, people are trained to become familiar with them. These tools concentrate on learning during and after doing a project. On the other hand, these tools also help teams to learn before doing projects by gaining feedback from other project teams. An example from one of the interviewees demonstrates how teams learn before, during and after doing:

Building petrol stations is a regular business activity in BP. There are different types of stations, but some petrol stations are standard. Therefore, lessons learnt from building the first petrol station are not only beneficial in terms of knowing what was done well, but also help to avoid the any mistakes when building the second one. After building ten petrol stations, they can do it better, faster and cheaper.

Experiences have provided a useful source of learning materials for individuals, teams and the organisation. Knowledge management tools are operated within a team after a day of work or the accomplishment of the whole project. This provides an opportunity to reflect on what people have learnt from the project from both an individual as well as a team perspective, and to share their learning with other individuals and teams. The on-going practice of using these tools has enabled not only individuals within the team but also the team as a whole to accumulate knowledge. Furthermore, a well-defined learning structure and the creation of a base for shared understanding also enforce the articulation of team knowledge.

No individual has a complete knowledge of a certain project, and without any one of the team members, the team knowledge must be regarded as incomplete. Kasl, Marsick and Dechant (1997) use the metaphor of a mosaic to describe team knowledge. This metaphor captures the dynamics, as well as the complexity, of team knowledge. The concept of team learning depicts not only its importance within the organisation, but also the essential role it plays in bridging individual and organisational learning. The articulation and accumulation of team knowledge indicates a higher level of knowledge than the sum of individual knowledge within the team. Reflecting on the argument above, no team has a complete knowledge of the organisation, but without one of the teams, the organisational knowledge must become incomplete. Team learning forms a critical level, which cannot be neglected, in terms of the understanding of organisational learning. Team knowledge represents a vital component in relation to the articulation and accumulation of organisational knowledge. Findings on organisational learning will be analysed in the following section.

4.5 Organisational learning

Extending the earlier discussion on team learning, this part of the paper expands the scope of learning and knowledge to the organisational level. After the accomplishment of certain cross-business unit projects, tool C is used. Representatives from business units gather together to discuss what they have learnt and what knowledge from the project can be articulated for subsequent use. What makes it more significant is that the learning activities have overcome the boundaries between teams and even business units. The creation and accumulation of knowledge at the organisational level indicates that the diminishing of boundaries between groups can facilitate organisational learning. Organisational learning in BP may not represent other perspectives of learning advocated in the literature, but it indicates very well how an organisation can learn from what it has done and from its past experiences and apply this learning to its future actions.

4.6 The interrelationships between various levels of learning

Without any doubt, the importance of individual and team learning contributed to organisational learning cannot be ignored. It is clear that team and organisational levels of learning cannot only be explained through each individual learning within the team or the organisation. Nor is an understanding of team learning sufficient to understand how an organisation learns as a whole. There are some degrees of difference between each level of learning, and these differences can possibly be identified through the knowledge that individuals, teams or the organisation articulate and accumulate. In BP each level of learning has some degree of overlap. The various levels also represent different concepts and features, however. This explains why a group of individual learning is not the same as genuine team learning. At the same time, a cluster of team learning within an organisation will not necessarily qualify as organisational learning. Hence, it is essential to clarify these three types of learning. The learning literature primarily concentrates on individual and organisational learning. Therefore, it is vital to explore the concept of team learning in order to have a complete understanding of organisational learning.

Organisational learning processes, according to Huber (1991), are composed of four phases: knowledge acquisition, information distribution, information interpretation and organisational memory. Although Huber's learning processes concept derived from an extensive review of literature, there are no empirical data to sustain his argument. Therefore, this research seeks to investigate the process of organisational learning through cross-examining previous works by providing empirical evidence. Moreover, this research not only examines the learning processes within an organisation, but also the learning processes taking place within a team. The unclear differentiation between knowledge and information in Huber's model has been avoided. (This differentiation has been explained in the previous discussion above). This research also seeks to identify the interrelationships between learning processes within the contexts of teams and organisations.

Findings from the empirical research suggest that team and organisational learning processes in BP can be generalised into a five-stage model (see Figure I). More detailed explanations are as follows:

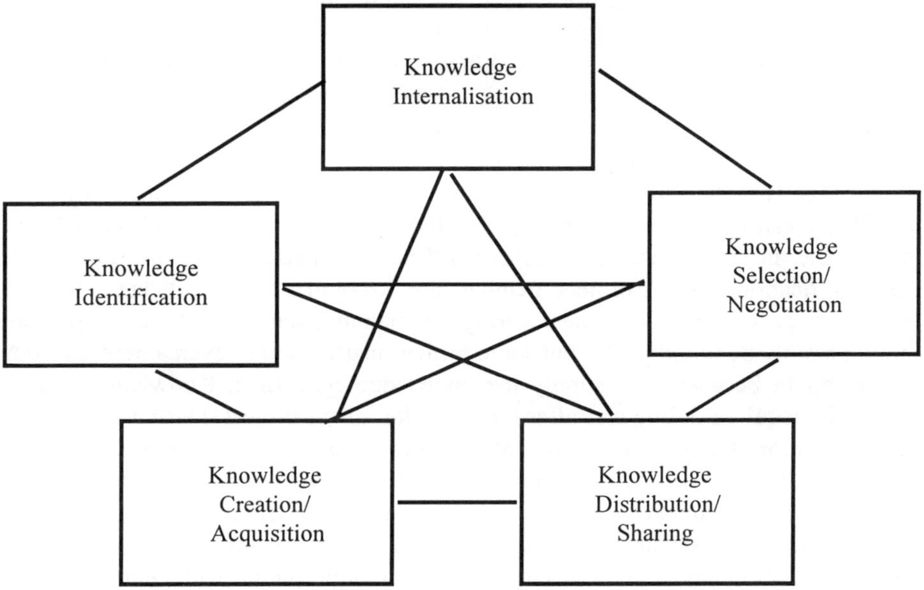

Figure I. This is a model depicting team and organisational learning processes.

Knowledge identification - This is the process of identifying the meaning of knowledge, what type of knowledge is required and the location of knowledge. To identify knowledge is vital for a team or an organisation, but this point is absent from Huber's work. The knowledge identification process plays an essential role in the forming of goals (Hackman 1990; Liebowitz and De Meuse 1982), as a socialisation process within the team or the organisation (Nonaka and Takeuchi 1995) and as an initial stage for a project (Wheelwright and Clark 1992).

By using tool B before doing a project, knowledge from previous projects can be drawn on. Targets for knowledge articulation during or after doing the project can be set up before doing the project. A pre-project meeting with people having relevant knowledge may contribute much to project planning. In the context of learning during doing, knowledge identification helps members to detect whether different kinds of knowledge are further required during the progress of the project. In addition, knowledge identification also helps to ensure that knowledge gained during the operational processes has followed the set target. After completing the project, knowledge identification facilitates the understanding of what knowledge has actually been learnt from the project, who has the knowledge and the difference between the target and the actual outcome.

Knowledge creation/acquisition - This is a process to determine whether knowledge can be acquired internally or externally, or whether no such knowledge is available, in which case the company will have to create it in-house. Following the identification process, project members are aware of what knowledge is needed in order to accomplish the project during the planning stage. Required knowledge might possibly be found in BP, but some further justification or even a new research project might be needed to supply the knowledge concerned. Knowledge is also created or acquired during and after the operation of the project. During the project processes, more knowledge is articulated by groups of learners, even though some knowledge is also created by reusing old knowledge. Problems which occur during the project also stimulate the input of knowledge either from the project team or from outsiders. This directly enhances the creation of knowledge. After the completion of the project, a team level or an organisation-wide knowledge is created through well-planned meetings and the application of the tools. During this stage, the outcome of knowledge creation is different from the outcomes of the 'learning before doing' and 'learning during doing' stages. It is vital to differentiate between the stages in order to know what knowledge is created by each of them.

Interactions between learners enable the conversion of both tacit and explicit knowledge. Four types of conversion, socialisation, externalisation, combination and internalisation, as suggested by Nonaka and Takeuchi (1995), can be found during the interaction processes, but knowledge creation in BP provides a very different picture from the spiral process depicted by the two authors. Four types of knowledge conversion interact in a more dynamic manner than that suggested by the image of a spiral. During the socialisation stage, knowledge is internalised and externalised spontaneously by individual learners. Moreover, during the combination process, individuals are both externalising part of their tacit knowledge and internalising tacit and explicit knowledge. This dynamic interaction indicates that knowledge is not created only through certain arrangements of knowledge conversion or in a spiral-like process. The team and organisational levels of knowledge creating processes in BP present a more dynamic picture (see Figure II).

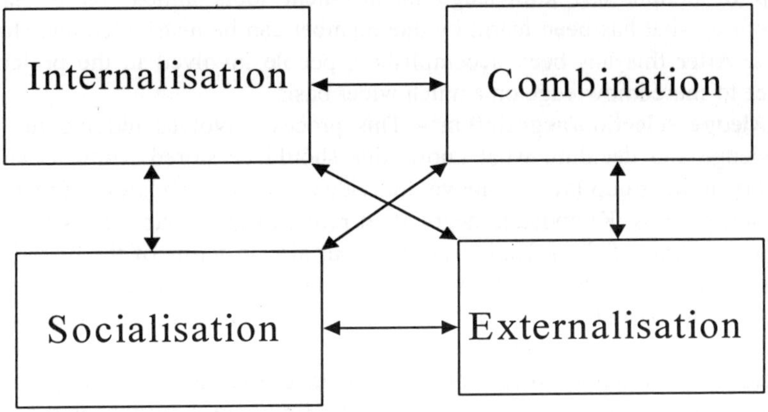

Figure II. This figure indicates the dynamic nature of knowledge creation in BP.

This helps to expand the possibilities for knowledge creation. The occurrence of knowledge creation is not only determined by the conversion of knowledge, but also depends on the interaction with other learning processes, namely knowledge identification, knowledge sharing/ distribution, knowledge selection/ negotiation and knowledge internalisation.

Knowledge sharing/ distribution - This is a process which follows when the required knowledge has been made available. Distribution and sharing have rather different meanings. Knowledge is distributed for wider use, whereas knowledge sharing refers to the common understanding among participants, as well as the shared interpretation by the individual, the team and the organisation.

The definition of interpretation provided by Daft and Weick (1984) only emphasises the giving of meaning to information and knowledge, not the development of a shared understanding within a team or organisational context. This paper uses the idea of knowledge sharing instead of Huber's 'information interpretation' to address the processes of both meaning-giving and sharing of understanding.

Interviewees indicate that in BP knowledge is distributed by e-mail, groupware and hard copy. Knowledge is shared when a group of people engage in the use of knowledge management tools. Before doing a project, knowledge is made available throughout the group by various methods. This occurs not only through the provision of reading material or documents, but also by the provision of training courses and coaching to ensure that people know exactly what to do and that tacit knowledge can be transferred. During the evolution or progress of the project, a group of people use knowledge management tools almost on a daily basis. Therefore, what has been learnt by one member can be distributed and shared with others. After this has been accomplished, people involved in the project have a chance to share knowledge on a much wider basis.

Knowledge selection/negotiation - This process involves judging the value of knowledge and deciding what knowledge should be stored within organisational memory at the group level or above. Individuals also absorb knowledge through the selection process. Knowledge negotiation refers to the group discussion activities, not only as part of the selecting process, but also in terms of the negotiation of a socially acceptable and commonly shared meaning of knowledge. Conflicting views may form the basis for exposing misunderstanding, different interpretations and the variety of assumptions held by individual members. It may seem arbitrary to address the negotiating process only at this stage because four other processes also contain negotiation activities. What is emphasised here, however, is the activity of the decision-making process under group negotiation. In the discussion below on the complex dynamics between learning processes, a clearer view on how the negotiation process has a simultaneous effect on other processes will be provided.

BP's three-stage learning model (learning before doing, learning during doing and learning after doing) can be applied to this process. Knowledge is selected before doing, which implies planning. During the planning stage, the knowledge selecting process is carried out by teams or across business units' discussion, as well as by negotiating the commonly accepted meaning of knowledge. Selection during doing a project refers to the choosing process to determine what knowledge will be beneficial so that more attention can be given to it. During this stage, individual members constantly reflect and exchange what they have learnt and negotiate an agreed interpretation to carry on the project. After doing a project, knowledge selection plays the role of 'gatekeeper' for organisational memory. Again, at this stage, individual members discuss what they have learnt within the team, the business unit or even outside their own business unit. They also negotiate and determine what knowledge needs to be codified in various forms such as documents or a database.

Knowledge internalisation - This is a process which explains knowledge as being located in various places. The term internalisation refers to absorption by individuals during the learning process. Nonaka and Takeuchi (1995): 96 state that "internalisation is a process of embodying explicit knowledge into tacit knowledge". Knowledge is learnt by individuals and internalised by individuals; yet knowledge could also be embedded within a group of people. From the empirical data, we find that this concept of internalisation can be expanded to an even broader organisational level. As Walsh and Ungson (1991) state, knowledge is stored in organisational memory in various forms such as documents, database and business processes. As soon as knowledge is internalised, it becomes the source of knowledge identification, provides criteria to judge knowledge acquisition or creation, serves as a basis for wider distribution, and becomes the object for negotiation and selection. According to the distinction suggested by Polanyi (1958), tacit knowledge is internalised by people, and explicit knowledge is recorded in various forms. It must be stressed that the team and organisation levels of knowledge contain both the tacit and explicit dimensions, and it is very easy to miss the wider picture if attention is only given to only one of the two dimensions.

4.7 Dynamic interactions between learning processes

After discussing the detail of each element of the learning processes, we now want to explain the interrelationship between each learning process. As shown in Figure I, each process simultaneously interacts with four other processes. In fact, each process has an impact on all the other processes. For instance, knowledge identification occurs not only during the beginning of learning processes, but also during the processes of creation/acquisition, sharing/distribution, selection/ negotiation and knowledge internalisation. In addition, during the process of knowledge creation or acquisition, this process also affects the other four processes. This can be explained as the phenomenon of "nonlinear dynamics" (Waldrop 1992). The story of building petrol stations provides evidence to explain the dynamic nature of learning within BP. During the process of building a petrol station, members may face some problems which they have not encountered before. In order to solve these problems, they need to acquire new knowledge or even create new devices to overcome these problems. At this stage, they not only what knowledge is needed to solve these problems, but also where they can possibly find this piece of knowledge. After this knowledge has been acquired or created, they distribute it to members involved within this project. Members not only exchange their opinions and negotiate a commonly acceptable interpretation about the knowledge, but also continuously identify the appropriateness of this knowledge along with the progress of the project. Even at the stage of knowledge internalisation, members still have to identify what knowledge will be useful for the next project. By the same token, during the process of knowledge creation/acquisition, members have to share what has been created or acquired in order to carry on with the project. Members also have to select and negotiate the usefulness of the newly created/acquired knowledge and to internalise the knowledge for applying this knowledge to their project.

4.8 Knowledge management

The importance of knowledge management to BP, according to the leader of the BP knowledge management team, is as follows:

Knowledge management has fuelled our transformation. We found out it was better to learn than be dead. Re-using knowledge was the fuel for innovation. We are great at putting knowledge into action, but we are not great at learning from others. We have the know-how; we just have to try it.

Project 1990 stressed the importance of cultural change. This project also emphasised the appreciation of information flows and information *per se*, as well as information sharing. The current focus in BP, according to an interview with BP chair Mr. John Browne in Harvard Business Review, has shifted from the appreciation of information flows to that of knowledge flows, and from information sharing to knowledge sharing.

The goal of a knowledge management project in BP, according to one of the interviewees, is "to re-use learnt knowledge so that business process may be optimised and that the correct decisions are made". In other words, it is to improve that the way the company works by learning from past experiences, and to do things faster, cheaper, safer and better. A high degree of repeated business activities, such as drilling wells, building petrol stations and so on, enables BP to manage their knowledge through structured learning. On-going learning activities at various levels not only enhance each individual's understanding, but also help the articulation of knowledge from day-to-day operations at the team and organisational levels.

Knowledge created during business processes exists in both tacit and explicit forms. To transfer all tacit knowledge employees possessed into explicit knowledge and codify it totally into organisational memory is virtually impossible. As most interviewees stress, "BP does not believe in 100 per cent IT solutions like other companies do". Therefore, they started from the perspective of people. Reasons to focus on people as the first step of implementing the Knowledge Management Project are not only because the knowledge people possess, but also because knowledge transferring is possible only when people want to share it with others. This point also reflects on the argument made by Tsoukas (1994) that knowledge is dispersed within the organisation. On the other hand, it also challenges the assumption commonly accepted by the knowledge management literature that knowledge is something solid and can be easily transferred. Through connecting people by various channels, they are able to ensure that the increase of knowledge sharing facilitates problem solving in day-to-day work as well as supporting decision-making.

Two interviewees indicate that "approaches to knowledge management can be

described on a continuum, the focus of people is located on one side and the focus of information technology on the other". Although the Knowledge Management Project starts from the 'perspective of people', it does not mean that the other side of the continuum is neglected. By providing advanced information technology and the integration of other communication channels, BP have overcome, though not yet completely solved, some problems which are caused by the global operation. One of the interviewees comments that "BP does not intend to move toward the side of information technology, but to find a balance between these two".
Like many other companies, BP has its own set of best practices which are made available for users throughout the company. The success stories, recorded in codified form or spread during both formal and informal interaction, have become one of the important assets for management. Some interviewees indicated that "knowledge learnt from failures is sometimes more important than knowledge learnt from successful experiences". This is not only because knowledge learnt from failures can provide insights in terms of difficulties and problems for a project. It is also about personal preferences in choosing what knowledge people would like to learn. The explanation provided by one of the interviewees is that "people listen to failures and challenge the best practice". Another comment from one of the interviewees can demonstrate this point.

In BP people believe that they can do even better in terms of solving the same problem compared to the success stories, and they are willing to spend an awful lot of time just to find a new way to demonstrate their abilities.

Another important factor to explain this phenomenon relates to BP's organisational culture. Some interviewees noted that "BP's culture encourages people to explore new ideas". In order to make sure that efforts from employees are not spent on "reinventing the wheel" (this term is used by most interviewees), but on some breakthrough thinking, the Knowledge Management Project also helps people to know what BP already knows.

4.9 IT Networked Organisation/Virtual Organisation

The components of the new BP organisation following "Project 1990" consist of about 90 business units based at the same level. The most significant impact on the organisational structure is the disappearance of middle management. This project has changed the previously hierarchical organisation into a flat structure. Layers of the reporting system have been reduced. To empower each business unit with its own authority to cope with its environment is also one of the purposes of this change project. Business units are able to determine their goals and targets in relation to their local knowledge. Another important project in the development of a common IT infrastructure in BP, according to one of the interviewees, is called "COE- Common Operating Environment". The COE project not only initiates the establishment of IT networking, but also promotes the idea and the use of IT networking. Business units are linked through the IT network and the linkage also extends to customers and suppliers. In addition, the virtual teamworking technology has provided BP with a rather unique way to integrate its global business processes across various geographic regions and time zones.

The benefits which result from the adoption of a virtual arrangement, as identified by various authors, include business efficiency (Grenier and Metes 1995), the revitalisation of the corporation (Davidow and Malone 1992), cost saving (Crainer 1996) and flexible working places and hours (Barnatt 1995). One of the interviewees also states that "the company's aim was to get the best human resources from different part of the organisation". Despite all of the promising benefits that virtual working can provide, there are a few problematic issues which also need to be taken into account. For instance, the frequency of face-to-face interaction has been limited due to the time difference and geographic distance which separate people. This has had a costly impact on the development of interpersonal relationships. One of the project consultants indicates that "Group Chair Mr. John Browne states that quality conversation not only comes from formal business talks, but also from informal chatting during coffee break in the corridor. In order to encourage informal conversation, as well as to overcome the physical barriers caused by the dispersion of staff, the idea of a "virtual corridor" was carried out by the establishment of BP Connect". By having "Connect" functioning as a virtual corridor, people are able to get to know each other and share knowledge through this channel. Since there are 90 business units in BP, as one interviewee say, "it is very difficult to have an overall view about what other people are doing and to know what knowledge they have".

The main purpose of the common IT infrastructure is to provide a more convenient and comprehensive communication channel for everyone in BP. For people in different time zones and in the same place at different times, intranet can be used. For people in different places at the same time, video-conferencing facilities can be used. For people in the same place and at the same time, a face-to-face meeting is the most favourable method. The result is that BP's new IT project has built a company-wide platform for people to communicate more easily.

Various opinions on the implementation of virtual arrangements, from individual member's perspectives, are illustrated by interviewees. Most of the interviewees agree that the virtual arrangement provides a more flexible way to arrange their working time and they are happy with this type of arrangement. Two interviewees emphasise that "it is great to have the freedom to choose where and when to work". Another interviewee describes that "I do not need to come to my office at night to have a meeting with Australia like before, I can just do it at home". The way to choose working at home or in the office, as to most interviewees explain, "depends on the tasks needed to be done". One of the interviewees illustrated: "for example writing a report is best to be alone, therefore, it is better to work at home, but if something requires the discussion with colleagues, it will be better in the office". One important factor pointed out by most interviewees is that " after staying at home for a long period, the productivity drops dramatically", and the reason is that they "feel isolated". On the other hand, different members have different working arrangements, and it is harder to find a time when all the members are in the same office. Social activities such as "away days" have been introduced. Members gather together somewhere outside the office and their homes to share what they have learnt. All of the interviewees stress the importance of the "away days" as well as other activities by which all the members can be together. Despite certain advantages virtual arrangement provided, it is also vital to balance this with social perspective of work. As one of the interviewee comments, "work for me is not only about money, but also the social activities work can provide". After discussing the virtual arrangement of individual's work, it is crucial to know how the impact of virtual arrangement on the issues of learning and knowledge management.

4.10 Knowledge management and learning in a virtual environment

Knowledge is located in various locations and forms. One focus of BP is to increase the sharing of tacit knowledge. As discussed earlier, tacit knowledge is mainly stored in our brains. The purpose of knowledge management is not to capture all the tacit knowledge and store it in codified forms. Rather, it is to let people learn from each other, and what BP provides is a directory by which people can find other people in different parts of BP and what knowledge they possess. This system in BP is called "Connect" (see earlier). It is like the Yellow Pages directory, containing a huge number of staff homepages including names, contact numbers, expertise areas and so on. When people are facing problems before, during or after doing a project, they can find others who have faced the same problems. Various communication channels enable one party to learn or get advice from another. The integration of various databases with "Connect" makes both tacit and explicit knowledge more accessible to everyone in BP. Individuals input what they have learnt into their homepages, and they are responsible for keeping those homepages constantly updated.

User feedback about "Connect" contains both positive and negative comments. Some argue that it is "easy to use" and "convenient and tied with business". On the other hand, others complain that sometimes they "can not find the person or thing they need", and that "new expertise areas should be added". One of the project consultants states that "these complaints will disappear as Connect becomes populated. It currently contains about a quarter of the IT-enabled population in BP". Inevitably, physical distance and time difference create difficulties in terms of learning at the individual, team and organisational levels. From our previous discussion of the nature of both tacit and explicit knowledge, and according to Nonaka and Takeuchi (1995), it is vital for individuals to learn tacit knowledge through observation. The difficulty in terms of articulating tacit knowledge is obvious, and it is clear that tacit knowledge cannot be simply transferred through e-mail or phone calls. The system "Connect" also helps individuals to find the people who may have the knowledge they can learn from. Moreover, the availability of virtual teamworking technology has facilitated individual's learning from others through observation. Since individual learners can identify the location of knowledge, further actions, such as travelling to learn or arranging training courses on how to solve certain problems, can always be arranged.

The learning process model on team and organisational learning, illustrated in Figure I, contains five processes: knowledge identification, knowledge creation/acquisition, knowledge sharing/distribution, knowledge selection/ negotiation and knowledge internalisation. The establishment of Connect and the availability of IT have facilitated the identifying processes both for teams and the organisation, and the geographic distance created by virtual arrangement can also be reduced. In terms of knowledge creation/acquisition, this process may cause more difficulties compared to other processes. The difficulty lies in the nature of tacit knowledge as well as the interaction of tacit knowledge with both tacit and explicit knowledge. Virtual teamworking technology, according to most interviewees, has "overcome most of the problems caused by the geographic distance". On the other hand, as two interviewees indicate, "for people at the different time in the same places or at different time and in different places", the process remains difficult. For knowledge sharing/distribution, approaches such as e-mail, telephone and the intranet are used, as is the practice of knowledge management tools. These tools combine with other communication channels, not only to ensure that knowledge is distributed to involved individuals as well as teams or other groups cross various geographic regions, but also to facilitate the achieving of shared meaning and understanding of certain knowledge. In terms of knowledge selection/negotiation, the dispersion of individuals and teams may mean that considerably more time is taken before consensus is reached than would be the case for individuals and teams located in the same place. On the other hand, the virtual arrangement may provide certain advantages by gathering more opinions from people in different regions to involve them in the selecting/negotiating process. The process of knowledge internalisation concentrates on embodying explicit knowledge into tacit knowledge. The impact of virtual arrangement on learners as individuals in the knowledge internalisation process may not be significant. At least, there is no direct evidence, but from team and organisational perspective the effect is obvious. Since team or organisational knowledge is internalised, the dispersion of members has caused some difficulties in terms of tracing the knowledge, especially when members change from one team to another or leave the job.

4.11 Difficulties Behind Learning and Knowledge Management in BP

It cannot be denied that there are a number of difficulties involved in BP's approach to learning and knowledge management. The main problems, which this research has identified, are as follows:

- Willingness - Most interviewees indicate that learning and knowledge management will not work if no one is willing to contribute what he or she knows in order to share it with others. Knowledge is still considered as power in BP, as in many other organisations. Despite the fact that BP's flat organisational structure facilitates knowledge sharing, the possession of knowledge is still a key advantage in fighting for resources. Furthermore, a few interviewees stated that "each new project is always perceived as something extra after work". Updating homepages, recording lessons learnt for sharing, solving other people's problems and learning something new are only some of the extra efforts which individuals need to devote their time to in order to support the company-wide learning and knowledge management programme. One interviewee illustrated this point by saying that "people in BP can appreciate the value of learning and knowledge management and agree that these are good things to do, but it does not necessarily mean that they will do it". People are motivated by rewards and mutually beneficial incentives, which means that someone might help another with a problem if he or she can help them at the same time. To promote the benefits of learning and knowledge management, as well as to advertise the concept of knowledge ownership, BP have tried to encourage people to take the first step. Only when people feel the direct benefits, however, are they are more willing to dedicate their efforts in order to gain greater benefits in return.

- The measurement of learning and knowledge management outcomes - Undoubtedly, both learning and knowledge management contribute significantly to operational efficiency and to the improvement of performance. However, both are difficult to quantify and measure. BP, like many other companies, requires a rigid process for planning and outcome appraisal and appraises performance in quantified figures. It is difficult and virtually impossible to isolate contributing factors and to determine what proportionate contributions are made by learning and knowledge management to a project. As one interviewee says, "it is hard to know how much a nut contributes to a washing machine." Alternatively, as another interviewee say, "if knowledge and learning is intangible, there is no need to measure it at all".

- Culture, sub-culture and trust - "Project 1990" has made a significant impact on BP's organisational culture. Interviewees state that "BP is more open than before, but culture is still a big issue in BP". National cultural differences and sub-cultures inherited from the operational backgrounds have led to very different perceptions about knowledge. This also creates major difficulties in terms of developing trust. As one interviewee put it, "learning how to appreciate different cultures and turning the differences into useful resources is the key lesson for BP to learn in order to overcome the cultural barrier". Trust is hard to develop and maintain, and it is even harder in a virtual organisation (Handy 1995). Both learning and knowledge management require trust to let knowledge flow more smoothly through the organisation and across organisational boundaries.

- The dilemma- Early discussion of the concept of knowledge management in BP indicates that knowledge is created according to experiences and learnt knowledge. Knowledge is accumulated time after time, and is reused. When asking interviewees about the relationship between knowledge management and innovation, most interviewees responded that "knowledge management enables innovation through managing knowledge-creating activities". Dramatically different to most of the responses, one interviewee states that "knowledge management in BP might kill innovation, and it is always hard to figure out when to use old knowledge and when to invent new things". The profound understanding of knowledge has provided a valuable piece of advice: that knowledge might be helpful for an organisation, but, on the other hand, it may also become an obstacle to innovation. Most literature on knowledge management emphasises the potential benefits which knowledge can provide, and gives knowledge highest credibility without considering its potential negative effects. Therefore, an awareness of both the positive and negative values of knowledge should provide a first step in applying knowledge to business processes. This point also reflects the essence of 'unlearning', as well as the dilemma organisations may face when to apply existing knowledge and when to discard it. As Hedberg (1982): 3 argues, "understanding involves both learning new knowledge and discarding obsolete and misleading knowledge, and the discarding activity – unlearning - is as important a part of understanding as is adding new knowledge".

5 Conclusion

This paper explored the concept of learning processes at the team level, and examined its relationships with individual and organisational learning in the context of virtual settings. A learning process model was generated and the dynamic interrelationships between each learning process revealed. Evidence gathered from a case study of BP suggested that learning at each level has its own character and each level of learning is crucial to the articulation and accumulation of organisational knowledge. The analysis also showed that a clear understanding of the team learning concept is essential and a holistic approach to examining the learning concept is desperately needed.

This paper also attempted to bridge the concept of learning and knowledge management. Since most learning literature has assumed that knowledge is just something existing within the learning processes, and little research demonstrates the role of knowledge in relation to learning. On the other hand, the knowledge management literature commonly takes knowledge as something already existing within organisations without illustrating the essential role of learning for articulating knowledge. This analysis not only illustrated the close link between these two concepts, but also the important need to cross-examine both concepts in order to understand each field.

The importance of both learning and knowledge management concept is notable. It is also vital to examine both concepts within the context of virtual settings. Organisations are gradually 'going virtual', but existing research has only promoted the advantages that the virtual setting can deliver without carefully examining its impact. For an organisation to learn and to manage its knowledge is difficult. It will be even harder for it to do that within a virtual setting. Our understanding of these areas is still limited and more effort to integratively examine these three areas is surely needed.

6 Acknowledgements

We thank Dr. Nick Milton for his critical review of previous version of the paper, and other members of Knowledge Management Team in BP for the time and efforts devoted to this research.

References

1. Adler, Niclas and Peter Docherty. 1998. "Bringing Business into Sociotechnical Theory and Practice." *Human Relations* 51(3):319-45.
2. Allday, Deborah. 1998. *Spinning Straw into Gold- Managing Intellectual Capital Effectively*. London: The Institute of Management.

3. Anderson, Robert and Karolyn Snyder. 1989. "Team Training." *Training and Development Journal* 43(2):58-61.
4. Antal, Ariane, Meinolf Dierkes, and Kartin Hahner. 1997. "Business Perception of Contextual Changes." *Business and Society* 36(4):387-407.
5. Antonacopoulou, Elena. "Mathophobia and Philomathia: A New Perspectives and Its Implications for Management Education." *The 1995 Annual Meeting of the American Academy of Management* (Vancouver) Vancouver.
6. Argyris, Chris. 1982. *Reasoning, Learning, and Action, Individual and Organisational*. London: Jossey-Bass.
7. Argyris, Chris and Donald Schon. 1978. *Organisational Learning, a Theory of Action Perspective*. London: Assison-Wesley.
8. Barkema, Harry, John Bell, and Johannes Pennings. 1996. "Foreign Entry, Cultural Barriers, and Learning." *Strategic Management Journal* 17(2):151-66.
9. Barnett, William and Morten Hansen. 1996. "The Red Queen in Organizational Evolution." *Strategic Management Journal* 17:139-57.
10. Barrie, John and Wayne Pace. 1997. "Competence, Efficiency, and Organizational Learning." *Human Resource Development Quarterly* 8(4):335-42.
11. Bartlett, Christopher and Sumantra Ghoshal. 1998. "Beyond Strategic Planning to Organization Learning: Lifeblood of the Individualized Corporation." *Strategy & Leadership* 26(1):34-39.
12. Beer, Michael and Russell Eisentat. 1996. "Developing an Organization Capable of Implementing Strategy and Learning." *Human Relations* 49(5):597-619.
13. Bell, D. 1973. *The Coming of Post-industrial Society: A Venture in Social Forecasting.* New York: Basic Books.
14. Blacker, Frank. 1995. "Knowledge, Knowledge Work and Organisations: An Overview and Interpretation." *Organisation Studies* 16(6):1021-46.
15. Boisot, M., D. Griffiths, and V. Moles. 1997. "The Dilemma of Competence: Differentiation Versus Integration in the Pursuit of Learning." *Strategic Learning and Knowledge Management*, editors R. Sanchez and A. Heene. Chichester: John Wiley & Sons .
16. Bolk, H, P. van Elswijk, T. Melis, and A. van Praag. 1997. "Only People Learn: Organisations and Systems Can Adapt." *Computers in Industry* 33(2,3):209-16.
17. Bradley, Keith. 26Jul. 1996. "The Value of Intellectual Capital." *Financial Times* :11.
18. Brooking, Annie. 1996. *Intellectual Capital*. London: International Thomson Business Press.
19. Brooks, Ann. 1994. "Power and the Production of Knowledge: Collective Team Learning in Work Organizations." *Human Resource Development Quarterly* 5(3):213-35.

20. Burgoyne, J., D. Brown, A. Hindle, and M. Mumford. 1997. "A Multi-Disciplinary Identification of Issues Associated With "Contracting" in Market-Oriented Health Service Reforms." *British Journal of Management* 8(1):39-49.
21. Burgoyne, J., V.E. Hodgson. 1983. "National Learning and Managerial Action: A Phenomenological Study in the Field Setting." Journal of Management Studies 20(3): 387-399.
22. Cangelosi, Vincent and William Dill. 1965. "Organisational Learning: Observation Toward a Theory." *Administrative Science Quarterly* (10):175-203.
23. Chatterji, Deb. 1996. "Accessing External Sources of Technology." *Research Technology Management* 39(2):48-56.
24. Cohen, Michael and Lee Sproull. 1996. *Organizational Learning*. London: Sage.
25. Collins, H. 1993. "The Structure of Knowledge." *Social Research* 60(1):95-116.
26. Cook, Scott and Dvora Yanow. 1996. "Culture and Organizational Learning." *Organizational Learning*, editors Michael Cohen and Lee Sproull. London: Sage.
27. Crainer, Stuart. 1996. "Virtual Development." *Management Training* :15-16.
28. Crossan, Mary. 1997. "Improvise to Innovation." *Ivey Business Quarterly* 62(1):36-42.
29. Cyert, Richard and James March. 1992. *A Behavioral Theory of the Firm*. Cambridge, Massachusetts: Blackwell Publishers Inc.
30. Daft, R. and K. Weick. 1984. "Toward a Model of Organizations As Interpretation Systems." *Academy of Management Review* 9:284-95.
31. Davenport, T. and L. Prusak. 1998. *Working Knowledge: How Organizations Manage What They Know*. Boston: Harvard Business School Press.
32. Davidow, William H. and Michael Malone. 1992. *The Virtual Corporation, Structuring and Revitalizing the Corporation for the 21st Century*. New York: Harper Business.
33. Doz, Yves. 1996. "The Evolution of Cooperation in Strategic Alliances: Initial Conditions or Learning Processes? " *Strategic Management Journal* 17:55-78.
34. Drucker, Peter. 1968. *The Age of Discontinuity: Guidelines to Our Changing Society.* New York: Harper & Row.
35. ———. 1993. *Post-Capitalist Society*. Oxford: Butterworth-Heinemann.
36. Easterby-Smith, Mark. 1997. "Disciplines of Organizational Learning: Contributions and Critiques." *Human Relations* 50(9):1085-113.
37. Edvinsson, Leif and Patrick Sullivan. 1996. "Developing a Model for Managing Intellectual Capital." *European Management Journal* 14(4):356-64.
38. Egidi, Massimo and Alessandro Narduzzo. 1997. "The Emergence of Path-Dependent Behaviors in Cooperative Contexts." *International Journal of Industrial Organization* 15(6):677-709.

39. Epstein, Marc and Marie-Josee Roy. 1997. "Using ISO for Improved Organizational Learning and Environmental Management." *Environmental Quarterly Management* 7(1):21-30.
40. Fiol, C. Marlene and Marjorie A Lyles. 1985. "Organisational Learning." *Academy of Management Review* 10(4):803-13.
41. Fisher, Kimball and Mareen Fisher. 1998. *The Distributed Mind: Achieving High Performance Through the Collective Intelligence of Knowledge Work Teams*. New York: Amacom.
42. Garvin, David. 1993. "Building a Learning Organisation." *Harvard Business Review* 74(4):78-91.
43. Gephart, Martha, Victoria Marsick, Mark van Buren, and Michelle Spiro. 1996. "Learning Organizations Come Alive." *Training & Development* 50(12):34-44.
44. Giunipero, Larry. 1997. "Organizational Change and Survival Skills for Materiel Managers." *Hospital Materiel Management Quarterly* 18(3):36-44.
45. Gnyawali, Devi, Alice Stewart, and John Grant. "Creation and Utilization of Organisational Knowledge: An Empirical Study of the Roles of Organisational Learning on Strategic Decision Making." *The 1997 Annual Meetings of the Academy of Management* (Chicago) Chicago.
46. Goh, Swee and Gregory Richards. 1997. "Benchmarking the Learning Capability of Organizations." *European Management Journal* 15(5):575-83.
47. Greiner, Ray and George Metes. 1995. *Going Virtual: Moving Your Organisation into the 21st Century*. New Jersey: Prentice-Hall.
48. Hackman, J. 1990. *Groups That Work (and Those That Don't): Creating Conditions for Effective Teamwork*. California: Sage.
49. Hall, Richard. 1997. "Complex Systems, Complex Learning, and Competence Building." *Strategic Learning and Knowledge Management*, editors Ron Sanchez and Aime Heene. Chichester: John Wiley & Sons Ltd.
50. Hamblen, Matt. 1998. "NetMeeting Cuts Dow Travel Expenses." *Computerworld* 32(10):20-21.
51. Handy, Charles. 1995. "Trust and the Virtual Organisation." *Harvard Business Review* 73(3):40-50.
52. Hedberg, Bo. 1982. "How Organizations Learn and Unlearn." *Handbook of Organisation Behaviour*, vol. 1, editors P. Nystrom and W. Starbuck. New York: Oxford University Press.
53. Hodgson, Geoffrey. 1998. "Competence and Contract in the Theory of the Firm." *Journal of Economic Behavior & Organization* 35(2):179-201.
54. Howells, Jeremy. 1996. "Tacit Knowledge, Innovation and Technology Transfer." *Technology Analysis & Strategic Management* 8(2):91.
55. Huang, Chun-Ming, Sue Newell, and Elena Antonacopoulou. "Team learning processes: The impact of virtual organisations." *The 16th Labour Process Conference* (Manchester) Manchester.

56. Huber, George. 1991. "Organisational Learning: The Contributing Processes and the Literatures." *Organisation Science* 2(1):88-115.
57. Hult, Tomas and O. Ferell. 1997. "Global Organizational Learning Capacity in Purchasing: Construct and Measurement." *Journal of Business Research* 40(2):97-111.
58. Jones, Alan and Chris Hendry. 1992. "How Learning Organizations Are Created." *Management Development Review* 5(4):10-15.
59. ———. 1992. *The Learning Organisation: A Review of Literature and Practice*. Coventry: Human Resource Development Partnership.
60. Kamoche, Ken. 1996. "Strategic Human Resource Management Within a Resource Capability View of the Firm." *Journal of Management Studies* 33(2):213-33.
61. Kasl, Elizabeth, Victoria Marsick, and Kathleen Dechant. 1997. "Teams As Learners: A Research-Based Model of Team Learning." *The Journal of Applied Behaviour Science* 33(2):227-46.
62. Keys, Bernard, Robert Wells, and Trey Denton. 1998. "Japanese Managerial and Organizational Learning." *Thunderbird International Business Review* 40(2):119-39.
63. Khanna, Tarun, Ranjay Gulati, and Nitin Nohria. 1998. "The Dynamics of Learning Alliances: Competition, Cooperation, and Relative Scope." *Strategic Management Journal* 19(3):193-210.
64. Kim, Rowon. 1996. "Learning-Induced Control Model to Allocate Managerial Resources for Production Technology Development." *International Journal of Production Economics* 43(2,3):267-82.
65. Klavans, Richard and David Deeds. 1997. "Competence Building in Biotechnology Start-Ups: The Role of Scientific Discovery, Technical Development, and Absorptive Capacity." *Strategic Learning and Knowledge Management*, editors Ron Sanchez and Aime Heene. Chichester: John Wiley & Sons Ltd.
66. Kogut, Bruce and Udo Zander. 1993. "Knowledge of the Firm and the Evolutionary Theory of the Multinational Corporation." *Journal of International Business Studies* 24(4):625-46.
67. ———. 1996. "What Firms Do? Coordination, Identity, and Learning." *Organisation Science* 7(5):502-18.
68. Kolb, D. 1984. *Experiential Learning*. New Jersey: Prentice-Hall.
69. Kolb, D. and R. Fry. 1975. "Towards an Applied Theory of Experiential Learning." *Theories of Group Processes*, editor C. Cooper. London: John Wiley & Sons.
70. Lei, David, John Slocum Jr., and Robert Pitts. 1997. "Building Cooperative Advantage: Managing Strategic Alliances to Promote Organizational Learning." *Journal of World Business* 32(3):203-23.

71. Levitt, Barbara and James March. 1988. "Organisational Learning." *Annual Review of Sociology* 14:319-40.
72. Liebowitz, S. and Kenneth De Meuse. 1982. "The Applications of Team Building." *Human Relations* 35(1):1-18.
73. Liedtka, Jeanne, Mark Haskins, John Rosenblum, and Jack Weber. 1997. "The Generative Cycle: Linking Knowledge and Relationships." *Sloan Management Review* 39(1):47-58.
74. Luxhoj, James, Jens Riis, and Brian Stensballe. 1996. "A Hybrid Econometric-Neural Network Modeling Approach for Sales Forecasting." *International Journal of Production Economics* 43(2,3):175-92.
75. Makridakis, Spyros. 1997. "Metastrategy: Learning and Avoiding Past Mistakes." *Long Range Planning* 30(1):129-35.
76. March, James and Herbert Simon. 1958. *Organizations*. New York: John Wiley.
77. March, James. 1991. "Exploration and Exploitation in Organisational Learning." *Organisation Science* 2(1):71-86.
78. March, James, Lee Sproull, and Michael Tamuz. 1991. "Learning From Samples of One or Fewer." *Organisation Science* 2(1):1-13.
79. Marsick, Victoria and Karen Watkins. 1997. "Lessons From Informal and Incidental Learning." *Management Learning: Integrating Perspectives in Theory and Practice*, editors John Burgoyne and Michael Reynolds. London: Sage.
80. Martinez, Michelle. 1998. "The Collective Power of Employee Knowledge." *HR Magazine* 43(2):88-94.
81. McBain, Richard and Mitchell Kusy. 1997. "On Team Effectiveness, Team Learning and Learning in Organizations." *Manager Update* 9(2):20-31.
82. McKenna, Stephen. 1992. "A Culture Instrument: Driving Organizational Learning." *Leadership & Organization Development Journal* 13(6):24-29.
83. McLaughlin, Curtis and Arnold Kaluzny. 1997. "Total Quality Management Issues in Managed Care." *Journal of Health Care Finance* 24(1):10-16.
84. Medcof, John. 1997. "Why Too Many Alliances End in Divorce." *Long Range Planning* 30(5):718-32.
85. Meyers, Patricia and David Wilemon. 1989. "Learning in New Technology Development Teams." *The Journal of Product Innovation Management* 6(2):79-88.
86. Miles, R. and W. Randolph. 1980. "Influence of Organizational Learning Styles on Early Development." *The Organizational Life Cycle*, editors J. Kimberly and R. Miles. San Francisco: Jossey-Bass.
87. Morgan, G. and Smircich, L. 1980. "The Case for Qualitative Research." Academy of Management Review 5: 491-500.
88. Nelson, R. and S. Winter. 1982. *An Evolutionary Theory of Economic Change*. Cambridge: Harvard University Press.

89. Nonaka, Ikujiro. 1994. "A Dynamic Theory of Organisational Knowledge Creation." *Organisation Science* 5(1):14-37.
90. Nonaka, Ikujiro and Hirotaka Takeuchi. 1995. *The Knowledge-Creating Company, How Japanese Companies Create the Dynamics of Innovation.* Oxford : Oxford University Press.
91. Parkinson, Stephen. 1995. "Marketing." *Manager Update* 7(2):12-22.
92. Polanyi, Michael. 1958. *Personal Knowledge.* Chicago: University of Chicago Press.
93. Powell, Walter, Kenneth Koput, and Laurel Smith-Doerr. 1996. "Interorganizational Collaboration and the Locus of Innovation: Networks of Learning in Biotechnology." *Administrative Science Quarterly* 41(1):116-45.
94. Probert, Philip. 1996. "How to Create Intellectual Capital Responding to the Learning Challenge of Change Management." *Competency* 4(2):21-24.
95. Purser, Ronald, William Pasmore, and Ramkrishnan Tenkasi. 1992. "The Influence of Deliberations on Learning in New Product Development Teams." *Journal of Engineering and Technology Management* 9(1):1-27.
96. Quinn, James. 1993. "Managing the Intelligent Enterprise Knowledge and Service Based Strategies." *Planning Review* 21(5): 13-16.
97. Raelin, Joseph. 1997. "A Model of Work-Based Learning." *Organization Science* 8(6): 563-78.
98. Ritcher, Frank and Kai Vettel. 1995. "Successful Joint Ventures in Japan: Transferring Knowledge Through Organizational Learning." *Long Range Planning* 28(3): 37-45.
99. Roos, Johan, Goran Roos, Lief Edvinsson, and Nicola Diago. 1997. *Intellectual Capital- Navigating in the Business Landscape.* London: Macmillan Press Ltd.
100. Sackmann, Sonja. 1991. *Cultural Knowledge in Organisations: Exploring the Collective Mind.* California: Sage.
101. Sanchez, Ron and Aime Heene. 1997. "A Competence Perspective on Strategic Learning and Knowledge Management." *Strategic Learning and Knowledge Management*, editors Ron Sanchez and Aime Heene. Chichester: Sage.
102. Sanchez, Ron. 1997. "Managing Articulated Knowledge in Competence-Based Competition." *Strategic Learning and Knowledge Management*, editors Ron Sanchez and Aime Heene. Chichester: John Wiley & Sons Ltd.
103. Sashkin, Marshall and Steve Franklin. 1993. "Anticipatory Team Learning: What Is It and How Does It Happen." *The Journal of Management Development* 12(6):34-43.
104. Schein, Edgar. 1992. *Organisational Cultural and Leadership.* San Francisco: Jossey-Bass.
105. ———.1993. "On Dialogue, Culture, and Organisational Learning." *Organisational Dynamics* 22(2):40-51.

106. ———. 1996. "Culture: The Missing Concept in Organization Studies." *Administrative Science Quarterly* 41(2):229-40.
107. ———. 1996. "Three Cultures of Management: the Key to Organisational Learning." *Sloan Management Review* 38(1):9-20.
108. Senge, Peter M. 1992. *The Fifth Discipline, the Art and Practice of the Learning Organisation*. London: Century Business.
109. Simonin, Bernard. 1997. "The Importance of Collaborative Know-how: An Empirical Test of the Learning Organization." *Academy of Management Journal* 40(5):1150-1174.
110. Sinkula, James, William Baker, and Thomas Noordewier. 1997. "A Framework for Market-Based Organizational Learning: Linking Values, Knowledge, and Behavior." *Academy of Marketing Science Journal* 25(4):305-18.
111. Snelbecker, Glenn. 1985. *Learning Theory, Instructional Theory, and Psychoeducational Design*. London: University Press of America.
112. Solomon, Charlene. 1998. "Sharing Information Across Borders and Time Zones." *Workforce* 3(2):12-18.
113. Sparrow, John. 1998. *Knowledge in Organisations: Access to Thinking at Work*. London: Sage.
114. Stamps, David. 1998. "Learning Ecologies." *Training* 35(1):32-38.
115. Starbuck, William. 1992. "Learning by Knowledge Intensive Firms." *Journal of Management Studies* 29(6):713-40.
116. Stein, Eric and Vladimir Zwass. 1995. "Actualizing Organizational Memory With Information Systems." *Information Systems Research* 6(2):85-117.
117. Stein, Robert and Gifford Pinchot. 1995. "Building an Intelligent Organization." *Association Management* 47(11):32-40.
118. Steward, Thomas. 1989. *Intellectual Capital: The New Wealth of Organisations*. London: Macmillan Press.
119. Suttling, Bob and Dave Wood. 1997. "Managing the Virtual Team." *Telecommunications* 31(12):61-64.
120. Sveiby, Karl. 1997. *The New Organisational Wealth: Managing and Measuring Knowledge Based Assets*. San Francisco: Berrett-Koehler Publishing.
121. Tsoukas, H. 1994. “From Social Engineering to Reflective Action in Organisational Behaviour”. New Thinking in Organisational Behaviour, editor H. Tsoukas. Oxford: Butterworth Heineman.
122. Toffler, A. 1990. *Powershift: Knowledge, Wealth and Violence at the Edge of 21st Century*. New York: Bantam Books.
123. Trussler, Simon. 1998. "The Rules of the Game." *The Journal of Business Strategy* 19(1):16-19.

124. Tuomi, Ilkka. "The Communicative View on Organisational Memory: Power and Ambiguity in Knowledge Creation Systems." *The 29th Annual Hawaii International Conference on System Science* (Hawaii). Hawaii.
125. Upton, David and Bowon Kim. 1998. "Alternative Methods of Learning and Process Improvement in Manufacturing." *Journal of Operations Management* 16(1):1-20.
126. Verganti, Roberto. 1997. "Leveraging on Systemic Learning to Manage the Early Phases of Product Innovation Projects." *R&D Management* 27(4):377-92.
127. van Maanen, John. 1983. Qualitative Methodology. California: Sage.
128. Vicari, Salvatore, Georg von Krogh, and Johan Roos. 1996. "Knowledge Creation Through Cooperative Experimentation." *Managing Knowledge- Perspectives on Cooperation and Competition*, editors Georg von Krogh and Johan Roos. London: Sage.
129. von Krogh, Georg and Johan Roos. 1996. "Arguments on Knowledge and Competence." *Managing Knowledge- Perspectives on Cooperation and Competition*, editors Goerg von Krogh and Johan Roos. London: Sage.
130. Waldrop, Mitchell. 1992. *Complexity- The Emerging Science at the Edge of Order and Chaos*. Middlesex: Penguin.
131. Walsh, James. 1995. "Managerial and Organisational Cognition: Notes From a Trip Down Memory Lane." *Organisation Science* 6(3):280-321.
132. Walsh, James and Gerardo Ungson. 1991. "Organisational Memory." *Academy of Management Review* 16(1):57-91.
133. Warkentin, Merrill, Lutfus Sayeed, and Ross Hightower. 1997. "Virtual Teams Versus Face-to-face Teams: An Exploratory Study of a Web-Based Conference System." *Decision Sciences* 28(4):975-96.
134. Wathne, Kenneth, Johan Roos, and Georg von Krogh. 1996. "Towards a Theory of Knowledge Transfer in a Cooperative Context." *Managing Knowledge- Perspectives on Cooperation and Competition*, editors Georg von Krogh and Johan Roos. London: Sage.
135. Wheelwright, Steven C. and Kim Clark. 1992. *Revolutionizing Product Development: Quantum Leaps in Speed, Efficiency, and Quality*. New York: The Free Press.
136. Willman, Paul. 1996. "Protecting Know-how." *Business Strategy Review1996* 7(1):9-13.
137. Wilson, D. and R. Rosenfeld. 1990. *Managing Organisations: Texts, Readings and Cases*. London: McGraw Hill.
138. Zuboff, S. 1988. *In The age of the Smart Machine: The Future of Work and Power*. New York: Basic Books.

KNOWLEDGE AND POWER IN INFORMATION TECHNOLOGY SERVICES: A CASE STUDY

ILAN OSHRI

Warwick Business School, University of Warwick, Coventry CV4 7AL, England

E-mail: phd97io@wbs.warwick.ac.uk

This study explores the organisational power of the Information Technology services at the University of Warwick, following the growing dependency of organisations on IT services. The study examines the power relations between administrative departments and identifies the factors which shape power distribution in an organisation, using interviews with managers, questionnaires and documentary research. The uniqueness of the University, in terms of the combination of academic and administrative management and decision-making mechanisms managed by committees, guides the author to analyse the research findings with an elaborated theory consisting of aspects of power and decision-making. The study's conclusions point to a relatively low organisational power for the Information Technology services, and ambiguous findings regarding the influence of specific factors on the power relations at the University. The elements that profoundly influence the organisational power of the Information Technology services are: (i) the domination of the business dialogue in the decision-making process at the University, (ii) the managers' perception that the Information Technology services are service providers and not policy makers, and (iii) the existence of latent conflict, exercised by influential players. The analysis of the influential factors led to the development of a subjective model which links different categories of factors with the players' current and prospective power. However, the study's conclusions need further investigation in different higher educational institutions in order to create a comparative frame of study which will assist in the understanding of the sources of power in the administrative section of universities.

1. Introduction

From the early days of computing, people have raised a variety of questions pertaining to the societal implications of the increasing pervasiveness of the computer [9]. According to Bloomfield and Coombs [3], the focus of study was on the relations between knowledge and power in organisations, an area which has been studied along the years in various ways. Some of the studies [17,10] have emphasised the process of centralisation of power in organisations, in which information providers, such as technical experts, would lose power to information gatherers who tend to be in positions closer to top management. Another area of study has concentrated on the influence that information technology has on the power of management, especially middle management, by causing their role to shrink and eventually disappear [15,13].

The influence of information technology on the power of computing occupations in organisations has been investigated with contradicting evidence presented regarding its 'organisational power' (a term that was developed by Child et al. [5] which will be elaborated in the next chapter). Robin Fincham [8] suggests that computing occupations are in a state of transition between a technical service and something more central to organisational power and decision-making. The increasing strategic position of IT occupations and the growing dependence of organisations on information technology has made the IT occupations central to decision-making and corporate power. Furthermore, Scarbrough and Corbett [22] suggest that experts - persons who possess knowledge and information - might exercise their potential power over non-experts, using their knowledge to advantage in the decision-making process. On the other hand, Boynton and Zmud [4] claim that restructuring of the IT industries has meant a growing 'market' for the supply of IT systems and services. They suggest that "IT departments are rather like some local government or public sector agency caught up in a process of competitive tendering"[p. 28]. The traditional situation of a centralised systems development function having a monopoly over the supply of an organisation's IT services is being transformed into a market of information, through which individuals become more adept in computing techniques as systems become more user-friendly.

This paper aims to investigate power relations between Information Technology services and other administrative departments in organisations, in an attempt to contribute to existing research in this area.

2. Views of Power in organisations

The theoretical framework builds on the theory of power in organisations, in which the three views of power as outlined by Lukes [16] will be presented.

2.1 Perceptions of Power

The theme of power in institutions is studied in various ways. Studies relate to power as the outcome of individual charisma and abilities as well as the outcome of an organisation's structure or decision-making processes. Power can be exercised in different manners and by different means, and we will suggest different ways of understanding the notion of power.

Lukes [16] investigates the theory and practice of power in politics and organisations. In his book (ibid.) 'Power: a radical view', he reviews two previous perceptions of power and develops his own point of view of power and its sources.

The one-dimensional view is the 'pluralist' view of power. In this view, power is described as follows: *A* has power over *B* to the extent that *A* can get *B* to do something that *B* would not otherwise do. One can see that there is a difference here between potential and actual power and between its possession and whether it is exercised. This approach to power refers to *A*'s capacity to exercise power and also specifies a successful attempt. One can have the capacity of power but the crucial question is whether it is exercised and to what degree of effectiveness. As Dahl [7] writes, a: "rough test of a person's overt or covert influence is the frequency with which he successfully initiates an important policy over the opposition of others, or vetoes policies initiated by others, or initiates a policy where no opposition appears"[p. 66]. Conflict, according to that point of view, is assumed to be crucial in providing an experimental test of power attributes. Without it, the exercise of power will not be evident. Conflict is between preferences, which are assumed to be consciously made, exhibited in actions, and thus to be discovered by observing people's behaviour. Thus, the one-dimensional view of power involves a focus on behaviour in the making of decisions on issues over which there is an observable conflict of interests.

Power can have two faces. The first is, as described above, the pluralist view. The meaning of the second face is, as Bachrach and Baratz [1] claim: "a set of predominant values, beliefs, rituals, and institutional procedures that operate systematically and consistently to the benefit of certain persons and groups at the expense of others. Those who benefit are placed in a preferred position to defend and promote their vested interests"[p. 43].

The two-dimensional view of power involves a qualified critique of the behavioural focus of the first view which allows consideration of the ways in which decisions are prevented from being taken on potential issues over which there is an observable conflict of interests.

The three-dimensional view, developed by Lukes [16], points to those areas that were outside the social dialogue of the previous two views. It suggests that the previous views are too committed to behaviourism - that is, to the study of overt, actual behaviour, of which concrete decisions in situations of conflict are seen as paradigmatic. Lukes pointed out that "trying to assimilate all cases of exclusion of potential issues from the political agenda to the paradigm of a decision gives a misleading picture of the ways by which individuals and groups can succeed in excluding potential issues from the political process"[p. 21]. In this sense, power takes simple forms: through the control of information, through mass media and through the process of socialisation. As Lukes described: "the elite seems to have possessed that most indispensable of all characteristics in a dominant group - the sense, shared not only by themselves but by the populace, that their claim to govern was legitimate"[p.23]. Dahl develops Lukes' ideas and claims that leaders do not merely respond to the preferences of constituents but also shape preferences. Moreover, the bias of the system is not sustained simply by a series of individually chosen acts but also, and most importantly, by the socially structured and cultural behaviour of groups, and practices of institutions, which may indeed be manifested by individuals' inaction.

Lukes's criticism of the two previous views is based on the assumption that power should be associated with actual, observable conflict. Lukes claims that this insistence on actual conflict as essential to power is not sustained in reality, for at least two reasons. First, two types of power may not involve such conflict: namely, manipulation and authority. Second, it is highly unsatisfactory to presume that power is only exercised in situations of such conflict. As Lukes concludes: "*A* may exercise power over *B* by getting him to do what he does not want to do, but *A* also exercises power over him by influencing, shaping or determining his very wants" [p.23].

The forgoing discussion has described three views of power in organisations and ways to measure power in organisations and political systems. It is obvious that the first two views of power are easier to measure and one can find these forms of power by observing the outcomes of decision or non-decision processes. But, unlike these forms of power, the three-dimensional view, which is also called the radical view of power, is very difficult to measure. Moreover, it is difficult to identify its existence. The theory, developed by Lukes, claims that this form of power exists due to a socialisation process and other social influences, usually developed over the course of time. This requires a longitudinal examination of people's attitudes and behaviour, linked to symbols of power.

Concluding the views of power, a view by Lukes creates a comprehensive framework which will enable this paper to confront the different forms of power influencing the power relations in organisations.

2.2 Definitions

The studies by Scarbrough and Corbett [22], Child et al. [5] and Fincham [8] discuss the notion of 'organisational power' but not all of them confront the need to define it. From their explanations, one can define it as: the ability to advance the interests of individuals or specific group(s) by promoting control over the resources of the organisation. In these studies, the scholars mentioned above refer to two ways in which control can be promoted. First, by controlling the specific group's role or activities in the organisation, usually in terms of knowledge and expertise. For example, a group may start to restrict other group activities, by setting the agenda for them, or for the entire organisation. Second, by influencing the structure and processes of the organisation. For this paper, both interpretations are suitable. This paper will seek those activities influencing the agenda, structure and processes of the organisation and will be within the decision-making processes over policies, budgeting, project launching and so on. Nevertheless, one should not be limited to this wording; the exercise of power can be the prevention of changes in organisation, as well.

The assumption that organisations are political arenas [2,20] assists in the understanding of the term 'power relations'. One can understand 'power relations' as a result of the development of organisational power by individual(s) or group(s) within an organisation. The extent to which each group succeeds or fails in its attempt to control the resources of the organisation, will define its position with relation to other individuals or groups. This picture of power relations defines the ability of individuals or groups to change their control over the organisation's resources. In turn, the position of individuals or groups in these power relations will define their quality of employment, by increasing their resources in terms of personal benefits and departmental development [5].

3. **Analysis**

In this chapter, the paper will establish the evidence focusing on the relatively weak organisational power of the Information Technology services and their weak image of influence in the picture of power relations at the University. The analysis of the findings will be presented in two stages.

First, the findings of the interviews, questionnaires and the documentary research of the case study conducted in Warwick University will be presented, summarising the opinions of the interviewees and respondents in this paper into four main perceptions regarding the power position of Information Technology services of Warwick University. Second, following this section, an analysis of these perceptions in the context of the theoretical framework of organisational power and power relations will be presented, concentrating on three main criteria of power measurement: image of power possession, control of the resource allocation process, and control of the agenda. The paper will also examine the factors that shape the power relations at the administrative section of Warwick University.

3.1 Findings

The information gathered during the interviews, questionnaires and documentary research endorses four main perceptions which define the power relations between the administrative departments at the University and shape the organisational power of the Information Technology services.

The four perceptions are as follows:

3.1.1 The image of the Information Technology services as a non-influential department: the view of the Information Technology services' managers.

This opinion was expressed by almost all of the Information Technology managers who were interviewed. The view emerges from two findings expressed by the Information Technology managers: (i) their image as service providers and not as policy shapers, (ii) their inability to control or significantly influence the processes of decision-making. Let us elaborate the first reason.

The evidence from the interviews suggests that managers from the Computer Services department (CS) conceptualise their role as service providers, though they would like to see more involvement at the level of strategic players. The CS's Mission Statement in the University Calendar expresses the department's ambition to be part of the IT policy designers of the University. However, the evidence suggests that this is not the case.

A senior manager from the CS department described the current influence the department has as one which evolved from the need to provide services to other administrative departments, in an environment that aspires to cut expenses but which encourages technological progress. He believes that the CS's responsibility in the University includes more than providing services. He sees the role of the CS department as an analyst of the University's IT needs and a promoter of solutions and alternatives. However, he feels that financial constraints, introduced by top management, inhibited the development of a more influential role for the Information Technology services.

Managers from Management Information Systems (MIS) also understand their role as service providers. A senior manager from MIS explained that their job is to provide technical support to all projects in Senate House. She thinks that MIS is responsible for providing any technical support which might be required, but is not responsible to analyse the needs of other administrative departments.

Managers from the Information Systems Support Unit (ISSU) present another view. They understand their role as designers of the IT policy in WBS. However, their case is totally different from that of the other Information Technology services; the main reason for this is the fact that the Unit only interacts with WBS. Therefore, their ability to influence and shape the IT policy of the Business School is much easier to achieve. In this sense, the ISSU managers feel that they can define their goals, and usually achieve them without much interruption. Managers from CS and MIS expressed the same view about the ISSU's image of power, though they pointed out that the Unit is not that influential, especially when it involves interactions with the Information Services Policy Committee.

Summarising, from the evidence above one understands that the Information Technology services conceptualise their role as service providers. Therefore, because their main role is presently seen as providing technical services, the result is that the personnel of the Information Technology services could easily undermine their potential position as strategic players.

The second reason for the low image of power of the Information Technology services is their inability to control or significantly influence the processes of decision-making at the University. This opinion was expressed by managers from the Information Technology services. Here, the paper will present two cases involving decision-making processes, which elaborate this point.

The first case was the process of purchasing new computing systems for the computer rooms of several faculties in the campus at the end of 1995. In this case, CS was expected to purchase the computers that the faculties preferred, according to their experience with similar systems. However, CS created a process in which the technical benchmark and the financial examinations were made without the participation of the faculties. Finally, only three supply companies reached the final decision, in which representatives from several faculties (engineering, social studies, WBS and others) participated in the process, and Research Machine (RM) was chosen by them to be the supplier of hundreds of computing systems over the next few years.

This case illustrates a manipulation process, performed by CS managers, who exercise their technological knowledge in the evaluation process of the computing companies, and influence the decision-making process. However, the final decision was taken by the faculties, based purely on financial measurements.

The second case was the implementation of a new payroll system at Senate House. In this case, MIS, and the Finance and Personnel departments discussed different options for payroll systems during 1994 and 1995. From interviews with people who were involved in the process, it emerges that the MIS department supported the implementation of BMS's system. The reason for this was the knowledge and expertise held by the MIS personnel regarding previous BMS systems. However, the final decision to install a different system, the Peterborough, was driven by the Finance department, following its preference to contract with well-established companies.

Some interviewees presented this case as a symbol of synergy among the three departments and their representatives. However, the level of influence that was presented by the MIS department was very low and it emerges from the perception of the MIS managers that their role is simple as service providers.

In addition to the evidence above, the questionnaires provide further information pointing to the same conclusion. Several questions aimed to measure the level of influence of departments on the decision-making process at the University. The findings of these questions are as follows:

The Information Technology services respondents are less likely to make suggestions that concern other areas (finance, personnel and general administration) in comparison to other administrative departments. The Information Technology services respondents feel they have a relatively low influence on the outcomes of the decision-making processes. Respondents from the Information Technology services also feel they have less influence on financial and general decisions, than other administrative departments'. Therefore, the evidence shows that MIS and CS, in these cases and according to their response in the questionnaires, do not possess significant influence over decisions, and conceptualise their position as non-influential within the University.

The case of the ISSU is different. Though, there is no specific case study to demonstrate the influence of the Unit on decision-making processes, the overall picture, as illustrated by managers from the Information Technology services points to relatively high influence. Managers from the ISSU suggest that this situation is a result of a trust-based relationship and useful communication channels between the Finance and Resources Committee of the WBS and the Unit. However, the Head of the ISSU expressed his opinion that ISSU has similar problems to other administrative departments in influencing decisions taken by the Information Service Policy Committee.

As a conclusion, the managers of the Information Technology services realise that their non-influential image evolved as a result of their service providers role. Furthermore, the domination of financial arguments in the decision-making process, even when it comes to technical proposals, contributes to their inability to influence the outcomes of decision-making processes.

3.1.2 The image of the Information Technology services as a non-influential department: the view of the administrative departments' managers.

This view was expressed by managers from Finance, Personnel, Retail Services and Registrar offices (administrative departments).

Three main explanations were presented by the managers to support their view. First, these managers conceptualise the Information Technology services as service providers and not as strategic players. Second, the growing understanding of individuals of the IT and computing fields undermines the IT experts' authority. Third, the influence that top management has on the decision-making processes.

Evidence regarding the first explanation suggests that managers from Finance, Personnel Management and Retail Services comprehend the role of the Information Technology services as service providers. None of these managers thought that representatives from the Information Technology services should have significant influence over decisions concerning the resources allocation of the University. One manager pointed out that the Director of Information Services participates only in three committees: the Information Services Policy Committee, the Joint Council and Senate Strategy Committee and the Library Policy Committee. Comparing the Director of Information Services' participation in committees with that of the Finance officers, the indication is that the Finance representatives have twice as much participation than the IT representatives, with significant influence in the important Finance and General Purposes Committee [24].

Managers from the Finance and Retail Services departments manifest a different picture regarding their ability to influence decisions. These managers expressed confidence in the committees' support. In particular, the senior manager from Retail Services presented confidence that his initiatives will be promoted in the relevant committees. He based his opinion on the fact that one of the factors that the committee is concerned with is the contribution of proposals to the University's business performance. Thus, because the Retail Services is a business-oriented, he believes that it is much easier to convince the committee to support its proposals, rather than to invest in non-profitable areas.

Concluding the first explanation, the evidence shows that managers from the administrative departments conceptualise the role of the Information Technology services as service providers. The implications of this position have been discussed above. Their evaluation relies on the insignificant participation of computing managers in committees.

The second explanation is based on the views of managers from Retail Services and the Finance department. These mangers claimed that computing expertise has become common knowledge. They suggested that computing experts possess the knowledge to install and maintain the computing systems, but do not control the information inside the systems. They believe that the link between expertise and the ability to influence decisions is dependent on the possession of specific information, regarding the decision-making process. Foe example, many of the administrative managers have the know-how to operate computers, and many of them have discovered personal interest in computing systems as a hobby or as a part of their work. Therefore, it seems to these managers that they would be able to deal with computing challenges relatively better than computing experts could confront financial issues.

A recent case in the Retail Services department illustrates this view. A senior manager from Retail Services was involved in the installation of a computerised telecommunication system at Claycroft residences at Warwick University during 1995.

At that time, the proposal was discussed with the former IT Director. His opinion was that it is technically impossible to implement this kind of system. However, because the financial opportunity of this project captivated the Earning Income Group from the Finance office, they figured it would be worthwhile to explore the technological and financial issues by themselves. The Director of Retail Services launched research into similar systems implemented in other universities, and reached the conclusion that the project was possible, and even profitable. The Finance and General Purposes Committee supported the project, and now, after a year, it is considered a great success, although it has had problems.

Such situations could undermine the technological authority of IT experts. They stress the ability of individuals to influence technological decisions even though they are not experts in this field.

The third explanation relates to the work of committees at the University and to the influence that top management has in the decision-making processes. The group which is called top management, consists of senior managers from the Finance and the Registrar offices.

Considering the complexity of this explanation, the paper needs to elaborate the ways by which the committees at Warwick University work. Following this, the paper will suggest evidence pointing to the domination of top management in the decision-making process.

The process of decision-making at the University is organised around the work of committees [24] according to the University of Warwick Calendar, though it is not the actual way that decisions take place in the University's day-to-day reality. Two sources of evidence describe the decision-making process at the University.

First, the University of Warwick Calendar, in which the structure, regulations and participants of different committees are described in a comprehensive way.

Second, through interviews with managers at the University that pertain to the work of the committees. The gap between the rhetoric of the Calendar and the reality of decision-making at the University is significant.

The University of Warwick Calendar [24] describes the work of the committees as a mechanism of rational decision-making. The members of the committees suggest their motions in advance to the review of the members. A discussion is developed over the motions and the members consider the various aspects that the University is concerned with, such as the academic benefits, the financial implications, the priority of the actions and financial sources. The members of the committee also examine alternatives that may have advantages over the discussed proposal, and then decide on the best action that provides for the needs of the University.

Conversations with almost all the interviewees, including senior managers, suggest a rather different decision-making process at the University. According to these interviewees, committees serve as rubber stamps, with decisions actually taken before the committee gather to discuss proposals. The members of the committees are advised in advance about the contents of the proposals and the person or the department which is promoting a specific proposal usually assesses the support it will achieve in the committee before any discussion takes place.

In cases when it turns out that the proposal will not be supported, then the proposal will not be presented to be a part of the committee's agenda. The managers promote their proposal in different ways. It can be done through the social relationships between Directors of departments, and also by creating an ad hoc coalition in order to promote one initiative and to block another one. Usually the main issue in this coalition is based around the budgeting of the proposals.

Manager from Information Technology services described this scenario in a sharp way, suggesting that certain departments will support other departments' initiatives when it is clear that they will gain a dividend.

According to almost all of the interviewees, the persons who influence the decision-making processes in the important committees are: senior managers from the Registrar's office and those from the Finance office. To put it bluntly, proposals that are supported by the Registrar or by senior managers from the Finance office will be approved. It should be stated that all the interviewees except one indicated the Registrar as the most influential person in the administrative section of the University.

The evidence suggests that there is an influential group who possesses a relatively high level of influence, both outside and inside the work of the committees. This group determines the issues to be discussed in the committees and actually sets the agenda of the administrative departments at the University.

The interviewees assess the influence of the Information Technology services on decision-making processes as relatively low, which means that managers from the Information Technology services possess a low level of influence on the important committees. There are two main reasons which were expressed by the interviewees which might explain this phenomenon.

First, some of the interviewees suggested that key people in the committees are not familiar with computing systems, and therefore will probably not promote the development of IT strategic policy at the University. Managers specified that decision-makers do not always conceive of the updating of computing services as a crucial process which will contribute to the success of the University.

Also, the communication between IT experts and members of the committees is problematic as well, especially when IT experts need to explain technical issues, a process that can cause some of the members to 'switch-off'.

The second reason refers to the type of dialogue that the members of the influential group have developed in the last years: many of the interviewees claimed that financial issues have became the primary issue in discussions in the committees.

In recent years', tight resources have created the need to cut expenses in almost all the departments and urged top management to be more careful about expenses. Lately, the University has cut the budgets of several administrative departments. Another change in the financial policy of the University is the direction of more capital to administrative departments that generate profits and less to departments that generate services such as the Information Technology services.

This situation is a result of a process, determined by influential players in top management, in which the discussion is organised around issues that are already approved in a process prior to the work of the committees, usually these issues are regarded as 'safe'. Themes such as investment in IT infrastructure and the creation of an IT strategy at the University are considered as controversial and therefore will be exterior to the business dialogue [12].

This has resulted in less influence of the Information Technology services over decisions, especially financial decisions.

The result of the questionnaires strengthens the evidence above. Managers from the administrative departments think that Information Technology services have a relatively high influence on technological decisions at the University.

However, these managers refer a relatively low influence of the Information Technology services on financial decisions or decisions in general.

3.1.3 Different views of power relations among administrative departments and managers.

Two different understandings of the power relations and the ways that power is exercised at the University emerge from the interviews. The power perception gap is between two groups: one group is formed of the Information Technology services' managers and mid-level managers (e.g., officers and supervisors from the administrative departments) and the second group is the senior managers (the Directors of administrative departments and the Registrar).

The first group conceptualises (with the exception of the senior manager from CS) the power relations between administrative departments as based on tough competition over restricted resources, meaning power relations which are based on conflict. These managers interpret the interactions between administrative departments and the committees over the resource allocation process as fundamental conflict, in which each department tries to increase its resources.

Opposing this view, the second group (other administrative departments and Directors at the University) thinks that the power relations are based on cooperation between administrative departments and individuals. Their perception supports the view that the University has one common goal.

Although, this common goal was expressed in different ways by each of the Directors and the Registrar, all of them shared the perception that the parties in the University are working together to achieve this target, which can be business success, more students or academic achievements. According to their view, one of the expressions of the common effort to achieve 'University success' is the wide and diverse participation of many administrative departments in the committees work. Also, the committee regulations lead to an open process, and demonstrate the shared effort of the parties to achieve the goal. This view is a unitary view.

The profound gap between the radical view of the computing managers and mid-level managers, and the unitary view of the senior managers, provokes the following conclusion. The conflict between the Information Technology personnel and the senior managers is covert or latent and senior mangers seem not take aware of its existence.

Various factors influencing the organisational power.

Analysing the responses from the interviews and questionnaires reveals that there are more than ten different factors which are considered to be sources of the organisational power of a department (or individuals). The list of factors includes: more staff, bigger budget, better communication skills of the representatives in the committees, charisma of the Head of department, better communication with other administrative departments, more involvement in the committees' work, size of the department, high level of expertise possessed by the department, more success of the department, accessibility to influential persons and consultation with other administrative departments.

The variety of factors raises the question whether it will be possible to understand which of the factors is more important and which of the factors could best bestow relatively higher organisational power?

Another issue that needs discussion is the relation between the exercise of these factors and the ability to improve the organisational power of the department. The fact that this paper encountered many factors, some of which are totally different from the others, implies that the way to understand the influence of these factors on power should be from the point of view of the player. This means an analysis which takes into account a power of the player in the present, and the player's attitude regarding his/her prospective power. These factors, therefore, should be viewed as the obstacles, and potential aids, to developing from the present power position to the prospective power position.

3.2 The analysis of Organisational Power and the power relations

The evidence of the study will be interpreted in this chapter into three main conclusions.

First, that the power relations between administrative departments at the University have not changed in the last three years. It will be argued that the committees, as decision-making mechanisms, preserve power within the closed circle of the decision-makers - the senior managers from influential offices.

Secondly, the Information Technology services possess low levels of influence over the decision-making process at the University, which can be interpreted as a relatively low level of organisational power. It will be argued that the top management exercises its power by dominating the business dialogue in the decision-making processes and excluding technological issues.

Thirdly, the factors which shape the organisational power of the administrative departments are subjective, players indicate those factors which reflect their position in the power relations. They also evaluate those factors that can change their future power. Based on the study's findings, a Subjective model will be presented, suggesting explanations of the link of power with the factors indicated by the interviewees.

3.2.1 Power relations at Warwick University

The power relations between administrative departments at Warwick University are highly dependent on the University's committees system and its implications. The literature which deals with the structure of committees and their influences shows that committees are considered as a source of power in which decisions are taken place.

This determination emerges from the understanding that there are two central decisions in organisations in which committees are involved: designing policies and determining budgets. With control over these two decisions, the committees manage the organisation. However, the way that committees work at Warwick University indicates that Barber's view is inappropriate. The findings of this study suggest that the committee is a mechanism for perpetuation of the power of the influential players in the top management of the University.

The paper identifies in the structure and activities of the committees at Warwick University two elements that facilitate the perpetuation of power.

First, the evidence pointing to a small group of people who participate and chair the main committees. This group consisting of persons from the Finance and Registrar offices, defines the nature of the dialogue that will take place in the committees' discussions. The relationship between the domination of the social dialogue in societies and the exercise of power on individuals was developed by Michel Foucault and elaborated by Cousins and Hussain [6]. The dialogue that appears to take place both inside and outside the committees is a business dialogue. The essence of this dialogue is the evaluation of any action in terms of profit or loss to the University's future resources, with considerations regarding the possible effects on the academic departments and student life.

The evidence suggests that discussions that approve proposals within the committees have to be negotiated according to the codes of a business dialogue. The influential group, which determines the codes of the dialogue, actually sets the agenda at the University and controls the type of issues to be discussed. In the last years, since the government started to cut the Universities' resources, the resource allocation process has become more business-oriented than ever before. Therefore, it would be best for proposals from technical departments to be negotiated by stressing the business aspects of the initiatives.

Top management possesses power and exercises it according to Lukes' view, namely the radical view of power. It seems that the attempts of the IT experts to develop an IT strategy at the University is restricted by top management's explanations that resources are not available. Thus, the demands of the Information Technology services for change in the resource allocation process are oppressed before these issues can be properly voiced. Top management exercises its control over the political agenda of the administrative departments, a process which distinguishes the radical view from the other two views of power [25]. The conflict, in this case, has two faces. The people from the Information Technology services conceptualise the situation as an overt conflict, while top management seems not to realise its existence.

The extent to which the exercised power influences the behaviour of the computing experts can be observed from the evidence presented above. The IT managers agreed that utilisation of technological knowledge may damage their ability to promote proposals rather than granting them a competitive advantage over non-expert persons. The Information Technology services already concentrate on the business aspects of their initiatives in cases when they need to promote proposals. Two pieces of evidence strengthen the analysis of power in accordance with the radical view.

In the first case, the CS managers used technological expertise to promote their interests during the upgrading process of the computing systems last year. In the second case, the discussions concerning the payroll system involved both technological and financial arguments. However, in both cases, when it came to the final decision, the IT experts adopted the codes of the business dialogue, evaluating for cost-minimisation of the deal and the financial background of the winning company. These cases show that the business codes have become part of the IT and computing experts dialogue in their participation in the decision-making process.

These examples, together with the managers' expressions in the interviews, strengthen the idea that there is a process in which the top management exercises power over the administrative departments and dominates the business dialogue in the political agenda. This process secures power with the top management and inhibits attempts to change the stabilised power relations in the administrative section of the University.

Before continuing to the next point, one must stress that the way the top management (the Finance and Registrar offices) is presented above does not mean the domination of this group by dictatorial methods of management but, rather, an academic view on the perception of these people, very influential within the management of the University, who take the responsibility to manage the University in times of tight resources and greater competition.

3.2.2 Organisational power of the Information Technology services

The organisational power of the Information Technology services will be analysed by the definition, presented above: the ability to advance the interests of individuals or specific groups, by gaining control over the resources of the organisation. From analysing the organisational power of the Information Technology services, the paper suggests three factors, emerging from the research evidence, that point to one simple conclusion: the Information Technology services lack the ability to influence the resource allocation process at the University. The question is: why?

The elements which explain the relatively low organisational power of the Information Technology services are connected, although each of them has a different effect on the organisational power.

The first and most important element is the managers' perceptions, both IT and administrative, that the Information Technology services are non-influential departments at Warwick University. The Information Technology services are not part of the "profit generator" departments but, rather, a prospective benefit generator which depends on heavy investment in the present. They are not considered as policy designers and their participation in the committees is very limited. The Director of IT is the only representative in the top management committees, and in the last several months this position was not filled. This locates the personnel of the Information Technology services outside the closed circle of the decision-makers.

Even the more active approach taken by the CS managers has not located them as influential players. One cannot ignore their influence on the determination of practical computing solutions; however, their ability to control resources is very restricted. The ISSU may be the only computing unit that enjoys a relatively high appreciation as a policy designer. However, this influence can not be projected to other areas of the University, or to the other Information Technology services.

Second, is the domination of the business dialogue in the University's agenda. The idea of the business dialogue, as developed above, stresses the necessity of a business understanding for participants in the negotiations over resources at the University. The codes of the dialogue are not in favour of persons who come from the Information Technology field. The IT experts have to negotiate their proposals in the finance field, in which their knowledge is not sound, and the considerations they encounter are sometimes outside of their evaluation abilities.

Therefore, this situation puts them in a relatively poor position to negotiate their proposals, in comparison to Finance personnel. However, this conclusion cannot be generalised to all the Information Technology services' managers, some of them have acquired a financial education during their career or by business programmes for managers.

The analysis suggests that as long as the dominant dialogue is the business dialogue, it will be difficult for the computing personnel to possess a significant influence over the allocation of resources.

Third, the different views of the power relations between senior managers and the managers from the Information Technology services indicates the existence of conflict between the two sides. The evidence suggests that the senior managers see [employment] relations as based on cooperation between the administrative departments, in contrast to the computing managers that conceptualise their [employment] relations as based on a conflict over resources.

The evidence points to a relatively low organisational power for the Information Technology services. This conclusion evolves from the fact that the Information Technology services conceptualise their power relations as conflict relations because of their inability to control or influence the resource allocation process [21]. On the other hand, those players who control the resources and are able to allocate resources according to their beliefs, will probably not recognise the existence of conflict in their [employment] relations. The rationale of this argument leads to the inevitable conclusion that the Information Technology services possess a relatively low organisational power.

3.2.3 A Subjective view of power relations

The study's findings point to various factors that have an influence on the organisational power of the administrative departments. It is therefore necessary to explore models that could assist in the understanding of this complex picture. The main target of the study is to understand the relations between the players' power and the factors that shape this power. Surprisingly, the literature that deals with power relations in organisations [21,14,22] somehow skipped the development of a methodical model in this field.

This paper develops a model, based on the study's findings, which will assist in the understanding of the relations between power and its influential factors from the subjective point of view of the player. The rational assumption of the model is that players who possess a certain level of power will aspire to maintain this power and will seek opportunities to increase it. The greater influence the players have, the more the players will invest in maintaining this power and the less in improving their position. The more insignificant the influence which the players have, the more the players will invest in improving their position [21].

The matrix of the Subjective view consists of four combinations. There are two classifications of a player's understanding of the current power relations; these perceptions can be (i) conflict or (ii) harmony. There are also two classifications of a player's power expectation which can be (i) change or (ii) conservation.

		Player's perspective of the current power relations	
		Conflict	**Harmony**
player's prospective power	Change	**1 Quantitative Factors**	**3**
	Conservation	**4**	**2 Qualitative Factors**

Figure 1: The Subjective view of power relations

The model identifies two categories of factor: quantitative and qualitative.

1. Quantitative factors - these include issues of larger budget, more staff, more nominations of senior managers on board of directors, a bigger department, better location of the department and so on. A change in any of these factors means a process of reallocation of resources. This represents a situation in which the

players' view of power relations is based on conflict, which means that the players probably have no significant control over the process of resource allocation. The players' .expectation of power is change. Therefore, these players, such as Information Technology services, will try to improve their control over the resource allocation process by raising their demands for quantitative resources. Their aspiration is to change the allocation of the resources.

2. Qualitative factors - these include communication and consultation channels in the organisation. The emphasis is on improvement of the grievances process. This represents the combination of harmony with the preservation of power. In this situation, the players probably possess control over the resource allocation process and would like to maintain this situation. The utilisation of qualitative factors is likely to imply that these players would like to direct dialogue into issues that do not create dramatic changes in the allocation of resources. In this way, the players would be able to preserve their power and create alternative channels of dialogue, described as better communication and consultation.

3/4. Undefined situations - These two situations are difficult to analyse. The combinations of conflict with conservation, and harmony with change, seem more irrational and contradictory with the model's assumptions. Therefore, the model's assumption allows the players to locate their behaviour only in either the 1 or 2 rubrics. The other two rubrics can not be rationally analysed; the model can not properly raise explanations to the players' behaviour in these rubrics.

Clearly, this is one of the disadvantages of the model and further research in this field may elaborate the missing panels. Another point that should be stressed is the limitation that the model's assumptions create by defining rational behaviour as the only analysable behaviour. Players sometimes follow seemingly irrational strategies in order to overcome the barriers of a rational process. A strategy to exceed the department's budget in order to gain a greater amount in next year's discussions, without going through the process of negotiations, could be an example.

This kind of behaviour can not be analysed by the model, though it can classify it as part of rubric 1 (conflict/change). However, the model is dynamic in terms of providing explanations in different levels of organisational power, and enables one to predict the category of factors that players will utilise according to their position and expectations.

4. Conclusions

The analysis framework of the paper is based on the analysis of the sources of power which was developed by Lukes [25] and it enables measurement of the 'organisational power' in the complex environment.

This paper, based on interviews with a range of managers from various departments at the University, and on analysis of questionnaires and official documents, leads to the inevitable conclusion of the relatively low organisational power of the Information Technology services.

The main reasons can be summarised into two main areas: (i) the structure of the University and (ii) department position.

Under the category of the structure of the University, the paper analysed the decision-making process at the University; in particular, the centrality of the committees in this process. This analysis led to the understanding that there is group of individuals from the top management of the University who significantly influence the decision-making process. This group acts as a 'gate keeper' in the dialogue that takes place during decision-making over the allocation of resources. The consequence of the domination of business dialogue is the exclusion of other aspects, from different areas, from the discussions; for example, technological issues. This results in less influence for the Information Technology services, and points to their inability to control the allocation of resources.

The position of the Information Technology services generally can be summed up as service providers. The Information Technology services do not generate profits and conceptualise their activities as technically supporting others initiatives. Therefore, their participation in the decision-makers group is limited, and in many cases they depend on the allocation of resources from other administrative departments.

These two categories can also be described as external and internal reasons that determine the relatively low level of organisational power of the Information Technology services.

The discussion over the factors that shape the organisational power shows that it is difficult to point to a specific factor or factors that embody the ability to change one's organisational power.

This conclusion led to the development of the Subjective model. The variety of factors compelled the exercise of a methodical model which enables the examination of the influence of various factors over the power relations in an organisation. The innovation in this model is the concentration on the players' points of view - on the way they conceptualise their power both in the present and in the future.

Following the conclusions of this paper, three comments should be addressed toward the foundations of power studies and investigations of power in higher education institutions and in institutions in general.

Firstly, the paper's conclusions refer mainly to the radical view of power developed by Lukes). The paper explored the ways in which potential issues, particularly technological issues, were kept outside the political agenda.

This theory provides satisfactory explanations to the relatively low organisational power of the Information Technology services and one might accept this approach as sufficient to explore power relations in different systems and organisations. However, according to the conclusions of this paper, we would like to explore Lukes' view on the existence of latent conflict. Lukes claims [16] that latent conflict consists of "a contradiction between the interests of those exercising power and the real interests of those they exclude"[p. 24-25]. He clarifies this point, claiming that "This conflict is latent in the sense that it is assumed that there would be a conflict of wants or preferences between those exercising power and those subject to it, were the later to become aware of their interests"[p. 24-25]. However, this study's evidence points to a different conclusion. The evidence shows that the Information Technology services personnel are mindful of their interests and conceptualise their [employment] relations as based on conflict while the senior mangers in the University understand the [employment] relations at the University as based on cooperation and synergy between players. The implication is that conflict can exist, at least in the reality of the players exposed to power upon them beyond their ability to react in return, as Lukes claims. But, this conflict has not became an observable conflict, as Lukes implies. Our conclusion is that the extent to which power is exercised by the powerful players will define their ability to eclipse the existence of the conflict.

Secondly, we would like to argue that recent views [12,11] which have explored the notion of social dialogue in organisations have not explored the implications which might emerge from the specialisation of occupations. The industrial world is going through a specialisation process in which each occupation has developed its own dialogue (language) and agenda.

The implication of the specialisation process on the development of dialogues in organisations may be as follows:

The Personnel staff use in their dialogue terms such as 'appraisal systems', 'structure of wages' and 'social benefits'. The Finance personnel use terms like 'gross profit', 'cash flow' and 'earning per share'. The computing personnel use the terms 'direct access memory', 'central processing unit' and 'floating point processor'.

This arena of different dialogues, also known as the 'Tower of Babel', may lead to the conclusion that the literature which deals with the influence of organisational dialogue ('talk') over power relations [12,11] needs to consider the possibility of having more than one dialogue in an organisation, without affecting the analysis of power in the organisation. The suggestion of this paper is that the effects of the social dialogue on the players' power will not be explored under the question: who determine the dialogue, but rather under the question: what is the contribution of each dialogue to the final decision?

Thirdly, this paper has explored the different forms of decision-making processes and power relations at Warwick University. The theoretical framework of this paper reviewed some of the main models in this area [25]. Looking at the complexity of Warwick University's management, one can consider its management to be a combination of all the models presented above.

The rational model is embodied in the formal structure of the committees as a decision-making mechanism. The 'organised anarchy' model can be found in the actual process of decision-making by the influential group of the University. And, finally, the interest groups pushing to gain more resources within a tight resources arena, represent the political model. It seems that what is needed in this paper, and perhaps also in studies of other universities is a composite model, synthesising elements from all the models mentioned above. The construction of such a model is not easy, but highly essential for the changing environment and increasingly complex role of universities. Considering that the paper's subject is not models of university management, I would like to outline only general ideas of the composite model, that can then be further developed.

The basis of the model is the political model [2], which meets today's political, social, technological and business conditions of universities. This means that we will consider universities as miniature political systems in which interest groups negotiate for resource allocation and exercise different levels of power. The three views of power (Lukes, 1974) will be the measurements of power, through which the sources of power can be analysed.

5. Summary

This paper has argued that the Information Technology services have not succeeded in translating their growing responsibilities around the campus into the position of an influential player at the University of Warwick.

Naisbitt [18] claims that "the new power is not money in the hands of the few, but information in the hands of the many". However, the information and knowledge that computing experts and IT experts possess has not necessarily empowered them. It seems that the search for the sources of power will continue as long as power exists.

References

1. Bachrach P. and Baratz M. S., Power and Theory, Theory of Practice, *Oxford University Press*, New York, (1970).
2. Baldrige J.V., Curtis D.V., Ecker G. and Riley G.L., Policy Making and Effective Leadership, *Jossey-Bass*, San Francisco,(1978).

3. Bloomfield B.P. and Coombs R., Information Technology, Control and Power: The Centralisation and Decentralisation Debate Revisited, *Journal of Management Studies,* 29(4).
4. Boynton A. C. and Zmud R. W., Information Technology Planning in the 1990's: Directions for Practice and Research, *MIS Quarterly*, 11 (2), (1987).
5. Child J., Loveridge R., and Spencer A., Microelectronics and the Quality of Employment in Services, in P. Marstrand (ed.), New Technology and the Future of Work and Skills, *Frances Pinter*, London, (1984).
6. Cousins M. and Hussain A., Michael Foucault: Theoretical traditions in the social sciences, *MaCmillan*, London, (1984).
7. Dahl R. A., Who governs? Democracy and power in an American City, *Yale University Press*, London,(1961).
8. Fincham R., Computing Occupations: Organisational Power, Work Transition and Collective Mobility, *New Technology, Work and Employment*, 9 (1), (1994).
9. Fleck J., Artificial Intelligence and Industrial Robots: an Automatic End for Utopian Thought?, in E. Mendelsohn and H. Nowotny (eds.), Nineteen Eighty-Four: Science Between Utopia and Dystopia, *Sociology of Sciences Yearbook,* Dordrecht, Reidel, (1984).
10. Gotlieb C.C. and Borodin A., Social Issues in Computing, *Academic Press,* London, (1973).
11. Gowler D. and Legge K., The Meaning of Management and the Management of Meaning: a View from Social Anthropology, In M. Earl(ed.) Perspectives on Management : A Analysis , *Oxford University Press*, Oxford, (1983).
12. Gronn P.C., Talks as the Work: the Accomplishment of School Administration, *Administrative Science Quarterly*, 28(1), (1983).
13. Hicks R.L., Developing the Top Management Group in a Total Systems Organisation, *Personnel Journal*, 50(9), (1971).
14. Kirkbride P. S., Power, in J.E. Hartley and G.M. Stephenson, Employment Relations, *Blackwell*, Oxford, (1992).
15. Leavitt H. and Whitler D., Manager in the 1980's, in H. Leavitt and L.R. Pondy, Readings in Managerial Psychology, *The University of Chicago Press*, Chicago, (1964).
16. Lukes S., Power: A radical View, *Macmillan Publishers Ltd*, London, (1974).
17. Mosco V., The Pay-per Society, *Garamond Press*, Toronto, (1989).
18. Naisbitt I., Mega Trends, *Warner*, New York, (1982).
19. Pettigrew A., The Politics of Organisational Decision-Making, *Tavistock,* London, (1973).
20. Pfeffer J., Managing with Power: Politics and Influence in Organisations, (1992).
21. Scarbrough H. and Corbett J.M., Technology and Organisation: Power Meaning and Design, *Routledge*, London, (1992).

22. The University of Warwick, The University of Warwick Calendar 1996-1997, *Coventry Prints*, Coventry, (1996).
23. Walford G., Restructuring Universities: Politics and Power in the Management of Change, *Croom Helm*, New York, (1987).

AN EXPLORATION OF PARADOXES IN MANAGERIAL COMPLEXITY: A POSTMODERN INTERPRETATION OF CORPORATE VISION

GERT VAN DER LINDEN

GP Group, 34b Quai des Sanbarbani, MC98000 Monaco
E-mail: vdlinden@gp-group.com

PAMELA PARKER

GP Group
E-mail: pparker@gp-group.com

This paper focuses on managerial and organizational complexities, studying paradoxes reflected in different management paradigms. It looks at these paradoxes both on an epistemological level as well as an applied level. The discussion shows that paradoxes offer an important vehicle for exhibiting that managerial and organizational complexities can no longer be characterized by linearity, causality and determinism. The paper then offers an extensive debate on the essence of postmodernism related to management, portraying a fundamental break with the classical paradigms. It then reflects on the implications of postmodern thinking to more applied fields, questioning whether this represents a paradox of postmodernism. Corporate vision is deconstructed as an example. This exercise looks into different anchors of corporate vision such as core philosophy, competencies and capabilities, and envisioned future, as well as issues of time and space, past and future. The result is that postmodern thinking is epistemological yet inherently pragmatic. Inspiration and application are integral part of it.

"We're still confused, but only on a higher level".

Introduction

Paradoxes are an interesting concept to use as a vehicle to look at different management paradigms, and especially to see how they deal with issues of complexity within the realm of a deeper understanding of managerial and organizational life. Too often complexity has been translated into "complicated". Too often scholars have had the tendency of translating "managing complexity" into "managing over complexity". More flexible organizational forms and the fragmenting impact that they exert on established structures and practices all contribute to the complexity - perhaps a better terminology is perplexities - from which managers cannot escape. Consequently, managers no longer hold a unique role or position in organizations; they are themselves subject and object in turbulent networks.

New economic drivers can be described in terms of technological discontinuities and different customer expectations, less protectionism and excess capacity, mergers and acquisitions, changes in industry structures and deregulation, shifts in intellectual and market leadership, global and multi-domestic competition, alliances, and the emergence of trade blocks. The diffusion of innovation in the marketplace, the importance of dramatic and radical innovations as well as the importance of knowledge and knowledge-building infrastructures, and the increasing emphasis on designing new products and moving them to the marketplace rapidly, are only a few of the many elements that portray the technological revolution. It is clear that economy, business, and management, which consist of these complexities, can no longer be characterized by linearity, absolute causality, and determinism. This is the essence of the paradoxes that will be presented.

Paradoxes are not defined in terms of a formal logic point, i.e. as an apparently unacceptable conclusion derived by apparently acceptable reasoning from apparently acceptable premises. Paradoxes portray a confusion or collapse of different logical levels. Even though contradictions are inferred, they aren't just referring to a simple contradiction. Paradoxes suppose *con-fusion*: the different levels cannot be confronted because the fusion doesn't make disappear the differentiation of logical levels. Paradoxes are a powerful vehicle since they portray an intellectual and philosophical context; they say something about the actor, action and the acting fields. It is in that movement and continuous transformation that paradoxes are expressed.

Paradox 1: "Whenever Mr. Weber is confronted with complexity, he reduces it to chaos"

Weber is one of the founding father of the bureaucratic approach to organizations, which is still used heuristically. Based on a particular form of power - the rational authority - it starts from the assumption that stability, coordination, integration, and control are pursued through path finding, problem solving, and the implementation of functionality. This bureaucratic organization paradigm is further characterized by the use of technocracies, generalizing typologies, dimensions and formulas while at the same time it positions the employee within a pyramidal hierarchical principle. And this hierarchy with its values and interests of a higher order clearly prevails to the individual. Organizations give rise to a crystalline body displaying well-defined symmetries, structural order, sharply regulated activities, and interactions in which managers hold a unique controlling position.

Although Weber's work emphasizes the process-character of organizational life, modern organizational analysis has seen his work re-written to particularly emphasize on the static issues of efficiency and administrative control. In other words, a manager is able to control the increasing complexity of organizational life by applying a template of functional rationality on emergent and disorderly patterns of social relations. He uses the tools provided by Weber's work in that problems are first isolated from its contexts and then analyzed separately. When the analysis fails to take into account the interconnectedness of events, the situation stemming from a pre-acknowledged and pre-defined issue, is defined as a new problem. Hence, individuals are seen as "rule following" - building analogies to known situations according to a set of rules.

This modernist perspective assumes that phenomena are reducible to a few fundamental interactions and rules, and that consequently models of control allow predictions about organizational activities to be made. Under well-defined conditions a system governed by a given set of laws will follow a unique course. A slight change in the causes will likewise produce a slight change in the effects.

The paradox "Whenever Mr. Weber is confronted with complexity, he reduces it to chaos" portrays the assumptions as defined within the bureaucratic approach to organizations. It accentuates the fact that within this framework, complexity is considered a drawback that has to be avoided. Complexity provokes such a panic-reaction as if it were complete chaos - a heightening breakdown and loss of predictability, certainty, and identity - that has to be immediately translated to a comprehensible simple and static context of linearity, determinism, and objective causality in order to rule over complexity.

The classic approach to strategy illustrates this reductionism. The complexities of industries are reduced in terms of maturity, continuity and stability so that a single prediction of an organization's future path can be described. As a consequence, the better the environmental analysis according to few dimensions, the better the course (strategy) can be defined and implemented.

Paradox 2: "The shift from static to dynamic is enhanced by static interventions in the dynamics"

Complexity arises due to a multitude of simple components interacting simultaneously. This complexity lies inside the myriad of possible ways in which the components of a system can interact (the internal determination of behavior) and to the control paradigm such information processing, problem solving, i.e. external determination. Complexity results from the great number of variables and parameters that blur some fundamental underlying ordering principle. This concept of complexity marked the beginning of modern complex systems thinking. Although it takes a more dynamic aspect into consideration, the paradigm still stresses mechanistic, causal, and externally controlled aspects of general systems and organizations.

Illustrative is the idea of "autopoiesis". Autopoiesis describes "living" and social systems as a network of processes of production which highlight both the aspect of autonomy or organizational closure of a system vis-a-vis its environment (autopoiesis and self-organizing properties of complex systems) and its cognitive activity (Maturana & Varela, 1980). In other words, it refers to the capability of a system to maintain its identity by keeping its organization and its internal coherence invariant. In an attempt to come up with a more elegant, but also more general definition, Zeleny (1981) refers to an autopoietic system as a unity realized through a closed organization of production processes generated through the interaction of their own products (components), and by a topological boundary that emerges as a result of the same constitutive processes.

Hence, the definition of the organization as a system is based upon the specific ways in which the system interacts with the environment and the ways in which the features of the environment are considered to be the most important factors used in explaining the dynamics of the adjustment of the organization (Van Gils, 1997). All that is required of the systems to be studied is that they be complex, dynamic, capable of "learning", and exhibit feedback and/ or feed-forward. Because the synthesis of information has to be created from situation to situation, the concept of learning is translated by how the system constructs models of itself, its environment, and the relation of itself to its environment. Information, and learning are intertwined within the system and cannot exist independently since they are co-created within the process of communication. Hence, learning understood as a change of structure within a closed system (and not as an adaptation to a changing environment), emerges locally and evolves locally.

The main epistemological consequence stems from taking into account special problems of universal or global theories, referring to a class of objects to which they themselves belong. Whereas classical epistemology looks for a set of conditions under which external observations yield identical results, this approach adds the issue of self-observation, including social systems as observed and as observing systems. This illustrates the attempt to include the dynamic aspect of systems. Within its perspective, however, complexity still seems to be based on metaphors more closely akin to the growth of a plant from a tiny seed, or the unfolding of a computer program from a few lines of code (Waldrop, 1992).

Attempts to model and simulate variable structure models, for example, are characterized by assumptions of structural change, i.e. models that dynamically adapt their structure during a simulation. Although the system constructs models of itself, its environment, and the relation of itself to its environment - i.e. "learns" - such models still feature a technocratic approach to structure and process. Concretely, by introducing events into the dynamic aspect of the model, it is enabled to evolve during a simulation (see for instance in this book Christodoulou, 1998). Whether these events are time-driven, or have a "sensitive" dependence on causal relationships between environmental conditions or/ and agent's conditions, linearity and determinism still prevail.

Another illustration of an attempt to simulate variable structure models is seen by the logic used in Human Resources Information Systems (HRIS). The most optimistic scenario deems this technology capable of allowing individuals to live better with dispersion and fragmentation, but it does not accommodate managerial practices as such. The attempted complexity of the technology or HRIS, coupled with the structural organization, in essence, detracts from the performance of people. By streamlining processes in a rigid-defined, technocratic and uniform structure, HRIS aim to create clear answers for the future. Hence, the logic on which HRIS are based accentuates structural and linear change under the umbrella of complexity (see Van der Linden & Parker, 1998).

Paradox 3: "The irreversibility of chaos is reversed in/ by its determinism"

The discovery that simple, non-linear systems with only a few degrees of freedom could show chaotic behavior gives us a new perspective in deterministic thinking. This perspective indicates the interest in the "true" complexity of systems amidst local chaos. Systems exist at the edge of chaos, finding order, but in a haphazard and uncertain way. The smallest interference might lead to wild, unpredictable changes and provoke non-proportional consequences. These changes accentuate a non-linearity that is in contrast to classic equilibrium models, and in which the dynamic character appears via positive feedback phenomena (see Baets, 1998). As such, the basic premises of chaos theory are relatively simple. Decision-making units referred to as agents (machines, IS, as well as workers) interact with other agents to trigger unpredictable and uncertain consequences.

Instead of pure computation, (cognitive) modeling uses concepts of calculation and bounded rationality in order to deal with the processes between the design and the choice of means and of ends, between controlled and controlling system. In other words, the chaotic framework still presupposes a deterministic model, purposely relying on a well-defined input to the system, a conveyance, and an output. The deterministic character of chaotic systems is important to understand the consequences of its deterministic nature, and to make sense out of those consequences and keep them manageable. Systems are not pre-determined - similar systems in similar environments will not undergo similar changes. There still is, however, determinism, i.e. environmental determinism in that starting from a different set of initial conditions, the same dynamic systems behave differently. Hence, the "problem" of prediction and control of the system's behavior is balanced against a sensitivity to the unavoidable perturbations of environmental origin present in the real world.

Further, the objective representation of social reality is a key-element. For example, it is assumed that knowledge can be developed by making systematic representations of the outside world and checking these representations by means of generally accepted methodological rules and points of departure. Learning, as such, is more than the accumulation of knowledge - it is the operationalization of representations and experiences in a framework (Baets, 1998).

Because the chaotic paradigm is less rule-based and more based on complex equations, this approach claims to be connectionist. The focus is more on behavior and meaning rather than form (on equations). An illustration can be found in the application of neural networks. Through parallel distributed processing, dynamic and complex equations, data that deal with perceptions, and observations that are contingent and interconnected, neural network techniques imitate brain-like and assimilated thinking (see Baets, 1998).

In practice, the chaotic paradigm simplifies non-linearity into a so-called "workable" uncertain and unpredictable linearity, although it offers a higher level of aggregation. The paradigm allows for different paths to be taken, but the environmental determinism leads, in essence, to different linearities. Interestingly, although a strong awareness for non-linearity, dynamic behavior, and disequilibrium exists, one can read the heritage of the Cartesian anxiety for paradox in which duality is brought back to a dilemma in which there is (in its practical application) no real room for non-linearity and "absolute" (fixed) causality. Simply, that Cartesian anxiety for paradox led us to the paradox of the chaotic paradigm.

Postmodernism

Postmodernism outlines a philosophy in which tensions between unity and diversity, contexts and change, history and time, chaos and uncertainty, non-linearity and instability flow into one. It re-conceptualizes meaning and causality - how we experience and explain the world. Put simply, postmodern epistemology suggests that our shared language constitutes the world and that we can only "know the world" through the particular forms of discourse language creates (*con-text*: with text). It is argued, however, that as language games are continually in flux, meaning is constantly slipping beyond grasp and can thus never be lodged within one term. Meanings are not to be seen as individual constructs or precepts that are then socially evaluated and corrected, but socially created. As a consequence, postmodernism is characterized by the lack of fixed and absolute causality and determinism. Does that mean that as a result everything would happen "by accident", as positivist scholars would then conclude?

Postmodernism approaches rationality - meanings - as states in which small, local, temporal, antagonistic "truths" are produced. It is a world without demarcation between subject and object, in which content refers to mental content(s). This mental content refers to the alliance between "truth" and "claims of truth". Truths have the connotation of stability, transcending the temporal, the moment, the contingency of the singular. Claims of truth are subject to the position an individual has within a network that is produced in knowledge and action. As such, "claims" refer to the local and temporal, to flux. Thus, the postmodernists place themselves in a unique philosophical position by rejecting irrationality while not seeking to develop a new rationality. This makes that there is no absolute rationality or irrationality. Experiencing is not based on a simple and distant information processing. This results in the absence of absolute determinism.

Individuals create “truths” that are produced, anchored, and embedded in networks connecting signifiers and meaningful nods (issues), in options and projects. These networks can be imagined as rich webs or texts of meanings, dynamic interactions between multiple agents and actors, perpetual novelty, knowledge creation, only having a temporal meaning and gathering that temporal meaning in relation to each other. If a “truth” is expressed as a clear point of view, as an argument or a crystalline controversy, the possible existence of other points of view, other projects, of re-definitions are always present. Even if something was proven by sending people to the moon, by signing a contract, by transforming an organization’s structure, there’s always other possible ways, non-materialized possibilities, and non-realized fragments existent.

Individuals will take particular points of view, will take a stand, and will approach claims as “truths”, mobilizing things in the world and society. As our language games are continually in flux, meaning is constantly slipping beyond our grasp and can thus never be lodged with one term.

Postmodernism, Organizations and Management

Turbulent interactions with and between different stakeholders of a corporation prove that the complexity of the business, organizational and management environments cannot be reflected by one or a couple of variables. The essence of complexity is defined by the lack of linearity, determinism, and fixed causalities, which is exactly embraced by postmodernism.

There is no rational, standard and clear separation between an organization and its environment. Boundaries are diffuse: internal networks overflow external networks and vice versa. Each and every individual working within an organization has his own perception of this organization and its environment, dealing in a different way with stakeholders, agents and actors. There are no objective topological boundaries. Things "out there" such as markets, employees, suppliers, and customers, are called into being by language, and not by pre-existing objects in a relation to the observer and to each other. In a similar way, processes are not and don’t become things.

Organizations and environments are realities that are complicated, turbulent, chaotic, antagonistic, complex, and ambiguous. Individual interpretations (meanings) of the organization shape exemplaric pictures that form the organization's identity. In postmodern terms, organizations arise as differentiated universes of relative, local, partial and temporarily processive entities. Postmodernism recognizes the fundamental instability of organizations, the sense in which language and action are never final but are only moves within a game that leads to further possibilities (Parker, 1992). This perspective also expresses the tension between the individual and management, between unity and diversity or, for that matter, unity as diversity, and the tension between idiosyncrasy and coherence, and coordination and integration. It also continuously disrupts the normative structures about the organized world: Answers to problems about organizational systems turn into problems about answers. Hence, the focus is on the production of organization rather than the organization of production (Cooper & Burrell, 1988; Van der Linden, 1993). In other words, the focus is on how the exemplaric and fragmented meanings, i.e. the differentiation, form unity in constant flux - the organization's identity.

Management takes place within these differentiated universes of complex, partial, and temporal contexts. Managers are themselves subject and object in/ of these turbulent networks, in which managers experience issues which are paradoxical and which cannot, in the end, be defined or resolved. For example, processes are not the answer per se as they do not necessary solve problems or provide information and knowledge. Success does not lie in creating unity, as it is left only as a possibility. Success has to be seen as a temporal consensus in disensus. Success breeds failure and vice versa. In managerial life the "and ... and" prevails over the "or … or". It is the "genius of the and" dominating the "tyranny of the or". The answer is not a choice between two elements. "Or" does not lie in two opposing extremes. Travelling between tensions and making use of those tensions allows the search for openings which make new choices possible. That is the "core" competency of the manager. By using and reconciling tensions, managers become "agents in realities".

Management has a clear rhetoric function. It produces new realities, new language games or meaningful (con)texts with their own particular grammar. One could say it is the manager's role to identify temporally shared elements within complexity, to identify unity in diversity and diversity in unity while attempting to translate dynamic interactions into the momentary. Management of meanings has to be approached as a strategic journey. The manager's interest in meanings and knowledge starts from a time perspective and takes advantage of the richness of complexity and flux. Knowledge is exactly concerned with what one learns in the process of translating (giving meaning to) the flow of his world and involvement in contexts. Hence, management is a mindset to be understood as a continuous movement without location. It is differing in space and deferring in time.

Paradoxes of Corporate Vision: a postmodern example

Prompted by these elements, it can be said that postmodernism offers a fundamental break with the classical paradigms, its discourse being essentially epistemological in nature. Also, the more recent issues in organization and management thinking make clear the importance and relevance of postmodernism, and have set an interesting agenda for debate. On one hand, postmodernism brings about a revolution that undermines the basic tenets of research and understanding of managerial and organizational complexities. On the other hand, postmodernism could be critiqued exactly because of its essentially epistemological nature. Although the factual world is beyond our grasp, as postmodernism claims it to be, it is time for postmodernist scholars to discuss the implications of postmodern thinking to more applied fields (Van der Linden & Parker, 1998). Doesn't this yet suppose another con-fusion, i.e. a paradox, in which the different levels cannot be confronted because the fusion doesn't make disappear the differentiation of logical levels?

Theories can be situated on the crossline between epistemology and a philosophical psychology. They are networks of intelligibility, networks of inventions and textures of meanings occuring within a chaos of turbulences. Therefore, theory and practice are inseparable: there is no language of understanding placed beyond the boundaries of potential, as Gergen (1992) puts it. Postmodernists are only interested with the potential for theories to offer new possibilities. Theories give opportunity to think about things as psychological events. They give rise to an intellectual experience for readers and authors. Readers become authors of their own reading while producing new meanings. Authors also become readers and thus authors again.

The issue of corporate vision offers a good example of a philosophical psychology. It also portrays tensions between as states in which small, local, temporal, antagonistic "truths" are produced and turbulences of time and space. It describes a world without demarcation between subject and object, in which content refers to mental content(s). Therefore, we want to offer the reader a postmodern interpretation of corporate vision (one within the diversity of many possibilities).

Vision (1)

Vision connotes a dream or an apparition, a reading of the future. Some, however, prefer the term foresight to vision, which is based more on deep insights into trends. The typical corporate vision, however, is a muddled stew of values, goals, purposes, philosophies, beliefs, aspirations, norms, strategies, practices, and descriptions … a usually boring, confusing, structurally unsound stream of words that evoke the response ... "True, but who cares?" (Collins & Porras, 1994). This would be a vision approached from a sequential, event-paced point of view as it constitutes the familiar order of things and a consecutive (mechanical) series of events and actions. This is reflected in organization's typical planning cycle approach in which new products are only to be created when a promising technology comes out. It is an opportunistic way of dealing with transition and change, which does not reflect any proactivity and instability (of markets for instance).

"Vision" is rather time-paced as opposed to event-paced. Vision gives meaning to a future, and "speaks" about that future. It has an essence of direction and pro-activity, it is a continuous transition. This continuous transition between present and future creates one of the main cruxes for managers in dealing with vision in that they find working with time very uncomfortable and difficult. When discussing vision, there is a continuous movement of discussion among ideas about the future, experiences of the past, and meanings of the present. Human time, however, is illusionary in the sense that thinking, cognition, and the process of assigning meanings is only loosely related to "watch" time. Individuals don't think in "real time", they don't experience time in "real time". Therefore, it is interesting to briefly expand on the relation between past, present, and future from a postmodern point of view.

Time

Levy (1990) showed that language modulates time as much as it is modulated by time. Temporality starts with the act of classification as every object of perception evokes a particular relation with and in time. But also the rhythm of things, the temporality of registration and interpretation, are key in this process. They all focus on the qualitative step of assigning meanings to temporality, continuous flux, and transition. Classification temporalizes or even freezes flux and gives meaning to it, before it loses itself again in the turbulence and intensity of its chaos (Serres, 1982).

We will try to understand this process according to the time of expression of an event, object, or actor (i.e. the observation) and the time of content (Eco, 1985). The time of expression describes the way the expression happens before our eyes, i.e. the time the observation takes, its construction time. The time of content is the starting point of particular forms of discourse created by language. Language creates text in which meaning is assigned to the sequence of things. It reflects a configurational order, it is the report of an interpretation.

The time of content portrays the fusion of intertextuality, the dialogue and connectionism between content of elements and signifiers. In other words, it reflects the rhythm of contexts. The unity of the discourse is to be found in a temporal diversity as there is no demarcation between object and subject in the text. Its author enables meaningful actors to talk and makes words into signifiers that can be read by others. The construction of meanings represents social networks. It also describes the individual's competency and knowledge of things as it is a mental content that includes truths and claims of truths, and how things are read, and re-read. Simply, the time of content stresses the individual's temporal sensitivity.

Hence, this perspective recognizes a multiplicity of (time) "subjectivities". Its experience is chaotic and dispersed, stable and unstable. It is Derrida's idea of "*différance*", a continuous movement without location, a power always out of reach and thus never really there, to differ in space and to defer in time (Derrida, 1981).

A tangible illustration of human time could be portrayed by movies, where content sometimes is condensed or prolonged in time, and it jumps back and forth. Simultaneously, the movie has its own way and time of expression, as it takes time to experience it. Those "directors" who work in the context of the future don't necessarily construct their plot based on a fictional past, but simply treat the future as a destiny, evolving in and of itself. Other directors, on the other hand, evolve their movie based on a nostalgic past to which everything attempted in the present must appeal. Time is indeed fictional. The humanization of time runs in and through language and text. It is the smallest variation that causes the feeling for language and its inherent time feeling.

As previously stated, time has an element that transcends the temporal, the moment, and the contingency of the singular, as well as an element of reference to the local and temporal, and to flux. Whether past, present, or future, a particular moment is namely just the reflection of a momentary context. This means that the future cannot be predicted based on a model of the past, nor will memories of the past shape present action in a deterministic and linear way. In conclusion, talking about vision is talking about human time which, in itself, is a complex issue. The relationship between past, present and future is not clear or fixed. Time is *"é-motion"*. Reality becomes a continuous journey in which the present becomes a hardly noticeable nuance. Time is an adventure in the sense that it's a continuous process of discovery and sense, but without having a regular and fixed movement. In essence, the present is the coordinated product of passed futures.

This is reflected in Foucault's idea of genealogy (see Thanem, 1998 elsewhere in this book). It portrays an historical framework for analysis, but genealogies are at the same time full of gaps, jumps, discontinuities, superficialities and eclecticism. It is historical yet not history. It is like the movies we discussed earlier. It is self-reflexive, recognizing the continuous process of discovery and sense, of diversity and hardly noticable nuances. Genealogy is what Thanem (1998) calls a journey through the past, looking for traces of the present in the past.

Vision too (2)

What are the repercussions for the development of a vision now? How can all these signifiers of time, turbulences, instabilities, and different universes, etc. be captured and fused in a vision?

First, a vision is to be authored by an observant community, a network of intelligibility, of meanings, and of con-texts. Community is a condition for individuality and differentiation while guaranteeing a platform for fusion and symbiosis. This collectivity is deeply rooted within ever-progressing flux and continuous processes of change, dynamics and turbulences. Participation could therefore be regarded as an opportunity to offer a platform for free dialogue, to allow interpretation of individual worlds from the standpoint of their own frame of understanding. Again, meanings are not to be seen as individual constructs or precepts that are then socially evaluated and corrected, but they are socially created.

Second, a vision is rhetorical as it produces new realities through shared language and meaningful (con)texts with their own particular grammar. Vision also is characterized by alliances between “truths” and “claims of truth”. It is stability, transcending the temporal, the moment, and the contingency of the singular. Vision is also subject to the position an organization has within networks.

Working on a vision is working on a language and working on oneself. By identifying with signifiers, it leaves the opportunity for those signifiers to "talk", to tell their "story". In other words, a vision is to be set in scene by constructing meanings that relate to time and situations. This "*mise-en-scène*" enables the fusion of expression, content, description, and interpretation with its own time sensitivity. Therefore, it could be understood as a technology characterized by three building stones. Vision provides the context to bring the dynamics between those three to life in a harmonized way.

A first anchor of vision is derived from Collins and Porras (1994) core ideology, which we prefer to call core philosophy. Core philosophy reflects the core values an organization truly and passionately holds and the purpose of a company. It echoes the soul of the organization, the "grammar of the company's body language". Collins and Porras define the core values as the organization's essential and enduring tenets; not to be confused with the specific cultural or operational practices; not to be compromised for financial gain or short-term expediency. The purpose is the organization's fundamental reason for existence beyond just making money - a perpetual guiding star on the horizon; not to be confused with specific goals or business strategies (Collins & Porras, 1994; 1996). This grammar creates a context that transcends the organization. On an abstract level it offers stability, continuity and consistency. It is a monument (Thanem, 1998). Core philosophy exists as an internal element of authenticity that is not pre-determined by the (prevailing) state of the environment. Its role is to guide and inspire, not to differentiate. The authenticity and consistency with which the philosophy is lived differentiates visionary companies from the rest (Collins & Porras, 1996). It transcends the temporal, the moment, and the contingency of the singular. On an applied and concrete level, however, it refers to the local and temporal and to flux and instability. As a result, the core philosophy is continuously in transition and transformation.

The second building stone of a vision is the organization's core competencies and capabilities. The concept of core competencies was first introduced by Prahalad and Hamel in 1990. Core competencies is the creative bundling of the unique skill of harmonizing multiple core technologies, an understanding of how to exploit technologies, and core capabilities that enables a company to provide a particular benefit to customers and gives a company a distinctive advantage (Hamel & Prahalad, 1994). We will, however, define it here as the combination of strategic resources with technologies underlying a company's myriad production [service] lines, and the associated rhetoric to understand, connect and exploit resources in a uniquely competitive way providing a particular benefit to customers. These core competencies are supported by capabilities. Capabilities are embedded in a company's practices and its grammar (i.e. core philosophy), emphasizing the "behavioral" aspect of strategy with a special focus on supporting activities and customer value.

This idea of competencies, referring to activities, resources, technologies, and supporting capabilities, focuses on the idea of the production of organization. Competencies result when resources proven to be durable, difficult to imitate and to substitute, are harmonized and creatively bundled with multiple technologies in meaningful (con)texts - with their own particular grammar. Core competencies should be well aligned with the core philosophy, and are often rooted in it. They transcend a single business and are a significant source of differentiation, reflecting the capacity in creating new businesses, core products and functionalities, and end products or services (Prahalad & Hamel, 1990). Core competencies form an organizing framework as well as a way of communicating. They also continuously disrupt the normative structures about the organized world. Answers to problems about organizational systems turn into problems about answers. At the same time, competencies encompass the value chain, product and service variety, and ever evolving opportunities and developments, also showing the limitations of each project.

Competencies and capabilities, in other words, offer an operational "*différance*", to differ in space and to defer in time (Derrida, 1981). They are the organization's "time of content", portraying its rhythm as well as its operational unity, found in temporal diversity and the fusion of content of strategic elements and signifiers. By describing the mental content of competencies, capabilities, and knowledge, it gives the "reader" insight in the organization's temporal sensitivity.

The vision's last anchor consists of the envisioned future, a strong drive for progress. Vision-level audacious goals are a means to stimulate progress and create momentum, serving as focal points of individual and group effort, while reflecting the core philosophy. It is the commitment to this daunting and huge 10 to 30 years challenge that engages people (Collins & Porras, 1994). Vision-level audacious goals are tangible, energizing, and highly focused. People "get it" right away as these audacious goals have a "close" meaning to everyone. Vision-level audacious goals describe discontinuities, yet a sense of direction and destiny. The focus is on the unarticulated and unserved, which captures the philosophy of searching for openings in order to make new choices possible. Vision-level audacious goals go beyond the pure tactical. Therefore, organizations should continuously set audacious goals to avoid the "we've arrived syndrome". Once you have reached the summit, it's time to move on to other mountains (Collins & Porras, 1996). This consensus is a mere state in the journey, while disensus is the goal.

Finally, vivid stories - narratives - are important to capture issues of timing, drivers, certainties, uncertainties and dynamics. It brings together all the other anchors, moves between them, and the fusion offers more value than the stand-alone anchors. Compelling story lines picture how time unfolds, given the dynamics, between now and a future date. They reflect a configurational order, the report of meanings assigned to things and their sequences. Stories "talk" through formulas as the descriptions and the dynamics they tell bundle the perspectives, concerns, signifiers, and assumptions from which individuals generate futures and pose visionary questions. As such they transcend individuals, authors, signifiers, i.e. the momentary. These vivid stories are not deductive or analytic, they are not metaphorical or conclusive. They continuously disrupt normative structures about the organized world. They don't have a universal finality as they stress a pluralism of small stories that talk about organizational lives and futures. Hence, vivid stories encompass "*é-motion*" and break "out-of-the-box". They embrace complexity, paradoxes, turbulences, instabilities, and uncertainty and sometimes elevate them to the status of competitive weapon.

Rather than being predictive, formal, abstract or stylistic, vivid stories treat visions as "enabling constraints". It is important, however, to ensure the pluralism of different stories - diversity and disensus - in order to evoke thinking and movement, and stimulate new strategic possibilities. Put into different scenarios, vivid stories enable "readers" to dance through dramatic interpretations of different universes of signifiers, actors, and plots. They guarantee a visionary "*différance*" as different scenarios portray different rhythms and social networks and different contexts and fusions of intertextuality within and between scenarios. It is unity as diversity. Authors have to look for bridges between and commonalities within different scenarios.

This whole process makes the "*mise-en-scène*" of a vision. It doesn't attempt to systematize, define, or impose a logic on events. Instead, the limitations of all projects in passed futures and futures are recognized.

Postduction: the last seduction

Vision as described here is a good illustration of the rhetoric function of management. It clearly is a strategic journey through networks of intelligibility, fully taking advantage of the richness of complexity, turbulences and flux. As such, it presents the manager "vision intelligence", a philosophical psychology, a mindset to be understood as a continuous movement without location.

Vision intelligence is time paced, but not real-time paced, it is long-term yet not really long-term. The networks of intelligibility are produced along a multiplicity of time subjectivities, i.e. jumps and gaps. It's not where they jump from or where they land, it's the movement in between that opens fields of potentiality and offers new possibilities. To make the analogy, it's not particles but wavicles, the waves of movement along different anchors. And these wavicles portray fields and boundaries of potentialities. In other words, although the factual world is beyond our grasp, postmodernism certainly offers an understanding of managerial and organizational complexities as it exactly shows how a philosophical psychology about a particular more applied field can be developed. Vision intelligence is about what one learns in the process of translating (giving meaning to) the flow of the managerial and organizational world and the manager's involvement in contexts. Essentially, the epistemological nature does not coincide with the attempt of application. The different levels don't have to be confronted since there's fundamentally no logical differentiation.

The exercise we presented here shows that postmodern thinking is epistemological yet inherently pragmatic. Inspiration and application are integral part of it. They are, however, not to be seen on a linear continuum or understood as the quantifiable consequence of one another. As a result, there's no "loss of theory" as positivists would claim. Hence, there's no paradox on this issue.

The philosophical psychology which we called "vision intelligence" in this particular illustration also reveals how well postmodern thinking describes the essence of the "new economy", i.e. the knowledge economy driven by intellectual capital and brainpower. This brings us to the conclusion that postmodern is not post-modern. It's just passed modern or past-modern. Is this maybe the fundamental modern paradox?

References

1. Baets, W. *Organizational learning and knowledge technologies in a dynamic environment.* (Kluwer Academic, Amsterdam, 1998).

2. Christodoulou, K. (1998) Modelling and simulation of variable structure systems: a way to handle complexity. In: *Complexity and Management. A collection of essays, vol. 1*, ed. Baets, W. (World Scientific, Singapore, 1998).

3. Collins, J. & Porras, J. *Built to last*. (HarperCollins, New York, 1994).

4. Collins, J. & Porras, J. Building your company's vision. *Harvard Business Review*, September-October, 65-77 (1996).

5. Cooper, R. and Burell, G. Modernism, postmodernism and organizational analysis: an introduction. *Organizational Studies*, 9, 1, 91-112 (1988).

6. Derrida, J. The Law of Genre. In : *On Narrative.* Ed. Mitchell, W. J. T. (University of Chicago Press, Chicago, 1981).

7. Eco, U. *Sugli specchi e altri saggi*. (Fabbri, Bompiani, Sonzogna, Etas, Mila, 1985).

8. Gergen, K. J. Organizational Theory in the Postmodern Era. In : *Rethinking Organization*. Eds. Reed, M. and Hughes, M. (Sage, London, 1992).

9. Hamel, G. and Prahalad, C. K. *Competing for the future*. (Harvard Business School Press, Boston, 1994).

10. Levy, P. *La machine univers. Création, cognition et culture informatique*. (La découverte, Paris, 1990).

11. Maturana, H. & Varela, F. *The tree of knowledge* (Dutch Translation). (Contact, Amsterdam, 1989).

12. Parker, M. Post-modern organizations or postmodern organization theory. *Organization Studies*, 13, 1, 1-17 (1992).

13. Prahalad, C. K. and Hamel, G. The core competence of the corporation. *Harvard Business Review* 68, 3, 79-91 (1990).

14. Serres, M. *Genese* (Grasset, Paris, 1982).

15. Thanem, T. Death, Organisation and complexity: Thanatocracy and monumentalism in organisational architectures. In: *Complexity and Management. A collection of essays*, vol. 1. Ed. Baets, W. (World Scientific, Singapore, 1998).

16. Van der Linden, G. *Een essay over en empirisch onderzoek naar betekenissen van loon en beloning. Casus van een postmoderne visie op bedrijfsbeheer. (A treatise and survey on meanings of wages and rewards. The case of a postmodern view on management)* (State University of Groningen, Groningen, 1993).

17. Van der Linden, G. and Parker, P. On paradoxes between human resources management, postmodernism, and HR information systems. *Journal of Accounting, Management and Information Technologies* (1998).

18. Van Gils, M. *Social constructionism, non-linearity, and the significance of language*. (University of Groningen, Groningen, 1997).

19. Waldrop, M. M. *Complexity*. (Penguin Books, London, 1992)..

20. Zeleny, M. *Autopoiesis: a theory of living organization*. (North Holland, New York, 1981).

DEATH, ORGANISATION AND COMPLEXITY: THANATOCRACY AND MONUMENTALISM IN ORGANISATIONAL ARCHITECTURES

TORKILD THANEM

Warwick Business School, University of Warwick, Coventry CV4 7AL, England
E-mail: Torkild.Thanem@warwick.ac.uk

This paper is a thinking exercise. It tries to think differently about complexity by highlighting the filthy netherside of organisation that so often is neglected and pushed into the dark shadowlands. More specifically, it tries to think complexity by looking at the filthy relationship between death and organisation as spelt out in the organisational architectures of the thanatorium and the monument, and in their corresponding discourses of thanatocracy (rule by death) and monumentalism. Two sides of the same coin, the relationship between the thanatorium and the monument is complex but often denied. Dissatisfied with this denial, the paper attempts to illustrate how the thanatorium hides in and beneath the monument. The structure of the paper is threefold and it starts out by throwing up some of the central features of complexity. It then relates complexity to filth and death by bringing in some ideas developed in contemporary French thought. Finally, it employs this framework to appreciate complexity in the filthy relationship between the thanatorium and the monument.

1 Preludium

We are finally reaching the last session of a summer school that, albeit has seemed to be granted with eternal life, soon is about to reach its end. Perhaps it is the irony of this event, then, that my paper attempts to deal with exactly such issues that relate to the end, to death and to murder. More specifically, I attempt to deal with issues of death and organisation, thanatocracy (rule by death) and monumentalism in a way that tries to respect and problematise the complexity of these topics.

But before I begin my visit into matters of complexity, I have a confession to make. I must confess that complexity is at the periphery of my project. Furthermore, my thinking on complexity may be both provocative and filthy to some of you. You may not at all like what I am going to say now. However, I still think complexity is too important to be neglected anymore. It is therefore time to bring complexity up from its underground existence. It is time to stop taking complexity for granted. It is time for me to say what I mean by complexity.

"Death is perhaps the last pornographic issue left." Oliviero Toscani [26], the photographer of Benetton's many provocative advertising campaigns, said this in an interview with *The Guardian* this winter. And I can only add that death is perhaps the last organisational issue left. Nevertheless, it is through death that I want to think complexity. I want to think complexity by looking at the relationship between thanatocracy and monumentalism. I want to examine how thanatocracy hides in monumentalism, how the thanatorium hides in the monument, how death hides in life. For organisation is as much about death as it is about life. Organisation is *both* life *and* death.

But in order to appreciate this complexity, I think it is crucial to focus on the Other side, the filthy and deadly netherside of organisation where the smelly and stinking rakfish is buried. I think it is necessary to relate organisation not to life, as is normally done, but to death, which is done just too rarely and never by anyone normal. So I will therefore focus on the lavatorial. I will cross the lavatory rim and see what grows underneath its glossy porcelain surface. I will bend over, look down and look up in order to see what grows in that place which we usually do not touch, except with rubber gloves and these plastic disinfectant containers that are shaped like ducks, swans or whatever.

Now, all this may just give me too much to think about in what is supposed to be a straightforward conference paper. All this may give you the impression that I am in a hurry, like a nomad, a gypsy or a tinker, never finding time to rest, never settling down in one place, but always haunted by something or by someone. And you are probably right, for I feel haunted and I feel as if I am spread out across too many places and too many pages. I feel haunted by the demand to settle down with one truth, the Truth. Of course this freaks me out. For there are many truths in the Truth. Yes, there are many Truths in the truth. And to be honest, I quite like it like that, even if that makes me both paranoid and schizophrenic. But never mind. Let me get down to it. Let me get down to the lavatory. Let me get down to think complexity.

2 Thinking complexity, thinking differently

Yes, let me think complexity. But let me think differently about complexity. And let me first enforce a difference between complexity and complicatedness. For there is a difference, despite the stream of silent attacks that continuously tries to colonise the One with the Other, the complex with the complicated. The *Oxford English Dictionary* defines complicatedness as the quality of being complicated. Further on, being complicated is to be folded together, with the additional meaning of being intricate, involved, and difficult to analyse or unravel. More specifically, the word 'complicated' stems from the two Latin words 'com-', which means 'together', and 'plica', which means 'fold'. Complicatedness is therefore the folding over of planes so that one plane overarches the other, as in a verticality of levels. Complexity, on the other hand, is defined as the quality or condition of being complex. Being complex is to be interwoven, and a complex is the consisting of various parts that are united or connected together. In other words, a complex is formed by a combination of different elements. Actually, the latter part of the word 'complexity' stems from the Latin word 'plex', which means web. A 'plex' is therefore something that is woven together, like a plait or a braid. In conclusion, complexity cuts itself off from complicatedness by suggesting the uniplanar and two-dimensional interconnection of diverse parts, rather than the vertical and three-dimensional folding over of elements.

Hence, the difference I want to draw between complicated systems and complex systems is similar to the difference that the early Gilles Deleuze proposed between the multiplicity of order and the multiplicity of organisation [14, p. 18]. Like the multiplicity of order, a complicated system is determinate, pre-formed and static. The form and behaviour of a complicated system is always set out from the very beginning by someone else, by an Other who stands on the outside and looks down on it from a hierarchical and privileged position far away and high above. This Other, this chief executioner folds itself over the complicated system and assigns to it an external purpose. S/he then organises and reproduces the complicated system in accordance to this fixed and unchangeable purpose. Hence, the complicated system will never change. It does not have the power to change and it does not even have the will to change. Instead, the chief executioner will ensure that the complicated system remains the same, just the way it always was; highly formalised, specialised and divided into isolated parts that relate to each other like a set of 'either/or' dichotomies. Complicated, isn't it?

A complex system, on the other hand, is like the multiplicity of organisation. It is indeterminate, creative and dynamic, and its "organization is always unforeseeable" [14, p. 18]. Organisation is unforeseeable because the complex system has an internal purpose, just like Herder and Kant's organism [25]. Its purpose is assigned by no one else than itself and it takes no orders from any divine power outside its own territory. It is self-referential and does not need to legitimate or justify itself and its organisation to anybody else. It does as it pleases and it is therefore impossible to know what it will do next.

Hence, organisation is unforeseeable because the complex system is self-organising. Self-organisation is the *modus operandi* of the complex system. It organises itself from within and from below rather than being organised hierarchically and rigidly from the outside and from above. The complex system has its own will to power and its own will to organise. The complex system is a rhizome [8] which follows no plan, but spontaneously performs a multitude of unpredictable connections that never can be fitted into a complicated formalisation.

The connections produced by a complex system signify the reproduction of the complex system. The complex system reproduces itself and it does not need anyone else in order to reproduce. Its own body is enough because any complex system is diverse and creative. It is heterogeneous, both-and, many things at once. It is a schizophrenic body that disturbs any and every dualistic boundaries and accepts no fixed limits. It is a hermaphrodite with both male and female genitals who reproduces by masturbating and fucking itself. It is a cyborg that creates actants and links whenever and wherever it needs them for its own internal purpose of self-reproduction, evolution and survival. A complex system therefore evolves creatively. A complex system performs creative evolution [4, 8] and it knows that it does.

Hence, complex systems are self-reflexive. They do not just know about themselves. They also know that they know. The system's knowledge of itself is therefore part of the system, which again influences the way in which it organises and reproduces itself. The system's knowledge of itself enables the system to change and evolve. But since the its knowledge of itself can never be the same at any two given points in time, no initial state can ever be the same, and the system can never repeat itself. The complex system is therefore non-linear [1]. Hence, the complex system is full of surprises and it is impossible to predict what it will do next. Moreover, it is the self-reflexivity of complex systems that makes them unpredictable. In other words, complex systems are systems whose future cannot be foreseen by any method of forecasting or by any soothsayer staring into a crystal ball.

But self-reflexivity also has to do with the possible, the virtual and the diverse. Since nothing is predictable, everything is possible. However, this is not to say that a complex system merely is a possible and virtual system. For the complex system is realised. The complex system is therefore a system of the possible and the virtual because it allows for (and allows us to think) many different possibilities. Hence, the complex system is a diverse system. And it is the diversity of the complex system that I seek to deal with in what is to f(ol)low. I attempt to be reflexive by reflecting upon and recognising diversity in complexity.

3 Complexity the French way: complexity as filth

In order to recognise diversity in complexity, I now want to think complexity the French way. I want to think complexity by looking through the lens of contemporary French thought. I want to think complexity as filth. For, as I want to show below, diversity is filthy. But this is not completely French. There is a German involved here somewhere, a German who preceded these French thinkers, a German who inspired them and inspires me in my thinking on complexity; in my thinking of complexity as diversity and filth. His name is Friedrich Nietzsche, and his notion of filth and impurity is to be found in his genealogy.

Nietzsche, Derrida and the death of primordial purity

Friedrich Nietzsche, in *On the Genealogy of Morality* [20], did not merely write about traditional genealogy, understood as a search for origin. More importantly, he changed and extended the concept of genealogy to become an alternative to the search for the primordial, pure and homogeneous origin. Genealogy for Nietzsche is never pure, chronological and linearly progressive. Rather, genealogy is about inconsistencies and discontinuities, paradoxes, ruptures and anachronisms. Genealogy is about the things that are dirty because they are not in their proper place. Genealogy in this sense is about filth. It is about the diversities that make the world so unpredictable and uncontrollable, so dirty and so scary, so complex.

Nietzschean genealogy is about filth because it, by dismissing any search for a pure origin, resists confining the Self and the Other to a binary dualism. The relationship between the Self and the Other is more complex. Drawing upon Jacques Derrida's [9] spectral notion of the Self and the Other,[1] the Self cannot dualistically reject the Other as filthy. In fact, that would force it to reject itself. For the Self does not merely rely on filth and difference in its relationship with the Other. The Self is itself filthy and can never exist as purely different from, superior to or without the Other. The Self changes and reproduces itself in relation to the Other and the Other changes and reproduces itself in relation to the Self. The Self becomes what it is in its interaction with the Other and the Other becomes what it is in its interaction with the Self. There is no Self without the Other and there is no Other without the Self. The Self and the Other depend on each other, reflect each other and are reflected in each other because the Self exists inside the Other and the Other exists inside the Self.

Hence, we are not only our own friends. We are also our own enemies. The Other is not only its own friend and our enemy, but also its own enemy and our friend. Although the Other threatens our survival by making us filthy, we also need the Other in order to survive. 'We have never been modern', Bruno Latour [16] says on the cover of his book. We have never been Human, I could add. And we have never been pure. For to be pure is to be Human and to be Human is to be modern. But this is not us. Instead, we have always been cyborgs. For it seems impossible to survive both with and without the micro-organisms, germs, bacteria and viruses that exist on the insides and outsides, the nethersides and surfaces of our bodies and turn us into filthy and heterogeneous cyborgs. It seems as impossible to survive without the Other as it does with it.

And even if we *could* predict the Other, control and produce it in our own narcissistic image, that would not make much sense. For the Other would then cease to be the Other and become the Self. Instead, we must therefore learn to live with the Other. Rather than trying to in-form and per-form, shape and rape, re-present and re-produce the Other, we should let the Other be what it is. We should allow for filth, diversity and difference, and let the Other organise and reproduce itself in line

---

[1] Derrida's [9] thinking on the Self and the Other is highly influenced by the Jewish philosopher Emmanuel Levinas' ethics of infinity and the Other [17, 18].

with its own internal purpose. We should let the Other fuck, but we should not fuck it. The Other should be allowed to fuck itself.

Foucault's passions for filthy limit experiences

Foucault [12] picks up Nietzsche's idea of genealogy and utilises it in his historical studies of institutional practices. Genealogy is a way of getting closer to and becoming passionate about the institutional practices under investigation [11]. Genealogy admits for subjectivity and closeness to the subject matter and helps you to get 'under the skin' of your topic.

But even though genealogy provides a historical framework of analysis, it is not a history in the normal sense. Genealogy is not proper History because genealogies are full of gaps and suffer from discontinuity, superficiality and eclecticism. They make big jumps between distant and seemingly unrelated epochs and pay no attention to chronology. Hence, genealogy shows no respect for history. Instead, genealogy bends history to make it groan.

Furthermore, genealogy is about understanding the present rather than the past. It is not a journey through the present, looking for traces of the past in the present. It is rather a journey through the past, looking for traces of the present in the past. Hence, genealogy does not write a history of the past in terms of the present. Rather, it writes a history of the present in terms of the past [11, 13], thereby highlighting the barbaric and pre-modern features of contemporary society that the modern project tries to hide, disguise and run away from. A genealogy is therefore a critical project and a project of complexity, exactly because it is self-reflexive and attempts to recognise diversity in the filthy netherside aspects of its own world. It tries to discover the Other in the Self, even if that forces it to invent the Other.

At the same time as genealogy creates gaps, it therefore fills gaps. Genealogies fill gaps by passionately examining this Other netherside. This is a netherside endowed with limit experiences that relate to sex, madness and death [19]. It is a netherside that is typically ignored in conventional histories because it cannot normally be linguistically appropriated [11]. As Bataille suggests [2], sex, madness and death are somehow outside language. Sex, madness and death produce silence because they cannot be fully grasped in writing. However, writing from a genealogical angle may still enable me to investigate exactly this kind of aspects that populate the netherside of organisation. Writing a genealogy may enable me to recognise the ghostly Other in the Human Self and the way they reproduce themselves and each other. Writing a genealogy may enable me to recognise complexity.

Genealogising the rhizome, or: DeleuzeGuattari forget their anti-genealogy

Adding to Foucault's genealogy, it seems fruitful to think through Deleuze and Guattari's [8] notion of the rhizome, briefly mentioned above. Although Deleuze and Guattari insist on the rhizome being an *anti*-genealogy, both Nietzschean and Foucauldian understandings of genealogy seem compatible with a rhizomatic perspective. Like the rhizome, and unlike traditional and fascist genealogy, Nietzsche's and Foucault's genealogies dismiss any search for the pure origin. Nietzschean and Foucauldian genealogy imply a root structure that does not have one sole source of origin out from which species evolve in a linearly progressive fashion. Rather, their genealogies are anti-hierarchical, non-linear and non-chronological.

Because it is self-reflexive, genealogy enables me to study phenomena that are related, though not in a direct and continuous way, as if always on the surface of things. Instead, genealogy appreciates the way in which phenomena and their relations embody both the Self and the Other, and shift rapidly between surfaces and undergrounds. Rhizomatically, various phenomena may appear, then suddenly disappear, for again and without notice to reappear in the most unexpected modes and contexts. The Other is always inside the Self and the Self is always inside the Other, but we never know which character - the ghost or the living being, the demon or the angel - who will charge the surface and show its face.

Death indeed seems to be an issue that operates in such rhizomatic ways. Death does not merely produce ghosts and zombies. Death is itself a ghost or a zombie that comes and goes as it pleases, when least expected. It pops up from its habitat down in the filthy netherside to the glossy surface where everything seemed to be just perfect.

Swimming around in shit: Bataille's search for filth

Death also plays a significant role in Georges Bataille's scatology. Hollier [15] defines scatology as the science of filth, which promotes the study of taboos, stigma and perversions. Like genealogy, scatology directs its attention to sex, madness and death, issues that are too filthy to be talked and written about in polite and civilised conversations. But scatology recognises the filth of these issues more explicitly than genealogy does. And it is important to be explicit here. It seems necessary to be more passionate than Foucault when examining topics of death, organisation and complexity. It therefore seems worthwhile to step backwards a few decades into the history of French thought, adopt Bataille's attitude and let myself write passionately about the filthy aspects of organisation that I think can be found, or at least invented, in an organisational Netherland of thanatoria and monuments.

4 Thanatocracy and monumentalism in organisational architectures

As stated above, thanatocracy means rule by death. The word is made up of two roots, both of them stemming from Greek. The first part of the word, 'thanato-', comes from the Greek word 'thanatos' (θάνατος) and means death. The second part of the word, '-cracy', comes from the Greek word 'kratos' (κράτος). 'Kratos' is a common word in academic and political language and can take on many various yet similar meanings, both as a noun and as a verb (e.g. state, nation, a fixed area, to fix, to hold, etc.). In this context, however, I let 'kratos' mean rule. And again, thanatocracy is rule by death.

Further on, thanatocracy is the discourse of the thanatorium. The thanatorium is the organisational architecture where one is received in order to be murdered. The thanatorium is therefore the opposite of the sanatorium, which is the place where one is received in order to be healed and prevented from dying. The sanatorium is typically a hospital or a clinic (both monuments of the capitalist welfare state) whilst the thanatorium is usually depicted as a slaughterhouse, an execution site or a death camp. However, and despite the frequent dressing up of the thanatorium as a sanatorium, judges of normality rarely allow us to associate the one with the other. Judges of normality ensure that they are kept distinct, never to be confused with each other. Judges of normality deny the possibility of any other relationship between the two than that which depicts them as mutually exclusive opposites.

Similarly, monumentalism is typically separated from thanatocracy. 'Monumentalism', the *New Oxford Dictionary of English* states, is construction, especially of buildings, on a grand scale. Further on, the same dictionary defines a 'monument' as a statue, building, or other structure erected to commemorate a famous or notable person or event. A monument can also be a building, structure or site that is of historical importance or interest. And it can be an outstanding, enduring, and memorable example of something. In Middle English, the dictionary reminds us, a monument specifically denoted a burial place, and the word itself stems from the Latin word 'monere', which means 'to remind'. The monument was raised in memory of the dead.

In this context, therefore, monumentalism is taken as the discourse of building architectural edifices in order to transcend time and space, and to mask the Other that occupies its filthy netherside. Hence, monumentalism can only with much difficulty be separated from thanatocracy. More often than we like to admit, thanatocracy and monumentalism come together. Just as the thanatorium hides in the monument of the sanatorium, thanatocracy hides as monumentalism. Further on, the abattoir hides in the museum, the Other hides in the Self, and the filthy netherside hides under the pure and glossy surface. In conclusion, complexity is hidden because we are only allowed to see one side of the coin - the upper side.

Therefore, "[w]o es war [Schlachthaüser] soll Museum werden" [15, p. xiv].[2] This translates as: "where there were abattoirs there shall be museums." And this tenet has indeed been implemented. Museums have been constructed upon sites that used to occupy slaughterhouses. This is for example the case with La Villette in Paris, an area that was famous for its slaughterhouses and meat market. La Villette is now occupied by a park which, like a postmodern museum, displays edifices of contemporary culture and art. More generally, thanatoria have been replaced by monuments. For example at Place de la Greve (also in Paris) there was previously a gibbet used for the execution of prisoners. The gibbet has been replaced by a famous monument, the Obelisque Concorde. Hence, where there was filth, death and murder, there is now purity, eternal life and immortality, manifested in monuments that are forever visible and unquestionable. For how can there ever be anything ugly in something as beautiful as a monument?

The modern abattoir is an ideal example of an organisational architecture based on the discourse of thanatocracy. With its assembly line, the modern abattoir is a thanatorium that enables murder to take place on a large scale and with high precision, speed and efficiency. Typically located outside the areas where the civilised urban population lives [21], its murderous netherside of filthy slaughter is hidden away amongst the rude and rural inhabitants of the untamed countryside. Despite the abattoir itself is located away from modern city life, it manifests itself in the fast food restaurant which we now find in most cities in the westernised world [22]. However, the fast food restaurant is still rarely thought of as a thanatorium. Rather, it is usually depicted as a source of food, thus helping us to survive. Further on, the fast food restaurant is constructed as a monument of contemporary society, shining and smiling at us from a distance with its golden neon arches. But the M does not just stand for McDonald's. The M stands for murder, or as *The Smiths* were singing in the 1980s: 'meat is murder'.

The car factory is another thanatorium with strong connotations to the abattoir. When Henry Ford introduced the assembly line system to his car factories (now a monument of modern industry) he simply reversed the (dis-)assembly line system that was already being applied in the Chicago slaughterhouses for mass production of dead cattle and food [24]. Whilst the abattoir separates parts and organs from each other, the assembly line puts parts together, yet still keeping them distinct, divided and separate. Both systems operate as labours of division. Both systems claim purity and hygiene whilst trying to hide the blood dripping from their hands. Both systems are monuments that try to cover up filth whilst pursuing a project of

---

[2] Italics omitted.

purity and hygiene – a project that tries to remove 'the hair in the soup' [10] and hide its thanatocratic netherside.

This monumentalist project of purity and hygiene has produced some very tragic outcomes, and the mass murder set in motion by the Holocaust during the Second World War may be the largest scale thanatocracy to date. The Holocaust operated along the same organisational principles as the abattoir and the assembly line. Organisational discourse played a significant part in constructing the Third Reich as a monument of modernity. Bureaucratic principles of administration were central tools used to further the Nazi ideology of racial purity and fight for the immortality and eternal survival of the German race and nation. And like the cemeteries and abattoirs of modern societies, the Nazi concentration camps were located outside the cities of Europe [21].

According to Zygmunt Bauman [3], the massacre of Jews, homosexuals, gypsies, blacks, communists and others that did not fit into Nazism's plans for racial purity, was not a result of German racism and nationalism. Rather, Bauman argues that it was the rationalised bureaucracy of Nazi Germany, initiated by the Department for Economic Affairs, which enabled the execution of this mass slaughter. The anatomical division of labour in the Holocaust organisation alienated those working there, thus inhibiting participants, passive spectators and victims from conceiving the full implications of their own and other peoples' actions.

However, alienation and rationalisation may not be the only mechanisms at work here. Gillian Rose [23], e.g., would insist on the importance of moral sentiments. Nevertheless, Bauman's [3] account brings new light to the understanding of the Holocaust and he helps us appreciate its complex diversity. Furthermore, his argument represents a more terrifying account than previous perspectives [12].[3] If the Holocaust, both the major monument and victory as well as the major thanatorium and catastrophe of modernity resulted from modern organisation and not from pre-modern barbaric sentiments specifically related to German people, the Holocaust can happen anytime and anywhere again. It may suddenly knock on our door to pay us a visit right in the middle of our long and peaceful rest. And we may not even realise that it is inside until we have invited it inside and it is too late. For it never looks quite the same. It always wears different disguises, as if history was a fancy dress party. One time s/he wore the SS uniform, another time s/he wore the Mao uniform, and the next time, maybe s/he will dress up as a UN official – in khaki and light blue.

---

[3] As Rose [23, p. 22] says, "Fackenheim's [12] Holocaust philosophy [for example] argues for the *uniqueness* of the Holocaust; Bauman's Holocaust sociology argues for its *normality* – its continuity with the main features of bureaucratic, technocratic rationalization, shored up by the scientism of sociological reflection."

Strong powers also seem to be at work in order to hide what happened during the Holocaust of the Second World War. Attention has for decades already been turned away from the Holocaust death camp as a thanatorium and towards the death camp as a monument and museum. The death camp with its guides and curators, its souvenir shops and bookshops now reminds us less of a thanatorium. The experience of the death camp as a museum yields a catharsis that cleans us from the filthy memories of Second World War massacres. And we are not only tempted to say the following. Actually, we often say it: 'The Holocaust? That was then, more than 50 years ago. It could never happen again.' But it could. And it becomes even more likely because it does not seem to scare us anymore. Rather than acting as a reminder, the death camp as museum helps us forget this tragic past and look into the future with optimism, a view that perfectly suits the modern project of linear progression.

Throughout this summer school we have a number of times found ourselves by the ticket machines of such monuments, be it the Alhambra palace in Granada or the Cathedral in Cordoba. They are both beautiful works of art and architecture. They are true monuments that bring the spectator to silence and make us forget that many people died in order to build these monuments. We therefore forget that thanatocracy is part of monumentalism. We forget that there is a thanatorium in every monument. For whereas the bodies that enabled its construction are all dead and no longer here, the monument is immortal and still visible across time and space. We do not anymore have to care about the dead bodies that built the monuments. They were turned into lifeless matter even centuries ago. But it is difficult to ignore and stop thinking about the monuments themselves that shine at us in their own glory of power and beauty. It is difficult not to be amazed, silenced and struck by the monuments that cover up the thanatoria.

Postludium

In concluding my session on death, organisation and complexity - complexity as filth - perhaps time is ripe for the articulation of a warning. For have we really tried to discuss managerial complexity at this summer school, or have we wanted to manage complexity? It may seem that the Summer School on Managerial Complexity has just turned itself into a monument that covers up its filthy netherside. Rather than appreciating complexity, the Summer School has attempted to solve and mask complexity. And it is as the solution to complexity that we shall forever remember the Summer School. For by publishing the proceedings of the Summer School as a book, the Summer School builds a monument of itself. Purchasable, readable and consumable by anybody anywhere in the world, the proceedings know no spatial and temporal boundaries. Hence, they ensure that we will never forget the Summer School and never forget that the Summer School solved complexity. But we might forget that the Summer School masked complexity by trying to manage it. We might forget that the Summer School covered up complexity and we might thereby forget complexity itself. In other words, we might forget the thanatorium and just remember its monumental surface.

And having returned from the Summer School, I now find myself writing from my desk inside another monument – the University of Warwick. Located at Gibbet Hill, the University is surrounded by beautiful parks, gardens and phallic architecture. And I am seduced to forget the thanatocratic netherside of this place. I am seduced to forget that there used to be a gibbet placed at Gibbet Hill [5]. I now start wondering. Is the gibbet still there? And who is next in the line? As I submit my paper and myself to the editorial board, I can only hear a distant voice whispering out a name. But I never knew this person.

References

1. Baets, W. J. *Organizational learning and knowledge technologies in a dynamic environment* (Kluwer, London, 1998).
2. Bataille, G. *Eroticism* (Calder & Boyars, London, 1962).
3. Bauman, Z. *Modernity and the Holocaust* (Polity Press, Cambridge, 1989).
4. Bergson, H. *Creative evolution* (Macmillan, London, 1911).
5. Burrell, G. *Pandemonium. Towards a retro-organization theory* (Sage, London, 1997).
6. Deleuze, G. *Bergsonism*, transl. by H. Tomlinson and B. Habberjam (Zone, New York, 1988).
7. Deleuze, G. and Guattari, F. *Anti-oedipus. Capitalism and schizophrenia*, transl. by R. Hurley, M. Seem and H. R. Lane (Athlone, London, 1984).
8. Deleuze, G and Guattari, F. *A thousand plateaus. Capitalism and schizophrenia*, translated by B. Massumi (Athlone, London, 1988).
9. Derrida, J. *Specters of Marx. The state of the debt, the work of mourning, and the New International* (Routledge, London, 1994).
10. Douglas, M. *Purity and danger. An analysis of concepts of pollution and taboo* (Routledge & Kegan Paul, London, 1966).
11. Dreyfus, H. L. and Rabinow, P. *Michel Foucault. Beyond structuralism and hermeneutics* (Harvester Wheatsheaf, London, 1982).
12. Fackenheim, E. The Holocaust and philosophy, *Journal of Philosophy* **LXXXII/10** (1985) pp. 506-514.
13. Foucault, M. *Discipline and punish. The birth of the prison*, transl. by A. Sheridan (Allen Lane, London, 1977).

14. Hardt, M. *Gilles Deleuze. An apprenticeship in philosophy* (UCL Press, London, 1993).
15. Hollier, D. *Against architecture. The writings of Georges Bataille*, transl. by B. Wing (MIT Press, London, 1989).
16. Latour, B. *We have never been modern* (Harvester Wheatsheaf, London, 1993).
17. Levinas, E. *Le temps et l'autre* (Presses Universitaires de France, Paris, 1983).
18. Levinas, E. *Ethics and infinity* (Duquesne University Press, Pittsburgh, 1985).
19. Miller, J. *The passion of Michel Foucault* (Simon & Schuster, New York, 1994).
20. Nietzsche, F. *On the genealogy of morality*, ed. by K. Ansell Pearson, transl. by C. Diethe (Cambridge University Press, Cambridge, 1994).
21. Pick, D. *War machine. The rationalisation of slaughter in the modern age* (Yale University Press, London, 1993).
22. Ritzer, G. *The McDonaldization of society* (Pine Forge Press, London, 1993).
23. Rose, G. *Judaism and modernity. Philosophical essays* (Blackwell, Oxford, 1993).
24. Rose, M. *Industrial behaviour. Theoretical development since Taylor* (Allen Lane, London, 1975).
25. Rosen, M. *On voluntary servitude. False consciousness and the theory of ideology* (Polity, Oxford, 1996).
26. Toscani, O. Death is perhaps the last pornographic issue left, *The Guardian Media Supplement* (Monday 2 February 1998) pp. 10-11.